MW01470352

GUY T. LOGAN
1226 WENDELL AVE., APT. 5
SCHENECTADY, NY 12308

GUY T. LOGAN
1224 WENDELL AVE., APT. 5
SCHENECTADY, NY 12308

*STRATEGIC
MANAGEMENT OF
HUMAN RESOURCES
IN
HEALTH SERVICES
ORGANIZATIONS*

0-8273-4240-3

Delmar Series in Health Services Administration

Stephen J. Williams, Sc.D., Series Editor

Introduction to Health Services, third edition
  Stephen J. Williams and Paul R. Torrens, Editors

Health Care Economics, third edition
  Paul J. Feldstein

Health Care Management: A Text in Organization Theory and Behavior, second edition
  Stephen M. Shortell and Arnold D. Kaluzny, Editors

Ambulatory Care Management, second edition
  Austin Ross, Stephen J. Williams, and Eldon L. Schafer, Editors

Health Politics and Policy, second edition
  Theodor J. Litman and Leonard S. Robins, Editors

Strategic Management of Human Resources in Health Services Organizations
  Myron D. Fottler, S. Robert Hernandez, and Charles L. Joiner, Editors

STRATEGIC MANAGEMENT OF HUMAN RESOURCES IN HEALTH SERVICES ORGANIZATIONS

Edited By

Myron D. Fottler, Ph.D.
Professor and Director, Ph.D. Program in Administration–Health Services
Department of Health Services Administration
School of Health-Related Professions
University of Alabama at Birmingham
Birmingham, Alabama

S. Robert Hernandez, Dr.P.H.D.
Associate Professor
Department of Health Services Administration
School of Health-Related Professions
University of Alabama at Birmingham
Birmingham, Alabama

Charles L. Joiner, Ph.D.
Senior Associate Dean and Professor
Department of Health Services Administration
School of Health-Related Professions
University of Alabama at Birmingham
Birmingham, Alabama

DELMAR PUBLISHERS INC.®

NOTICE TO THE READER

Publisher and author do not warrant or guarantee any of the products described herein or perform any independent analysis in connection with any of the product information contained herein. Publisher and author do not assume, and expressly disclaim, any obligation to obtain and include information other than that provided to them by the manufacturer.

The reader is expressly warned to consider and adopt all safety precautions that might be indicated by the activities described herein and to avoid all potential hazards. By following the instructions contained herein, the reader willingly assumes all risks in connection with such instructions.

The publisher and author make no representations or warranties of any kind, including but not limited to, the warranties of fitness for particular purpose or merchantability, nor are any such representations implied with respect to the material set forth herein, and the publisher and author take no repsonsibility with respect to such material. The publisher and author shall not be liable for any special, consequential or exemplary damages resulting, in whole or in part, from the readers' use of, or reliance upon, this material.

For information, address Delmar Publishers Inc.
2 Computer Drive West, Box 15-015
Albany, New York 12212

COPYRIGHT © 1988
BY DELMAR PUBLISHERS INC.

All rights reserved. No part of this work covered by the copyright hereon may be reproduced or used in any form or by any means—graphic, electronic, or mechanical, including photocopying, recording, taping, or information storage and retrieval systems—without written permission of the publisher.

Printed in the United States of America
Published simultaneously in Canada
By Nelson Canada
A Division of the Thomson Corporation

10 9 8 7 6 5 4 3 2 1

ISBN: 0-8273-4240-3

Contributors

James W. Begun, Ph.D.
Associate Professor
Department of Health Administration
Medical College of Virginia/Virginia
    Commonwealth University
Richmond, Virginia

Dan Fogel, Ph.D.
Director of Executive Education Program
A.B. Freeman School of Business
Tulane University
New Orleans, Louisiana

Myron D. Fottler, Ph.D.
Professor and Director, Ph.D. Program in
    Administration–Health Services
Department of Health Services
    Administration
School of Health-Related Professions
The University of Alabama at Birmingham
Birmingham, Alabama

Lisa Frey
Graduate Student
A.B. Freeman School of Business and
    The School of Public Health and
    Tropical Medicine
Tulane University
New Orleans, Louisiana

Bruce Fried, Ph.D.
Assistant Professor
Department of Health Administration
Faculty of Medicine
University of Toronto
Toronto, Ontario

Cynthia Carter Haddock, Ph.D.
Assistant Professor
Center for Health Services Education
    and Research
Department of Hospital and Health Care
    Administration
St. Louis University
St. Louis, Missouri

S. Robert Hernandez, Dr.P.H.
Associate Professor
Department of Health Services
    Administration
School of Health-Related Professions
The University of Alabama at Birmingham
Birmingham, Alabama

Geoffrey A. Hoare, Ph.D.
Assistant Professor
Department of Health Services
Graduate Program in Health Services
    Administration
University of Washington
Seattle, Washington

Charles L. Joiner, Ph.D.
Professor, Department of Health Services
    Administration
Senior Associate, Dean
School of Health-Related Professions
The University of Alabama at Birmingham
Birmingham, Alabama

Kerma N. Jones
Vice President, Human Resources
Inter-Mountain Health Care Hospital, Inc.
Salt Lake City, Utah

Ralph H. Kilmann, Ph.D.
Professor of Business Administration and
    Director, Program in Corporate
    Culture
Katz Graduate School of Business
University of Pittsburgh
Pittsburgh, Pennsylvania

Richard Kurz, Ph.D.
Associate Professor and Associate Director
Center for Health Services Education
    and Research
Department of Hospital and Health Care
    Administration
St. Louis University
St. Louis, Missouri

Jacqueline Landau, Ph.D.
Assistant Professor of Organizational
    Behavior
A.B. Freeman School of Business
Tulane University
New Orleans, Louisiana

Peggy Leatt, Ph.D.
Professor and Program Director
Health Administration Program
Faculty of Medicine
University of Toronto
Toronto, Ontario
Canada

Ronald C. Lippincott, Ph.D.
Assistant Professor
Department of Government and Public
    Administration
University of Baltimore
Baltimore, Maryland

Anne L. Martin, Ph.D.
Assistant Professor
Department of Health Administration
Duke University
Durham, North Carolina

Alice A. Mercer, Ph.D.
Adjunct Professor
Department of Political Science
Memphis State University
Memphis, Tennessee

Norman Metzger
Edmond A. Guggenheim Professor of
    Health Care Management
Department of Health Care Management
The Mount Sinai Medical Center
New York, New York

Howard L. Smith, Ph.D.
Professor
The Robert O. Anderson School and
    Graduate School of Management
The University of New Mexico
Albuquerque, New Mexico

Foreword

Health services organizations are increasingly stressed institutions in a rapidly changing, competitive environment. In the best of times, which some would characterize as an excess supply of paying clients operating in a reimbursement environment where costs were largely a pass-through item, the management of human resources had almost a soap opera quality about them. Passive organizations tended to operate as a family where the premium was on stability and cooperation. There were consistent upgradings of wage and salary programs, improvement of fringe benefits, and a steady march toward better and more sophisticated technology for optimal patient care. The health services industry added many categories of skilled professionals and gained the reputation of being a good paying, growth industry. More and more could be done for the user of the service.

Aggressive health services organizations were only slightly less paternalistic than the passive organizations. The aggressiveness was driven by the organization's mission. Referral teaching organizations were in continual search of staff talent to improve clinical care programs. Physicians and support staff were recruited in order to protect the referral role. Human resources problems were often played out in terms of internal resources allocation decisions. The aggressive institutions attempted to maintain a broad base of attractive services. A limiting factor was more often a shortage of skilled personnel rather than the ability to fund the program. Chief executives rarely made reputations as "strong human resources managers." Reputations were usually made around the issues of medical staff relations, ability to deal with the lay board, ability to build programs and facilities and the ability to maintain strong operating margins while gaining access to capital.

Higher education has also played a major role in the patterns of staffing health services organizations. The super-specialization and extended training period for physicians has been well publicized. Perhaps less well understood has been the explosive growth of two-year, baccalaureate, and graduate degree programs for associated health professionals.

Health managers no longer were faced with the challenge of providing excellent nursing services to work with the medical staff while organizing a support staff of non-degree workers. Associated health professionals mimicked the medical profession by advocating advanced degree education which soon became very specialized. Professional societies were formed and national associations championed the importance and appropriate functions of their particular members. Chief executive officers tended toward presiding over a loose coalition of skilled employees who perceived themselves as independent contractors. Professional associations and/or unions determined conditions of employment. Each new technology seemed to spawn a new breed of skilled support personnel.

While all of this fell short of Jefferson's lament about professionals being a conspiracy against the public, the development did create a certain sense of unease among managers. Managers did not always understand the roles of the associated health professionals. Job descriptions often originated with professional societies. Even more puzzling was how could the manager intervene to modify professional behavior. Medical specialities were understandably protective of "their technologists." A spate of cost containment efforts did little to moderate the trend. "After all," a statement often echoed around expense reduction efforts, "does it make any sense to cut corners on a service that makes money and enhances the quality of care?" In the absence of stronger countervailing forces to cap costs, the answer was usually, "not really."

The Accrediting Commission on Education of Health Services Administration has recognized the importance of personnel administration and it is a requirement for program accreditation to have a satisfactory amount of curriculum devoted to human resources management. However, it would be difficult to identify a graduate program that has established a national reputation through the development of an excellent human resources concentration. Programs tend to put their efforts on quantitative analysis, problem solving, asset management, public policy, or strategic planning. Students have not been inspired by personnel course offerings and have been known to characterize such efforts as organized common sense. From time to time certain elements emerge with heightened visibility and increased student interest. Negotiations, union management, incentive compensation and physician contracts come to mind as recent "hot topics."

Strategic Management of Human Resources in Health Services Organizations now provides a framework of analysis in which to examine these trends and their impact on the management of human resources in the health care industry. By demonstrating the links that exist among strategic decision making, organizational design and behavior, and human resources management, this book illustrates how hospital administrators and human resources managers can manage their staffs effectively in the current health care environment.

It is interesting to note that 17 of the 19 authors are from an academic setting. Thus, one would expect an analytic and conceptual framework to the problem at hand. What is that problem? What is the importance of the topic to today's manager? How will the insights of this book be of use in day to day management of health services organizations? Only the individual reader can make those judgments. The introductory chapters on strategy and organizational systems make the case for a new approach to the management of human resources.

After laying the groundwork for a new look at human resources through the strategy and organizational systems in Part One, the book examines structural systems, behavioral systems, and human resources process systems. These chapters attempt to address some of the key issues in the post-1982 health management era.

This book identifies the causes of the changes in the health care delivery system since 1982. The health services industry is characterized by professionals over whom the organization has little control; rapidly changing and very expensive technology;

reimbursement by the government at below cost or private managed care at a competitive low margin rate; health organizations which do not know their true costs; patients who are underinsured; patients who are expectant of improved outcomes and are willing to litigate if the failure of an improved outcome appears linked to an omission or act of commission on the part of the provider; and a surplus of providers in an intensely competitive and heavily regulated market.

All things considered, these are not the characteristics of a business one would elect to start in 1988. The investment community has taken note of this state of affairs and is demanding more evidence of project success while building in more downside protections. As the demand for certainty about investment opportunities increases, the managers project an industry with greater risk and less certainty. Managers at the operating level rarely have the luxury of relying on hard data and a long track record. What happened prior to 1982 is of little relevance to 1988 investment decisions. Staff to occupied bed ratios, operating margins and ability to carry debt are at the heart of the daily conclusions and decisions. The increased volatility of health care organizations have focused new attention on the driving forces of organizational behavior. Today's health care manager must regard every element of cost as a mortal enemy of the organization's well being.

It is in this context that the whole issue of human resources management needs to be reconsidered. It is no longer satisfactory for the human resources manager to return from periodic meetings of like professionals with advice on what needs to be done to keep up with neighboring, collegial institutions. Human resources management must challenge past assumptions and look at new ways to increase performance among existing staff. The emphasis should be on quantification of every meaningful ratio and indicator. Continuous monitoring of work load/staffing ratios is essential in a tightly managed health services organization. Leaner work forces may well reduce costly administrative overhead. Consideration should be given to combining departments and continuing the trend toward organization around the needs of the patient. Product line management holds great promise. The human resources manager should feel accountable for each position and actively search for reduction techniques. Size is no longer a sign of prestige. Rather, size is a function of the market the institution is able to serve. That some organizations will fail or need to combine seems a given. This book provides a tool to begin or continue the reevaluation process.

There is always a downside to this painful sort of self-examination. The same individuals who cut corners to inflate staffs may use the business jargon of the day to foolishly trim staff in an insensitive manner. Professionals who are relatively untouched by the economics of health cannot be expected to warmly endorse an approach that will make their lives less pleasant. Unless the problem is accurately defined and understood by all parties concerned, it is unlikely that any conclusions and solutions will be supported by the key parties in the organization. Many health professionals chose their careers because of the desire to participate in the diagnostic, therapeutic, and caring process. Hospital staffs neither requested nor participated in the current approach to health care financing. There are mortgage payments to meet and college educations to worry about. Human resources policy adjustments have the potential to be career-threatening to the individual. Health care organizations are fragile coalitions of many talented individuals. It is of little comfort to the individual that the entire industry is in turmoil and facing the same unpleasant forces.

Thus human resources management must retain those elements that have stitched together the wonderful health care organizations that exist today. To continue serving the public with a quality service requires a thoughtful reshaping of our human resources system. Reading this book is the first important step that health care managers can take in this direction.

JOHN H. WESTERMAN

President and Chief Executive Officer
The Hospital of the Good Samaritan
Los Angeles, California

Preface

Several competing management issues and philosophies seem to be working at cross purposes within the health services industry in recent years. On the one hand, there has been growing interest in increasing the participation of employees in decision making processes and improving their level of psychological involvement and commitment with the organization. This interest is evidenced by the popularity of management texts on so-called Japanese management, the implementation of quality circles in numerous hospitals, and the discussions of the value of corporate culture in assisting a health services organization in its quest for performance. These techniques are envisioned as assisting the organization in its quest for high performance.

Conversely, financial and economic pressures on the organization have caused these same institutions to become concerned about maintaining market share, improving their competitiveness, achieving internal efficiencies, and maximizing reimbursement. The drive for financial performance often forces health institutions to reduce their workforce through terminations, to curtail spending on employee development, to postpone compensation increases, and to make other short-term decisions that are not conducive to a quality work life for their employees.

Given this paradoxical situation, we feel there is need for an up-to-date text on the management of human resources within health services organizations and *Strategic Management of Human Resources in Health Services Organizations* is an attempt to address this important gap in the health services management literature. We hope that the book makes a unique contribution by articulating the links that exist among strategy, organizational design and behavior, and human resources management. We believe that a need exists for this articulation, given the labor intensive nature of the industry and the emphasis being placed upon productivity in the increasingly competitive environment.

The relationship between strategy and selected organizational systems is described in the first part. Our view of the relationships that exist among strategy, organizational design and behavior, and human resources management are illustrated in the model that is presented in Chapter One. Subsequent chapters describe processes for formation of strategy within health services organizations, the relationship that should exist between strategy and human resources management, and methods for strategic management of personnel at various stages of the organizational life cycle.

The influence of these strategic choices upon structural systems are explicated in Part Two. Then, the effect of these systems upon human resources management activities are detailed. The third part contains descriptions of the contributions that behavioral systems make to individual and organizational performance. Finally, Part Four provides information on the performance of selected human resources functions such as recruitment, training and development, compensation management, and negotiating and administering the labor relations contract.

Our purpose is to describe not only human resources functions within organizations but also to provide a model of major organizational components that shape the human resources options available for health services managers. We hope that this approach will be an important contribution to our industry and will be of interest to practitioners as well as students in health services administration programs who are interested in expanding their understanding of strategic planning, human resources management, and senior management within hospitals, hospital systems, and other health services delivery organizations.

Given the diverse topics covered in our text, we feel that the contributing authors provided the expertise that was necessary for this undertaking. The contributors' expertise in the prescribed areas was essential for the achievement of the objectives of the book.

MYRON D. FOTTLER
S. ROBERT HERNANDEZ
CHARLES L. JOINER

Acknowledgments

We would like to thank a number of persons who contributed to the development of this book. Steve O'Conner, Bill Reddick, Patrick Sobezak, and Mei-Ling Tseng, Ph.D. candidates in Health Services Administration at University of Alabama at Birmingham, provided valuable research assistance as did Kimberly McCarthy and David Gray, masters students in Health Administration at UAB. Contributions and valuable assistance also were supplied by Katherine Howard, Jeri Beck, Sueann Middleton, and William Beavers.

Linda Godwin, Gayla Hulslander, and Patricia Washington provided typing of various drafts of the chapter manuscripts, administrative support, and assistance for the project. Without their help our other responsibilities could not have been fulfilled.

Steve Williams, Series Editor of the Delmar Series in Health Services, was extremely helpful in seeing that this manuscript was initiated, after the idea was first proposed at an American Public Health Association meeting long ago.

A note of thanks to our wives, Carol, Joy, and Gloria, for support of this research and publication endeavor. Dr. Hernandez would like to express special thanks to his children, Susan and Robert, for their understanding and patience with the time required for this activity. Dr. Joiner desires to acknowledge his girls, Amy, Rebecca, Laura, and Ashley, for the important place they fill in his life and to express appreciation for their understanding and interest in this project.

Finally, we wish to thank administrative officials at the University of Alabama at Birmingham for providing a conducive work environment that allowed this book to be completed.

MYRON D. FOTTLER
S. ROBERT HERNANDEZ
CHARLES L. JOINER

Contents

PART ONE STRATEGY AND ORGANIZATIONAL SYSTEMS   1

1 **Strategic Management of Human Resources in Health Services Organizations**   3
 S. Robert Hernandez, Myron D. Fottler, and Charles L. Joiner

 Effect of Environment on Health Practitioners 4
 Strategic Management of Human Resources 6
 Organizations and Strategy 8
 Structural Systems 8
 Behavioral Systems 10
 Human Resources Process Systems 11
 Outcomes 13
 Summary 18

2 **Formulating Organizational Strategy**   20
 S. Robert Hernandez

 Planning Methods 20
 Organizational Strategy 25
 Formulating the Organizational Strategy 29
 Organizational Strategy and Human Resources Management 42
 Summary 47

3 **The Organizational Life Cycle and Strategic Human Resources Management**   51
 Myron D. Fottler and Howard L. Smith

 Organizational Life Cycles 51
 Reinstituting Growth 55
 Organizational Decline 58
 Strategic and Tactical Responses 65

Constraints in the Management of Decline 70
Human Resources Issues and the Organizational Life Cycle 72
Summary 79

PART TWO STRUCTURAL SYSTEMS — 83

4 Technology and Human Resources Management — 85
Peggy Leatt and Bruce Fried

The Theory of Technology in Organizations 87
The Influence of Technology on the Human Resources Functions 93
Guidelines for Assessing the Potential Impact of Technology on Human Resources Management 107
Summary 108

5 Structure for Human Resources Management — 112
James W. Begun and Ronald C. Lippincott

Conceptualizing Organizational Structure 113
Influence of Structure on Human Resources Management 116
Summary 137

6 Information Systems for Human Resources Management — 141
Anne L. Martin

Definitions 141
The History of Hospital and Human Resources Information Systems 142
A Model of Human Resources Information Systems 149
The Role of Human Resources Information Systems 152
Functions of Human Resources Information Systems 157
Implementation 165
Special Issues 174
Summary 176

PART THREE BEHAVIORAL SYSTEMS — 179

7 Commitment and Motivation of Professionals — 181
Alice Atkins Mercer

Professionals in Health Services Organizations 182
Commitment 190
Motivation 199
Summary 204

8	**Management of Corporate Culture** *Ralph H. Kilmann*	206

What Are Adaptive Cultures? 208
How Do Cultures Form? 209
How Are Cultures Maintained? 211
Assessing and Changing Cultural Norms 212
Summary 218

9	**Organizational Change, Transformational Leadership, and Leadership Development** *Richard Kurz, S. Robert Hernandez, and Cynthia Carter Haddock*	219

Organizational Change 220
Leadership and Change 221
Leadership Development 231
Summary 234

	PART FOUR HUMAN RESOURCES PROCESS SYSTEMS	239
10	**Recruitment** *Jacqueline Landau and Geoffrey A. Hoare*	241

New Recruiting Needs 241
The Recruitment Process 242
Recruitment Planning 243
How? Implementing Recruiting Plans 256
Evaluating the Recruitment Process 261
Summary 262

11	**Selection and Placement** *Jacqueline Landau, Dan Fogel, and Lisa Frey*	267

Designing the Selection Program 268
The Selection Process and Federal Regulations 275
Selection Instruments 276
Determining the Utility of the Selection Process 288
Summary 290

12	**Training and Development** *Howard L. Smith and Myron D. Fottler*	294

Definitions, Goals, and the Importance of Employee Development 295
Incentives for Upgrading Staff Development Programs 305

Components of Effective Staff Development Programs 308
Trends in Staff Development 313
A Strategic Posture 315
Summary 316

13 Performance Appraisal 319
Charles L. Joiner

Historical Development 321
Common Performance Appraisal Methods 322
Evaluation of Performance Appraisal 324
Strategic Role of Performance Appraisal 333
Management by Objectives (MBO): A Strategic Appraisal System 336
Extrinsic Rewards and Job Performance 344
Summary 345

14 Compensation Management 348
Kerma N. Jones and Charles L. Joiner

Strategic Planning and Compensation 349
Base Pay 356
Incentive Compensation Programs 356
Benefits 362
Summary 368

15 Preventive Labor-Management Relations 370
Charles L. Joiner

Developing an Employee Relations Philosophy and Strategy 371
Maintaining Nonunion Status 371
Labor Law History and Trends 372
Causes of Labor-Management Problems 373
A Preventive Management Program 378
Management Strategy for Reactions During Union Organizing Campaigns 383
Summary 384

16 Negotiating and Administering the Labor Relations Contract 386
Norman Metzger

Legal Definition of Bargaining 387
Mandatory Bargaining Subjects 388
Voluntary Bargaining Subjects 389
Important Considerations That Affect the Bargaining Milieu 390
Selecting a Negotiating Team 391
Strategies for Bargaining 392

The 1974 Health Care Amendments 395
Key Contract Clauses 397
Administering the Contract 402
Summary 407

17 New Developments in Human Resources Management: A Summary for the Future   409
Myron D. Fottler and S. Robert Hernandez

Environmental Change 409
Human Resources Implications 411
Integration of Strategic and Human Resources Planning 414
Enhancement of Employee Productivity 421
Multiskilled Health Practitioners 422
Summary 423

Appendix   Checklist for a Collective Bargaining Contract   426

Index   437

*STRATEGIC
MANAGEMENT OF
HUMAN RESOURCES
IN
HEALTH SERVICES
ORGANIZATIONS*

PART ONE
Strategy and Organizational Systems

CHAPTER 1

Strategic Management of Human Resources in Health Services Organizations

S. Robert Hernandez
Myron D. Fottler
Charles L. Joiner

Like most other service industries, the health care industry is very labor intensive. One reason for the reliance on an extensive work force is that it is not possible to produce a "service" and store it for subsequent consumption. The manufacture of the commodity that is purchased and the consumption of that commodity occur simultaneously. Thus, the interaction between consumers and health care professionals is an integral part of the provision of health services. Given the dependence on health care professionals for service delivery, the possibility of heterogeneity of service quality must be recognized, both within an employee as skills and competencies change over time and among employees as different individuals or representatives of different professions provide a service.

The intensive use of labor for service delivery and the possibility of variability in professional practice require that the attention of leaders in the industry be directed toward managing the performance of the persons involved in the delivery of health services. The effective management of people requires that health services executives understand the factors that influence the performance of individuals employed in their organizations. These factors include not only the traditional human resources management activities (i.e., recruitment and selection, training and development, appraisal, compensation, employee relations), but also environmental and other organizational factors that impinge on human resources activities.

The strategic management of human resources is ensuring that qualified personnel are available to staff the portfolio of business units that will be operated by the *organization*. This book explains and illustrates the methods and practices that can be used to increase the probability that competent personnel will be available to provide the services delivered by the organization and that these personnel will perform necessary tasks appropriately. Implementing these methods and practices means that requirements for positions must be determined, qualified persons must be recruited and selected, employees must be trained and developed to meet future organizational needs, and adequate rewards must be provided to attract and retain top performers.

Of course, these functions are performed within the context of the overall activities of the organization. They are influenced or constrained by the environment, the mission and strategies that are being pursued, the structure of the organization, and the behavioral systems indigenous to the institution. *To manage human resources strategically, health care executives must understand the relationships that exist among these important organizational components and the human resources functions so that appropriate methods can be selected to accomplish the objectives in service delivery desired by the organization.*

The next section presents an overview of fundamental changes occurring in the health services environment that affect the numbers, types, and roles of health practitioners. This material is followed by a model of the relationships that exist among strategy, selected organizational design features, and human resources management activities.

EFFECT OF ENVIRONMENT ON HEALTH PRACTITIONERS

Table 1.1 provides an overview of the estimated supply of selected health personnel from 1970 to 1984. Few industries have the diversity of personnel and the wide variation in educational preparation, technical skills, professional responsibilities, and professional values seen in health services. Preparation ranges from 6 to 8 weeks on-the-job training for nursing assistants to more than 10 years of postbaccalaureate education for some medical specialties. Health practitioners share a common duty to provide services to consumers, but diversity exists in the responsibilities of these groups.

The numbers of practitioners and their roles have evolved over time and are expected to undergo radical change because of three factors that have transformed the health services industry. These factors are scientific and technological change, patterns of utilization, and funding for health services.

Advances in scientific understanding and technology in health services have triggered significant change in the industry and in the responsibilities of practitioners. Individuals have become more specialized, since one person or profession cannot possess all the knowledge in a specific area of medicine. This has caused not only specialization of medical practice but also the creation of new types of health practitioners with up-to-date and unique skills to staff new technologies.

The evolution of medical knowledge, concentration of technological capability in the hospital, and improved reimbursement have led to increases in the absolute number of employees in nonfederal short-term general hospitals from 662,000 in 1950 to 3,110,000 in 1982 and in the relative number of hospital employees per patient day from 1.78 in 1950 to 4.08 in 1982 (American Hospital Association, 1986). Recent

TABLE 1.1. Estimated Active Supply of Selected Health Personnel and Practitioner-to-Population Ratios, 1970, 1975, 1980, and 1984

Health Occupation	Estimated Active Supply				Percent Change		
	1970	1975	1980	1984	1970–1984	1975–1984	1980–1984
Physicians	326,200	384,500	457,500	501,200[a]	53.6	30.4	9.6
Allopathic (MD)	314,200	370,400	440,400	481,500[a]	53.2	30.0	9.3
Osteopathic (DO)	12,000	14,000	17,140	19,700[a]	64.2	39.7	14.9
Podiatrists	7,100	7,300	8,900	9,700[a]	36.6	32.9	9.0
Dentists	102,220	112,020	126,240	137,950	35.0	23.1	9.3
Optometrists	18,400	19,900	22,400	23,600	28.3	18.6	5.4
Pharmacists	113,700	122,800	143,800	157,000	38.1	27.9	9.2
Veterinarians	25,900	31,100	36,000	42,600	64.5	37.0	18.3
Registered nurses	750,000	961,000	1,272,900	1,453,000	93.8	51.3	14.2
	Practitioners per 100,000 Population						
Physicians	156.0	174.4	197.0	210.7[a]	35.1	20.8	7.0
Allopathic (DO)	150.0	167.9	189.5	202.2[a]	34.8	20.4	6.7
Osteopathic (DO)	6.0	6.5	7.5	8.5[a]	41.7	30.8	13.3
Podiatrists	3.5	3.4	4.0	4.2[a]	20.0	23.5	5.0
Dentists	49.5	51.6	55.2	58.0	17.2	12.4	5.1
Optometrists	8.9	9.2	9.8	9.9	11.2	7.6	1.0
Pharmacists	54.4	56.6	63.0	66.0	21.3	16.6	4.8
Veterinarians	12.5	14.3	15.8	18.0	44.0	25.9	13.9
Registered nurses	366	449	560	613	67.5	36.5	9.5

SOURCE: Data from the appropriate table in the individual chapters for the respective occupations courtesy of the U.S. Department of Health and Human Services, *Fifth report to the President and Congress on the status of health personnel in the United States*, DHHS Pub. No. (HRS-P-OD-86-1). Hyattsville, MD: U.S. Department of Health and Human Services, March 1986.

[a] 1983 data.

changes in utilization of services have occurred because of a number of factors described elsewhere (Aday and Shortell, 1988). Hospital admissions and patient days per 1000 population appear to have peaked in the 1970s after years of increase. The absolute number of hospital admissions and the average daily census for nonfederal short-term general hospitals have declined every year from 1981 to 1985 (American Hospital Association, 1986).

The number of physician visits for all sources and places has fluctuated from 4.6 per person in 1964 to 4.9 in 1976, returning to 4.6 in 1981 (U.S. Department of Health and Human Services, 1983, p. 137). From 1982 to 1983 the net income of surgeons increased 10 percent while net income of general/family practitioners declined. The expected physician increase to nearly 700,000 by the year 2000 is expected to foster increased competition in medical practice (U.S. Department of Health and Human Services, 1986).

Increasing emphasis has been placed on cost reductions and control of expenditures by major purchasers of health services. Since October 1, 1983, 5200 hospitals have been operating under the prospective payment system (PPS) for Medicare-eligible patients. This system, numerous other organized arrangements that encourage efficient use of health services, increased competition, and industry maturity have led to changes in job opportunities and entry of different types of individuals into training for the health professions. Employment in nonfederal short-term general hospitals dropped 3.4 percent from 1982 to 1985 (American Hospital Association, 1986). Within most health disciplines, enrollment in educational programs has either leveled off or declined in recent years (U.S. Department of Health and Human Services, 1986).

Certainly, more extensive discussions are available on the health services industry (Arthur Andersen, 1985; Starr, 1982; Williams and Torrens, 1988); the health professions labor force (Sorkin, 1977; U.S. Department of Health and Human Services, 1986), and manpower planning (Edwards et al., 1983). This brief review suggests that the labor force expanded and became specialized over the past 40 years in response to increased utilization of services. Recent emphasis on cost control and efficient operations appears to be having a negative effect on demand for services, employment opportunities, and number of persons seeking career preparation in the health professions.

Environmental shifts that influence demand for services and trends in the health labor force must be considered during human resources planning. It cannot be assumed that skilled personnel will be freely available in the labor market when services are started or expanded; thus, adequate time for recruitment and training must be allowed. When services are downsized or terminated, use of tactics such as attrition or employee retraining may reduce the probability that layoffs and terminations will be required. Attention to external trends and proper planning provides for the maintenance of the organizational staff at near optimum levels of size, skills, and experience. It also reduces problems from personnel shortages or surpluses such as reductions in quality, inefficiencies, inability to deliver services, and practitioner dissatisfaction.

STRATEGIC MANAGEMENT OF HUMAN RESOURCES

An organization that is managed strategically ensures that the functional and operational administrative systems are linked to the strategic and tactical decision-making activities of the firm. The planning, control, and management systems must be joined for the organization to be able to ensure that the plans developed by senior management are executed as intended.

As illustrated in Figure 1.1, organizations are complex entities that require constant interaction with the environment. If they are to remain viable, health services organizations must adapt their strategies to external changes. The internal components of the organization are then affected by these changes in that shifts in the organization's strategy potentially necessitate modifications in the internal structural systems, behavioral systems, and human resources process systems. There must be harmony, in turn, among these systems. The characteristics, performance levels, and the amount of coherence in operating practices among these systems influence the outcomes achieved in terms of organizational-level and employee-level measures of performance.

The strategic management of human resources involves attention to the effect of environmental and internal components on the human resources process system. Because of the critical role of health professionals in delivering services in this labor-intensive industry, a major concern of health services managers should be the development of personnel policies and practices that are closely related to, influenced by, and supportive of the strategic thrust of their organizations. In addition, managers must ensure that the human resources functions are linked to the other internal design

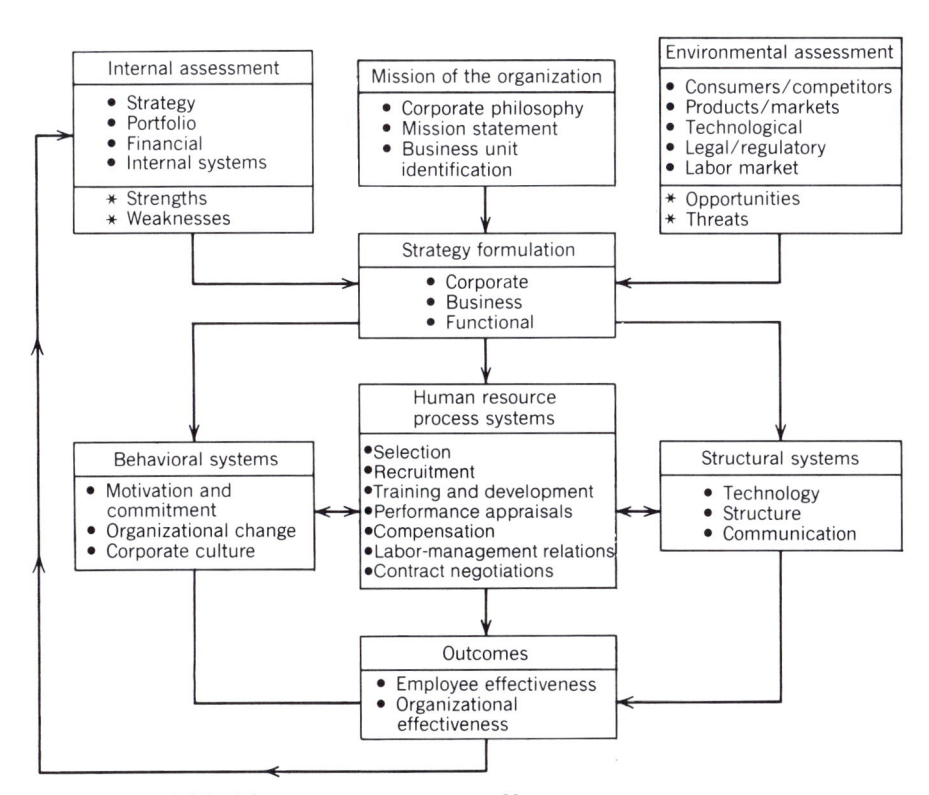

Figure 1.1. Model of the strategic management of human resources.

features of the organization. The sections that follow discuss these relationships in more detail.

ORGANIZATIONS AND STRATEGY

Organizations, either explicitly or implicitly, pursue a strategy in their operations. Deciding on a strategy means to determine the products or services that will be produced and the markets to which the chosen services will be offered. Once these selections have been made, the methods to be used to compete in the chosen market must be identified. The methods adopted are based on internal resources available, or potentially available, for use by managers.

As illustrated in Figure 1.1, strategies should be based on consideration of environmental conditions and organizational capabilities. To be in position to take advantage of opportunities that are anticipated to occur, as well as to parry potential threats from changed conditions or competitor initiatives, managers must have detailed knowledge of the current and future operating environment. Cognizance of internal strengths and weaknesses allows management to develop plans based on an accurate assessment of the firm's ability to perform in the marketplace at the desired level. A more detailed discussion of approaches for determining organization strategy is presented in Chapter 2.

STRUCTURAL SYSTEMS

The structure of an organization has a profound effect on the activities of employees. Structural concerns include deciding how task responsibilities will be allocated among employees and determining how coordination among the tasks will be achieved (Mintzberg, 1979), and these involve two features. One is the formal allocation of work roles to individuals. The second is use of administrative mechanisms to control and integrate work activity.

Implementation of strategies occurs through specific programs or services by the assignment of work to individuals, the control of worker activities so that standards of performance are achieved, and the coordination of tasks among workers for efficient operation. Thus, organizational structure should be determined by the strategy being pursued by the organization, since different strategies require work routines, control methods, and coordination patterns of distinctive types.

The several components of an organization's structure have a strong influence on human resources systems. One of these closely associated with task responsibilities of personnel is the technology applied by the institution. Another element of the structure used to control and coordinate work within or across units is the information systems that are required to assemble and organize data on human resources. A third component consists of the more traditional structural attributes of an institution such as formalization of roles and behavior, centralization of decision making, and complexity of the institution.

Technology Within Health Services Organizations

The numerous methods of conceptualizing technology will be discussed later in the text. In general, technology is conceived to include not only the equipment, supplies, and other physical materials required to provide a given service, but also the *knowledge* that must be used in the provision of the service or in the operation of the equipment.

Organizational strategy has significant implications for the technology that will be available to a health services organization. As discussed in Chapter 4, managers exercise strategic choice in determining the product lines their organizations will offer. The decision to provide selected medical services or products and to exclude others prescribes the medical technology that an institution must use. In addition, the strategic alternatives available to management in the short run are governed by the organization's current technology, including the knowledge level of its personnel.

Once a strategy, and therefore a technological thrust, has been chosen by an organization, it has an immediate effect on personnel practices of the institution. Since technology also includes the knowledge base that is required to deliver a service, an entire array of activities such as job design, human resources training and development activities, performance appraisal and compensation systems, and related activities must be orchestrated to ensure that service delivery occurs as planned. The relationships between organizational technology and human resource functions are presented in more detail in Chapter 4.

Structure for Human Resources Management

As previously stated, the methods used to allocate tasks within an organization and to achieve coordination among the tasks are structural concerns. Different strategies (e.g., markets served, methods for serving those markets) lead to different structures. Numerous authors (Chandler, 1962; Galbraith and Nathanson, 1978) have noted the changes that occur in structural features of organizations as firms attempt to serve different markets or to use new methods and practices in the strategies that they implement to gain a competitive edge.

The internal design or structure of an organization has a profound influence on the human resources systems. In turn, job design is an aspect of personnel practices that is influenced directly by structural decisions. Levels of formalization, internal complexity, and the amount of central authority used by organizations determine the appropriate design of work in the organization. Other personnel functions such as recruitment strategies, approaches to management development, and related activities also are affected by internal organizational design. The exact relationships among these factors are discussed in Chapter 5.

Information Systems

Appropriately designed management information systems provide data required to support strategic planning and management decision making. Information concerning human resources that can be used for planning as well as operational purposes is a crucial element of such a system. External planning depends on accurate data on future numbers of professionals that will be available for the jobs required by the organization. Internal planning requires significant detail on such matters as current productivity, employee skills, work demands, and related areas. Information systems also should be designed to ensure that managers have access to the data required to support the strategies the organization is pursuing.

Human resources information systems support personnel activities by providing managers with statistics on the future need for personnel as well as information on current levels of performance and skills of employees. This information can be used to identify types of personnel to be recruited in the future and to design compensation systems and related activities. The contributions of information systems to personnel activities are discussed in Chapter 6.

BEHAVIORAL SYSTEMS

The actual delivery of health services, the provision of most support functions, the development of plans, and the coordination and control of work occur through the behavior of people and the processes by which they interact. Thus, the ability to achieve an organizational strategy depends on the nexus between organizational behavioral systems and the mission and objectives of the institution.

The actions and behaviors of individuals within the organization constitute a major focus of behavioral systems. If management is sensitive to the personal and professional requisites of these individuals, their needs and expectations from participation in the organization may be translated into commitment to the organization and motivation to accomplish tasks desired by the institution.

Additional emphasis of behavioral systems is on group attributes and the quality of leadership available for guiding the institution through major change. Another significant element of behavioral systems is the corporate culture, that is, the characteristic day-to-day internal environment that is experienced and shared by those working within the organization. The approaches to leadership and organizational change are a third component of organizational behavioral systems that must be handled properly and linked to the mission of the firm if human resources are to be directed strategically by management.

Professional Commitment and Motivation

The essential building block of the health services industry is the quality and nature of that patient-professional interaction. Despite the existence of a number of standardized practices and protocols for delivery of some services, much professional behavior is governed by unwritten norms and reliance on individual judgment. Nurses and other professional groups perform tasks based on training received outside the hospital. Moreover, many professionals, such as medical staff physicians, are not employees of the institution but are autonomous individuals who are not regulated by the normal control structures that apply to other classifications of workers.

Thus, the ability of health services managers to obtain the psychological involvement of professional employees and the medical staff with the mission, goals, and objectives of the organization will be an important determinant of the success that the organization will enjoy. Chapter 7 discusses the commitment and motivation of professionals.

Additionally, a person's performance is viewed as a function of ability times motivation (Lawler, 1973). The behavioral outcomes desired by managers will be influenced by the desire of the employees and medical staff to perform the tasks required by the organization. Understanding the influence of commitment and motivation on individual behavior will assist a manager in discerning the range of actions that are available.

Corporate Culture

The system of informal customs and rules that transmit the behaviors expected of employees in most situations constitutes organizational culture. The quality of the relationship between the culture of an organization and its strategy has significant implications for the potential performance of that organization. Strong cultures do an excellent job of transmitting the values and beliefs of the organization to the individual. Thus, individuals are able to determine "what they are to do" in the vast number of situations for which no formal rules or guidelines exist.

Corporate culture should complement the human resources management process systems. The shared values and expectations that are reinforced by culture should blend with the systems that have been established for selecting organizational members, developing their skills, and rewarding desirable behavior. It is important that management assess cultural norms and plan interventions as required. Methods for managing the corporate culture are discussed in Chapter 8.

Organizational Change, Transformational Leadership, and Leadership Development

The importance of effective leadership in determining the overall mission and direction of an organization and guiding the institution through major change cannot be overstated. Whereas contingency theory posits that appropriate organizational strategy and design are determined by environmental conditions, senior management must be cognizant of those environmental conditions and determine the response the organization should initiate. In addition, the implementation of organizational transformation and change is the responsibility of senior executives.

To survive, organizations must constantly adapt. This is demonstrated by the radical changes occurring in the health services industry over the past 5 to 10 years. Managers must provide leadership to enable their organizations to make the transformations necessary in a changing environment. In Chapter 9, the approaches that are available for renewing the organization and for developing future leaders are discussed.

HUMAN RESOURCES PROCESS SYSTEMS

The processes used for human resources management may be viewed as a cycle of related activities. Employee recruitment, selection and placement, training and development, performance appraisal, compensation, and labor relations may be envisioned as a continuum of tasks and responsibilities that flow logically from one to another. ==These functions must be performed to ensure that the necessary human resources are available to support the strategic thrust of the organization.==

Although each of these areas is important and must be performed competently for the organization to function, the importance of the activity to sustaining excellence varies depending on the strategies chosen. Some strategic initiatives emphasize selection and placement activities, while others require health services managers to concentrate on refining the appraisal system. The relationships of selected strategic activities to human resources functions within the organization are discussed in Chapter 2, and the variance of these relationships over the organizational life cycle is further detailed in Chapter 3. The association between other organizational activities and human resources functions as well as the contribution of these functions to organizational performance are discussed in the sections that follow.

Recruitment

Health services organizations must have a constant influx of candidates for potential employment. New employee positions are required as market areas are expanded or services are initiated. Recruitment occurs even in the face of limited growth or decline in service capacity, because individuals with specialized skills or training who leave the organization must be replaced and because services or technologies that have been revised or modified must be staffed.

The recruitment of personnel plays an important role in helping the organization to adapt and remain competitive. Employees who have recently finished professional training are the source of information on new methods and techniques in service delivery that allow the organization to remain competitive in its traditional services. In addition, it may be necessary to recruit outside the institution if personnel with the managerial or professional skills needed to implement new strategic thrusts are not available internally.

As discussed earlier, the health services industry employs a wide variety of workers. Thus, the sources of applicants and types of method used to expand the applicant pool vary depending on the occupational classification being considered. A description of recruitment methods is provided in Chapter 10.

Selection and Placement

A responsibility highly correlated with recruitment is employee selection and placement. Once the applicant pool has been filled, methods must be in place to choose the persons who will be most qualified for each job. Successful implementation of organizational strategy requires that management have a thorough understanding of the jobs that must be performed and the qualifications of individuals to fill these jobs.

Selection of competent personnel for new or unique positions created by initiation of services may be problematic. For example, an institution initiating an innovative, nontraditional service may have difficulty identifying the activities that must be performed by individuals to be employed. In addition, it may have little experience in evaluating candidates for new positions.

A number of steps are critical in the selection and placement process. Job analysis must be conducted to determine the tasks to be performed and the qualifications required to perform them. Then, criteria for predicting employee effectiveness are required, with adequate attention given to validity and reliability of the selection instrument. These activities are examined in Chapter 11.

Training and Development

Investment in the existing human capital of a health services organization through a well-managed training and development activity offers potential for significantly enhancing the ability of the enterprise to achieve its objectives. Indeed, improvement in the skills and abilities of current employees will contribute to sustained levels of performance because the technological change occurring in the health services industry requires the constant updating of the knowledge base of health professionals. The increased competency of the health services organization staff will provide significant benefits to the enterprise no matter what strategy is being pursued.

The changing environment of the health services industry itself ensures that the training and development of current staff members will contribute to organizational performance. Institutions are required to develop innovative responses to the competitive reformations taking place today. The providers of health services within organizations need to be informed of the factors causing these changes, as well as the role they can potentially play in helping management cope with change. Information on competitive shifts and marketing alternatives conveyed to current employees and medical staff will help these professionals to understand the rationale for fundamental changes that are occurring. An improved understanding should make it easier to obtain their support and advice on implementation issues.

A third factor relating the strategic thrust of the enterprise to staff training and development is that the initiation of unique strategies and tactics may call for programs to be implemented by current employees who do not possess the necessary skills. For example, strategies to become a low-cost provider (Porter, 1980) may dictate the use of internal management control methods that are foreign to supervisory staff. Management must ensure that plans for strategic initiatives include an inventory of the skills required for program implementation and must design necessary developmental activities. These and other issues in training and development are discussed in Chapter 12.

Performance Appraisal

Performance appraisal is the systematic evaluation of an employee's work behavior on criteria measuring important job-related activities. To determine the extent to which work requirements and responsibilities are met, valid, reliable criteria must be developed for a job, and job behaviors must be measured.

Performance appraisal can serve multiple purposes for management. First, it can provide guidance for the selection of individuals for promotion. When advancement criteria have been determined, employees who meet or exceed those criteria can be identified as promotion opportunities arise. Second, it can be used as a method for determining increases in employee compensation. Levels of employee performance can be measured and comparisons can be made to improve the probability of equitable rewards being given for desired performance. Third, performance appraisal can serve as a tool for identifying areas in which personnel need training and development or additional counseling. Areas in which one or a number of employees exhibit consistently inadequate performance can be identified and remedial programs to improve performance developed.

If properly designed, an appraisal system complements the strategic planning of an organization because it translates the initiatives desired by the organization into specific behaviors required of individuals. Management must identify the key activities necessary for the achievement of the institutional mission and translate those activities into employee appraisal criteria. Performance appraisal methods are discussed in more detail in Chapter 13.

Compensation Management

The compensation system of an organization can influence the strategic direction of an organization and organizational effectiveness in a number of ways as identified by Lawler (1984). First, an enterprise's ability to attract and retain conscientious workers is partially determined by the kind and level of rewards provided. Second, a compensation system may foster the achievement of desired outcomes by motivating employees if a firm can link valued rewards to the performance of essential behaviors. Third, compensation systems contribute to the corporate culture that is perceived by organizational members as well as serving to reinforce the structural systems of the organization.

The linking of compensation systems to the mission and strategies of the organization occurs through several vehicles. An assessment of the current compensation systems and their influence on behaviors should be an input to strategy development. Once a strategy has been generated, the firm must focus on the type of human resources needed and the behaviors required to make the plan effective. Then, compensation systems should be designed to reward the behaviors necessary for obtaining desired outcomes. The coupling of compensation to the behaviors required to support diverse strategic thrusts is a major component of strategic human resources management.

A difficulty health services managers face in developing compensation systems is the heterogeneity of personnel employed within an institution. This heterogeneity suggests that unique methods for compensating different types of employees may need to be developed. Variability also must be faced by larger health services delivery systems across facility sites or, as the merger of larger systems occurs, across systems in compensation programs. Such diversity means that managers must be sensitive to the need for blending these potentially different approaches for employee remuneration into a logical framework that is supportive of the mission of the organization. The methods for development of compensation packages for rewarding the diverse activities of health services organizations are discussed in Chapter 14.

Labor Relations

Senior managers seek to retain flexibility in selecting among alternatives when determining organizational strategy. Preservation of administrative flexibility

(Thompson, 1967) is critical for long-term adaptability and survival for health services organizations, especially in today's volatile environment. Thus, development of a positive relationship between management and employees by practicing "preventive health" in labor-management relations is important for strategic management of human resources. This is because if groups of employees feel that management is not interested in their welfare, they may elect to have a union represent them and bargain for them.

Nonunion status allows management greater flexibility in exercising its prerogatives. Loss of flexibility accompanies union status when provisions of the labor-management contract serve as a barrier to the implementation of programs or options that are being considered by management.

For example, a hospital may have agreed with a union that workers whose jobs are terminated will be shifted to other comparable positions or that present workers will be hired first if new jobs are created within their job classification. If such a facility desires to cease offering obstetric services while initiating new inpatient services for psychiatric patients, there may be an impasse. It will be almost impossible to shift personnel from obstetrics to psychiatry without significant retraining, and the obvious alternative (i.e., for the hospital to hire new personnel for psychiatric services) may be blocked, at least temporarily, by the terms of the union agreement. Thus, the change in services desired by the hospital cannot be implemented without major costs and delays. Discussion of methods to create a positive human relations climate in the hospital is provided in Chapter 15.

Health services organizations that already have unions should consider the future strategic thrusts they wish to follow when they negotiate new collective bargaining agreements. Otherwise, an agreement may serve as an internal constraint that inhibits the performance of the organization. If the institution decides to follow a low cost strategy in the marketplace, the management team must be tough negotiators, granting few wage demands, holding the line on numbers of workers, and stressing productivity standards in an attempt to restrain costs.

Conversely, the negotiations of an organization wishing to compete on the basis of service quality should reflect that desire. Management should bargain to gain concessions that focus on quality of care such as use of employee assessment criteria that stress job performance and quality of care. Bonuses may be paid to employees for quality enhancement activities. The organization may emphasize funding for training and development for job improvement rather than providing larger pay increases. Less stress may be placed on reducing staffing ratios or controlling overtime pay.

These factors suggest the importance of integrating strategic considerations into the handling of labor relations and negotiating collective bargaining agreements. Approaches to labor relations and negotiations are discussed in more detail in Chapters 15 and 16.

OUTCOMES

The outcomes achieved by a health services organization depend on the environment of the organization, the mission and strategies being pursued, the internal structural systems, the behavioral systems, the human resources process systems, and the consistency of operating practices across these internal systems. The appropriate methods for organizing and relating these elements are determined by the specific outcomes desired by the managers of the health services organization and the other major stakeholders of the entity. Although numerous methods exist for conceptualizing organizational performance and organizational outcomes (Cameron and Whetten, 1983; Goodman and Pennings, 1977), the outcomes that might be useful for this discussion can be thought of as organizational outcomes and employee outcomes.

Organizational Outcomes

For long-term survival, a health services organization must have a balanced exchange relationship with the environment. An equitable relationship must exist because the exchange is mutually beneficial to the organization and to the elements of the environment with which it interacts. A number of outcome measures can be used to determine how well the organization is performing in the marketplace and how well the organization is producing a service that will be valued by consumers.

One subset of organizational outcomes that may be used to measure the effectiveness of the exchange relationship and the performance of the organization involves the economic viability of the entity. Growth, profitability, return on investment, and related measures are methods for determining the financial performance of the organization. An organization needs to perform well financially so that funds are available for maintenance of current operations, for investment in new service development, and for return to investors if the organization is investor-owned.

Another subset of outcomes involves the nature and level of service performance achieved. Quality of services is especially critical for the health industry, and it contains several dimensions. Technical quality is the appropriate management of a medical problem through medical and other health science technology. A second component of quality, called interpersonal, pertains to the psychological and social interaction between provider and patient. Amenities (e.g., appetizing food, quiet room) are a third element, which Donabedian (1980) perceives to be subsumed under the interpersonal element. He also asserts that an inseparable relationship exists between technical and interpersonal quality.

A third subset of organizational measures involves the efficiency with which services are provided, that is, the level of organizational resources required to produce units of service. Staffing ratios per occupied bed day or numbers of laboratory tests performed by discharge diagnosis are common standards used in hospitals. The ratio of the costs to units of service provided, such as the cost for types of physical therapy procedure or the cost per X-ray procedure, is another type of efficiency measure.

Employee Outcomes

There are other purposes and criteria for the existence of organizations and for determining their level of performance. An organization should provide its work force with job security, meaningful work, safe conditions of employment, equitable financial compensation, and a satisfactory quality of work life. Health services organizations will not be able to attract and retain the numbers, types, and quality of professionals required to deliver health services if the internal work environments are not suitable.

In addition, employees are a valuable constituent group whose concerns are important because of the complexity of the service industry.

A number of measures of the attitudes and psychological conditions of employees are available for use within the health services industry. Employee job satisfaction (Hackman and Oldham, 1975), commitment to the organization (Porter et al., 1974), motivation (Warr et al., 1979), levels of job stress (Maslach, 1976), and other constructs can be used. These measures provide management with an assessment of the quality of work life in the organization. Besides being important outcomes themselves, these factors appear to be associated with economic performance in that employees tend to remain in the organization and are likely to put forth more effort and to work more effectively if the quality of work life is improved (Macy and Mirvis, 1976).

Turnover and absenteeism are two measures of employee withdrawal from the organization. Price and Mueller (1981) have found that a number of organizational characteristics and practices cause professional employees to withdraw psychologically from hospitals and eventually to separate voluntarily from such institutions. The financial costs of recruiting replacement personnel can be quite high. In addition, the organization loses operating efficiency when new employees must be trained in organizational policies and procedures.

Strategic Management and Outcomes

As previously mentioned, the mission and objectives of the organization will be reflected in the outcomes that are stressed by management and in the strategies, general tactics, and human resources practices that are chosen. Management makes strategic decisions that, combined with the level of fit achieved among the internal organizational systems, determine the outcomes the institution can achieve. For example, almost all health services organizations need to earn some profit for continued viability. However, some organizations refrain from initiating new ventures that might be highly profitable if the ventures would not fit the overall mission of the organization for providing quality services needed by a defined population group.

Conversely, an organization may start some services that are acknowledged to be break-even propositions at best, because the services are viewed as critical to the mission of the institution and the needs of the target market. The concerns of such an organization would be reflected not only in the choice of services offered but also in the human resources approaches used and the outcome measures viewed as important. This organization would likely place more emphasis on assessment criteria for employee performance and nursing unit operations that stress the provision of quality care and less emphasis on criteria concerned with efficient use of supplies and the maintenance of staffing ratios. This selection of priorities does not mean that efficiency of operations is ignored, but that greater weight is placed on the former criteria. The outcome measures used to judge the institution should reflect those priorities.

Another institution may place greater emphasis on economic return, profitability, and efficiency of operations. While quality of care also is important, the driving force for becoming a low-cost provider causes management to make decisions that reflect that resolve. Maintenance or reduction of staffing levels is stressed. Prohibition of overtime is strictly enforced. Recruitment and selection criteria stress identification and selection of employees who will meet minimum criteria and expectations and, possibly, will accept lower pay levels.

In an organization striving to be efficient, less energy may be spent on "social maintenance" activities designed for employee needs and to keep them from leaving or unionizing. It can be anticipated that the outcomes in this situation will reflect, at least in the short run, higher economic return and lower measures of quality of work life.

SUMMARY

The intensive reliance on professionals for service delivery requires health services executives to focus attention on the strategic management of human resources in the delivery of services and to understand the factors that influence the performance of persons employed in their organizations. To assist managers in understanding these relationships, this chapter presented a model of the association that exists among strategy, selected organizational design features, and human resources management activities.

Since different strategies require work routines, control methods, and coordination patterns of distinctive types, the structure of a health services organization is determined by the strategy being pursued. Components of an organization's structure have a strong influence on human resources systems. The behavioral systems of the organization (i.e., commitment and motivation, culture, leadership) also influence the personnel functions performed by the organization.

The outcomes achieved by the organization are influenced by numerous factors. The mission determines the direction that is being taken by the organization and what it desires to achieve. The amount of integration of mission, structural systems, behavioral systems, and human resources systems defines the level of achievement that is possible. The remainder of the text addresses these issues in detail.

REFERENCES

Aday, L. A., and S. Shortell. "Indicators and Predictors of Health Services Utilization." In S. J. Williams and P. R. Torrens, Eds., *Introduction to Health Services,* 3rd ed. New York: Wiley, 1988.

American Hospital Association. *Hospital Statistics.* Chicago: American Hospital Association, 1986.

Arthur Andersen and American College of Hospital Administrators. *Health Care in the 1990s: Trends and Strategies.* Chicago: Arthur Andersen and American College of Hospital Administrators, 1984.

Cameron, K. S., and D. A. Whetten. *Organizational Effectiveness: A Comparison of Multiple Models.* New York: Academic Press, 1983.

Chandler, A. *Strategy and Structure.* Cambridge, MA: MIT Press, 1962.

Donabedian, A. *Explorations in Quality Assessment and Monitoring, Vol. 1: The Definition of Quality and Approaches to Its Assessment.* Ann Arbor, MI: Health Administration Press, 1980.

Edwards, J. E., C. Leek, R. Loveridge, R. Lumley, J. Mangan, and M. Silver. *Manpower Planning: Strategy and Techniques in an Organizational Context.* New York: Wiley, 1983.

Galbraith, J., and D. Nathanson. *Strategy Implementation: The Role of Structure and Process.* St. Paul, MN: West Publishing, 1978.

Goodman, P. S., J. M. Pennings, and associates. *New Perspectives on Organizational Effectiveness.* San Francisco: Jossey-Bass, 1977.

Hackman, J. R., and G. R. Oldham. "Development of the job diagnostic survey." *Journal of Applied Psychology,* 60:159–170 (1975).

Lawler, E. E. *Motivation in Work Organizations.* Pacific Grove, CA: Brooks/Cole, 1973.

Lawler, E. E. "The strategic design of reward systems." In C. Fombrun, N. M. Tichy, and M. A. Devanna, Eds. *Strategic Human Resource Management.* New York: Wiley, 1984, pp. 127–148.

Macy, B. A., and P. H. Mirvis. "A methodology for assessment of quality of work life and organizational effectiveness in behavioral-economic terms." *Administrative Science Quarterly,* 21:212–226 (June 1976).

Maslach, C. "Burned out." *Human Behavior,* 5:16–22 (September 1976).

Mintzberg, H. *The Structuring of Organizations.* Englewood Cliffs, NJ: Prentice-Hall, 1979.

Price, J., and C. W. Mueller. *Professional Turnover: The Case of Nurses.* New York: SP Medical and Scientific Books, 1981.

Porter, L. W., R. M. Steers, R. T. Mowday, and P. V. Boulian. "Organizational commitment, job satisfaction, and turnover among psychiatric technicians." *Journal of Applied Psychology,* 59:603–609 (1974).

Porter, M. E. *Competitive Strategy.* New York: Free Press, 1980.

Sorkin, A. L. *Health Manpower.* Lexington, MA: Lexington Books, 1977.

Starr, P. *The Social Transformation of American Medicine.* New York: Basic Books, 1982.

Thompson, J. D. *Organizations in Action.* New York: McGraw-Hill, 1967.

U.S. Department of Health and Human Services. *Health: United States and Prevention Profile, 1983.* DHHS Pub. No. PHS 84-1232. Hyattsville, MD: U.S. Department of Health and Human Services, December 1983.

U.S. Department of Health and Human Services. *Fifth Report to the President and Congress on the Status of Health Personnel in the United States.* DHHS Pub. No. HRS-P-OD-86-1. Hyattsville, MD: U.S. Department of Health and Human Services, March 1986.

CHAPTER 2

Formulating Organizational Strategy

S. Robert Hernandez

A critical function of health care management is the development of a plan for the future actions to be taken by the organization. Senior managers must identify the major tasks to be accomplished by the firm, assign responsibility for performance of those tasks, and monitor organizational actions to ensure that the tasks are executed satisfactorily. An adequate planning system allows a health care organization to achieve its objectives by identifying and responding to environmental change through deployment of internal resources in an efficient and effective manner.

The planning system or systems used by an organization vary based on the sophistication of management, the rates of environmental change, and the level of competition experienced within a geographic region or service sector. A planning system also must match the corporate culture of the firm as well as the complexity of the business in which the organization is competing. Major planning approaches available to health care managers are reviewed in the section that follows. Next, selected organizational strategies that might be used by health care institutions are presented and guidelines for formulating the organization's strategy are suggested. Selected relationships that should exist between possible business strategies and human resources functions conclude this chapter.

PLANNING METHODS

The methods used for the formulation of strategies and plans by organizations in the health services industry have evolved rather dramatically over the past several years. The major approaches to planning that have been used are **budgeting, long-range** planning, strategic planning, and strategic management.* Budgeting involves planning for anticipated revenues and costs for a given time period, usually annually. Long-range planning is concerned with the projection of organizational goals, objectives, programs, and budgets over an extended period. This approach requires forecasting of environmental trends based on historical data. Strategic planning involves an organization's choices of its mission, objectives, strategy, policies, programs, goals, and major resource allocations. This method is intended to define the strategy for the firm so that internal resources and skills are matched to the opportunities and risks created within the environment. Finally, strategic management is concerned with integration among administrative systems, organizational structure, and organizational culture for both strategic and operational decision making. This approach views strategic planning as one element of administrative functioning that must be blended with other management processes for an organization to function efficiently.

Each of these methods represents a fundamentally different approach for planning and organizing the activities of a health services organization. The evolution of each of these planning approaches in the health services industry will be reviewed, and implications for use of these techniques in strategy development will be discussed.

Budgeting

One of the first planning methods to be applied systematically in American industry, budgeting emerged more than 50 years ago to assist management in the difficult task of planning, coordinating, and controlling the numerous, often disparate activities undertaken by a firm (Hax and Majluf, 1984). Although the budget is not an operational plan and some have suggested that it is not planning (Marrus, 1984), it is the *expression* of the operational plan in dollars (Berman and Weeks, 1979). Budgeting converts management intentions for future actions of the organization into a financial format.

Although the budget has been available as a management tool at least since the 1930s, health services organizations have been slow to adopt budgeting as a management technique. One survey of hospitals in a midwestern state in the early 1960s found that only 23 percent of hospitals prepared income and expense budgets (McNerney et al., 1962). Even worse, only 3 percent used budgets appropriately by involving supervisory personnel in estimating expenses and in receiving reports of financial performance. Certainly, the use of the budget has expanded dramatically in the quarter-century since the McNerney study, possibly because of increased management awareness of the value of budgeting, or the mandatory requirement for budget development as a condition for participation in the Medicare program.

Budgeting has value for institutional planning because the process of budgeting requires that the hospital direct its attention toward the future and plan for it. Thus, it *raises the priority of planning* within an organization. Without annual attention to budget development, many institutions would not devote much time to a systematic consideration of future operations.

---

*Management by Objectives (see Chapter 13) is not a method for developing strategy but a system to plan, monitor, and control. Facility planning is seen as following strategy formulation. These approaches are not reviewed.

Additionally, budgeting *provides a structure* for the planning activity insofar as projected revenues, expenses, capital needs, and cash flows for organizational units must be identified. Thus, the hospital is forced to consider the unique units (responsibility centers) for the organization and the varying operating requirements that might be anticipated during the coming year. Performance can be meaningfully planned, evaluated, and controlled by the responsibility centers identified through the budgeting activity.

A typical process used in the decision-making process for budget development is illustrated in Figure 2.1. According to this simplified diagram, the board establishes financial goals for the organization and gives final approval to the budget. Senior management develops operational objectives and policies, establishes priorities for program development, plans for budget development including financial assumptions, and conducts the final administrative review once budgets have been developed. Line management must convert policy and operating objectives into programs by specifying resources required to accomplish approved projects.

Management must be sensitive to the importance of planning driving the budget process, rather than financial issues driving the process, if maximum benefit is to be obtained from this approach to planning. The budget will have little value if it becomes a "paper shuffling" exercise driven by finance as contrasted with a comprehensive institutional plan controlling projected revenues and expenses.

Budgeting will be inappropriately developed as a planning tool if the process does little more than use historical data for budget projections with only incremental change anticipated over past operations. One method to overcome this myopic concern for previous operating history as reflected in the incremental approach to budgeting is called zero-based budgeting (Dillon, 1979; Pyhrr, 1973). In this approach each unit must examine critically and justify each of the activities it proposes to operate during the projected budget period. Whereas traditional budgeting processes require only that a manager justify the increases planned over previous years, in the zero-based option each department defends its entire budget request each year.

Another concern arises if management uses budgeting as the only planning process for a hospital: that is, the focus is on only one aspect of hospital activities—financial

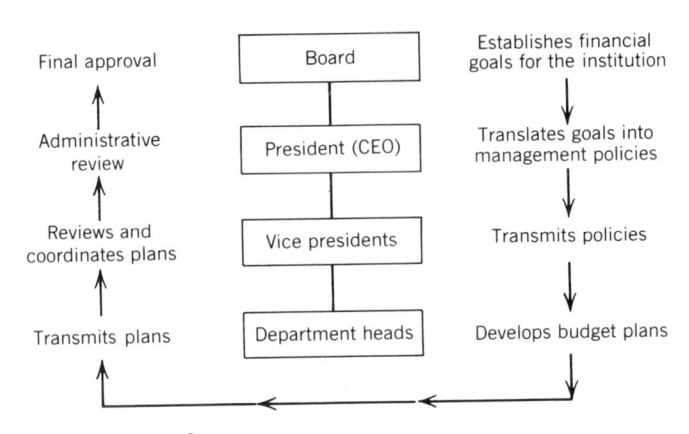

Figure 2.1. Budget preparation.

performance. It is important that management prevent preoccupation with short-term return on investment at the expense of the organization's long-term growth.

Long-Range Planning

The introduction of long-range planning has led to a substantial improvement in planning future operations throughout the business world. The use of this method followed the rapid growth of the U.S. economy in the post–World War II era. Corporations found that one-year budget projections were inadequate for identifying future operating plans in the rapidly expanding economy.

Long-range planning calls for an integrated approach to the attainment of corporate goals and objectives. Since the future is expected to be predictable through extrapolation of previous growth, study of historical operating trends is an important component to the development of the plan (Ansoff, 1984; Moore, 1970). Thus, the underlying assumption of the method is that establishment of challenging objectives based on projection of environmental trends will provide direction for future growth and development.

Descriptions of long-range planning techniques were relatively widespread in the hospital and health services literature by the mid-1970s. Perlin (1976) presented a model outline for a hospital long-range plan that included several steps. The characteristics of the population residing in the facility's service area with projections of future demographic changes were to be identified in one step. A planner then would forecast future community needs based on anticipated population changes and health delivery techniques, inventory current health resources and services, and conclude with a projection of future gaps or excesses in services based on a comparison of the first two factors. Historical utilization of an institution's programs and services would be chronicled. Then, based on an organization's mission and the gaps previously identified, a facility's plan to carry out future service programs would be developed. A long-range plan would conclude with an implementation plan that assigned short-range and long-range priorities to programs and services, and described physical plant requirements for plan accomplishment, manpower required, and financial resources needed. A typical flow of steps in long-range planning is illustrated in Figure 2.2.

This method represents a major improvement over the use of budgeting as a planning tool because it increases managerial awareness of and responsibility for planning. However, several limitations potentially detract from its use for charting future directions for an organization. One weakness is the assumption that the future is predictable from historical growth and utilization patterns. Complex environments that feature rapid changes in operating rules, conditions of demand, and levels of competition do not allow accurate extrapolations of the future from past operations.

A related, implicit assumption is that future performance can and should be an improvement on past operations. This assumption can lead managers to think they can control the future and achieve sustained growth when, in fact, external forces beyond management's control may influence organization operations. Overly optimistic objectives sometimes are attributable to this belief.

Moreover, long-range planning frequently is developed by support staff or external consulting firms using exhaustive arrays of operating and environmental data, but without adequate involvement of either senior or line management. Meager participation by senior managers results in a plan that may not address major issues critical for

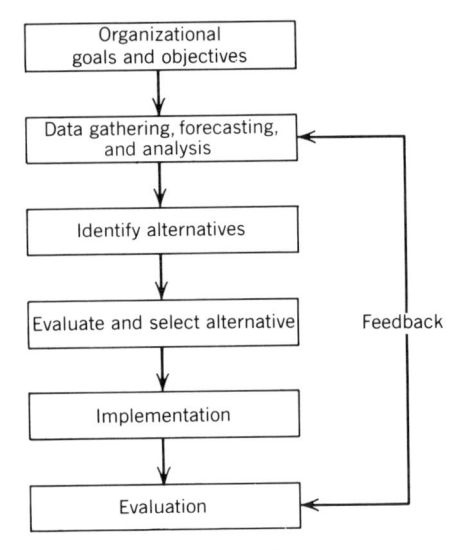

```
┌─────────────────────────┐
│   Organizational        │
│   goals and objectives  │
└───────────┬─────────────┘
            ▼
┌─────────────────────────┐
│ Data gathering, forecasting, │◄──┐
│      and analysis       │       │
└───────────┬─────────────┘       │
            ▼                      │
┌─────────────────────────┐       │
│   Identify alternatives │       │
└───────────┬─────────────┘       │
            ▼                      │
┌─────────────────────────┐       │
│ Evaluate and select alternative │  Feedback
└───────────┬─────────────┘       │
            ▼                      │
┌─────────────────────────┐       │
│     Implementation      │       │
└───────────┬─────────────┘       │
            ▼                      │
┌─────────────────────────┐       │
│       Evaluation        │◄──────┘
└─────────────────────────┘
```

Figure 2.2. Long-range planning.

the organization's survival, hence cannot reliably provide strategic direction for a firm. Such a document becomes a "dust gatherer" that sits on a shelf and is reviewed annually, at best.

Strategic Planning

Strategic planning emerged in American industry as a reaction to changes in the nature of competition and to increasing uncertainty in the operating environment. Organizations were confronting environmental fluctuations that had not been projected by their long-range planning systems, and discontinuities caused significant problems for the firms (Ansoff, 1977). Long-range planning is appropriate when a firm is operating as a single dominant business or product line in an industry experiencing high market growth, when the future is relatively predictable based on previous experience, and when there is low rivalry and/or a regulated environment in the industry. However, the changes in the U.S. economy required development of planning systems that systematically analyzed alternative possible futures and allowed firms in multiple markets to evaluate their options.

Given the rapid changes occurring in the health services industry described in Chapter 1, it is small wonder that many health services organizations are using strategic planning approaches in determining future directions. These techniques allow a complex organization to evaluate multiple market and service/product opportunities while considering the strengths of the organization itself and its competitors.

Strategic planning focuses on the market environment facing a firm, including future actions by competitors and consumers. A major contribution of strategic planning is the use of market segmentation and portfolio analysis, which allow management to assess the relative attractiveness and competitive strength of the business units of the organization.

Typically this approach to planning contains four steps (Ansoff, 1984). First, management conducts an *analysis of the anticipated results* of the firm's operations through projection of operating trends as well as identification of environmental threats and opportunities. These data allow management to project levels of performance likely to be attained if there are no changes in strategy. Then, *competitive analysis* identifies performance improvement that is possible from changes in the competitive strategies employed in current business units of the firm. Next, *portfolio analysis* allows management to establish priorities for the business units by comparing the anticipated future performance of all business areas. Since it can be anticipated that success among the subunits will be unequal, priorities for allocation of resources to business units can be based on this analysis. Finally, *diversification analysis* provides management with an opportunity to identify new business opportunities if deficiencies exist in the current portfolio of services. Improvement in health services organization performance is possible after each of these analyses, with implementation of management actions that are suggested as appropriate. Incremental performance increases are possible from improving competitive strategies in current businesses, more efficiently allocating resources among current portfolios of businesses, and diversifying into new areas.

Strategic Management

The planning systems previously described serve to identify future actions to be taken by an organization. As noted initially, senior management's responsibilities extend beyond planning future actions and identifying major tasks. They also must assign responsibility for performance of those actions and monitor organizational activity to ensure that the tasks are satisfactorily achieved. Outcomes of the planning process must be integrated into the organization for implementation and evaluation.

Strategic management is concerned with linking strategic planning activities with other internal management systems, the structure of the organization, and culture of the firm. This means that management must work to see that the outcomes of strategic planning become administrative reality. Additionally, if an organization is to develop a viable strategic plan, management systems must be considered during the planning process.

Ensuring that the strategic planning activities are at the nexus of action systems of the organization is a prime responsibility of senior management. The purpose of the text is to identify the relationships and techniques for achieving desired organizational strategies through strategic management of human resources within a health care organization. The remainder of this chapter is devoted to strategic planning and management processes that provide direction for the organization.

ORGANIZATIONAL STRATEGY

Many people do not distinguish between organizational strategy, operating tactics, and functional operating policy. Since, however, strategic planning activities should focus on *strategic* issues and *strategy* development rather than operating tactics, a clear delineation of what constitutes strategy is critical. A complex health services organization contains many operational units and can offer a vast array of products and services to a community. Thus the major subunits of an organization, which are called strategic business units and form the starting point for development of organization strategy, also must be identified. What constitutes strategy and how strategic business units are to be identified are the subjects of the sections that follow.

Types of Strategy

Although numerous management theorists and authors have definitions of what constitutes strategy,* Peter Drucker (1974) captured the topic best when he stated that thinking about the mission and strategy of an organization is concerned with answering the questions: What is our business, and what should it be? These questions translate into *identification of the business areas in which the organization will compete.*

Corporate-level Strategy

Health services organizations must determine the services they will provide and market areas that will be served by each one. The current portfolio of services and/or business areas must be reviewed, and what will be done with each of the services/businesses must be decided. This analysis is a corporate-level strategic decision that focuses on what areas will be *built, held, harvested, or divested.* Unique operating decisions concerning levels of investment, allocation of resources, functional policies, and related concerns are associated with each of these four strategic options that may be applied to a business unit operated by a firm.

Build. The decision by a health services organization to build a service or business area means that the institution plans to invest heavily in the service in an attempt to increase its market share of consumers receiving the service. This approach may be taken even though it will cause the institution to forego some short-term profits to build share. The management of a system of "emergicare" centers may decide to retain operating profits for development of new service delivery sites and to raise capital for growth by offering convertible debentures to a limited number of investors.

Hold. A decision to hold suggests that the institution wants to maintain its current market share. This strategy usually is associated with services or businesses that are expected to generate large cash flows which are to be diverted to other investment opportunities. Increased levels of investment are made only to the extent that there are increases in patient demand. Many hospitals providing inpatient acute care have implicitly adopted this strategy.

---

*For examples of strategy definitions see Aaker (1984), Andrews (1971), Hofer and Schendel (1978), and Steiner (1979).

Harvest. Organizations that attempt a harvesting strategy desire to increase short-term cash flow emanating from a service regardless of the consequences for that service. This strategy may be implemented by reducing staffing levels, cutting maintenance, allowing technical obsolescence, or adapting related tactics that decrease operating costs while maintaining prices. Eventually the service or business will be terminated or sold. A multi-institutional system may employ this strategy with a group of hospitals it owns in an attempt to reap short-term profits before disposing of the holdings to another system.

Divest. The decision to divest means that the organization will abandon the market. Resources that would have been consumed in service delivery will be diverted to other parts of the organization. For example, a multi-institutional system might sell a hospital in a geographic area it no longer wants to serve, or a community general hospital might terminate a major service such as obstetrics.

Business-level Strategy

A second, related strategy formulation issue involves the determination of *how the organization will compete in the delivery of the services to the markets that have been identified.* Managers must decide what general approach they will use to provide services to the desired areas. The answer to this question is a function of the *competitive advantage* the organization will attempt to achieve. Three basic strategies have been found to lead to sustained competitive advantage: overall cost leadership, differentiation, and focus (Hall, 1980; Porter, 1980, 1985).

Cost Leadership. To achieve overall cost leadership, a health care provider attempts to become the low-cost producer for an area. If a firm is able to provide services at a lower *cost* than its competitors and yet obtain reimbursement at *prices* that are comparable, it will receive above-average returns.

Whereas for industrial-sector firms, use of economies of scale through being the market share leader for a product is a common means of becoming the cost leader, numerous approaches may be used in the health services industry. Some investor-owned hospitals attempt to achieve cost leadership through management systems geared to control resource consumption (e.g., close monitoring of full-time equivalent employees per occupied bed). Health maintenance organizations attempt cost leadership through controlling utilization, especially hospital admissions. The methods for achieving cost leadership in health services depend on the type of organization providing the service, the consumers of the service, payer mix served by the institution, forms of reimbursement, and related factors.

Differentiation. An organization seeking to achieve differentiation desires to be perceived by consumers as offering a service that is unique on important dimensions. As a rule, a premium price can be charged for the unique service.

Achieving differentiation may be associated with *technical quality* of the medical care that is being delivered. The medical staff, the technical sophistication of the equipment provided, and related matters may be viewed by the community as superior. Differentiation also may be associated with offering *patient amenities* that are valued, such as valet parking in crowded inner-city hospital locations, immediate access to medical personnel in emergency rooms, or a reputation for having a responsive, caring nursing staff for inpatient acute care.

Focus. Organizations choosing to focus are those that target a narrow scope of competition within their community. The institution decides on a market segment or group to which it will provide a service or range of services. After the target market has been chosen, the institution may select a *cost focus* or a *differentiation focus* strategy.

One example of a differentiation focus strategy is initiation of pediatric emergency services by a community general hospital. Despite increasing competition in many communities for the emergency patient, identifying pediatric patients as the market to be served may give a hospital a competitive advantage in a saturated urgent care market. Targeted marketing strategies, special billing practices, separate entrances, unique employee uniforms, and related operating tactics support this strategy of focusing on the pediatric patient in need of urgent care and then differentiating the service from other emergency services available in the community.

Identifying Strategic Business Units

A fundamental concern for health care organizations is to identify their current business units by grouping related services into strategic business units (SBUs). Once the services provided by an organization have been logically grouped, it is possible to define the businesses in which the organization is competing. Businesses are usually defined by the services provided and/or the markets served (Abell and Hammond, 1979). One of the primary values of identifying SBUs is that an institution can identify the divergent markets it serves, develop logical strategies that will be successful in these unique markets, and assign responsibility for implementing the plan.

An SBU usually contains several related services or programs; but its exact composition will depend on the level of organization that is being examined. At the corporate level for Hospital Corporation of America or Humana, SBUs may be identified by groupings such as psychiatric hospital division, inpatient acute care hospital division, health insurance division, preferred provider arrangements, and related groupings. Conversely, a multispecialty medical group may define its SBUs based on clinical services such as pediatrics, obstetrics, and general surgery.

For acute care hospitals, data may be collected and operational management decisions reached on the basis of diagnosis-related groups (DRGs). However, it is *not* practical to plan strategically using DRGs as the unit for which strategies are developed, since it would be almost impossible for a community hospital to develop unique market approaches for more than 400 business units. An SBU for these organizations more realistically might contain a number of DRGs in "product groupings that have economic and marketplace significance" (Alfirevic et al., 1983, p. 13). It is possible that clinical service groupings will be appropriate units for SBU analysis by these hospitals.

Although there are a number of ways of identifying SBUs, an ideal SBU might have the following characteristics (Kotler, 1984).

- It is a single business or collection of related businesses.
- It has a distinct mission.

- It has its own competitors.
- It has a responsible manager.
- It consists of one or more program units and functional units.
- It can benefit from strategic planning.
- It can be planned independently of the other businesses.

Not only is it important that units designated as SBUs develop plans for the markets they serve, these SBUs should receive resources to implement the plan and should be held accountable for the success or failure of the plan. Some health organizations do not delegate the latter functions to the managers at the operating unit level and require senior management oversight of these responsibilities. The separation of these activities appears to be working well at present. In any case, SBU plan development, resource allocation, and accountability for performance must be done in conjunction with an overall strategy for approaching the marketplace that is developed by corporate planning.

FORMULATING THE ORGANIZATIONAL STRATEGY

A model process used to generate strategy for a health care organization is illustrated in Figure 2.3. Planning begins with a mission statement that has been articulated with sufficient detail to guide organizational decision-making; then external and internal analyses are conducted, the portfolio is reviewed, strategies are identified and selected, implementation is accomplished, and evaluation begins. The material provided in this section, though not presented at the level of detail required to guide the production of a strategic plan, allows the reader to understand the flow of data and the rationale that supports corporate strategy development. This understanding is beneficial for comprehending the relation of strategy to human resources functions. Detailed discussions of strategic planning approaches are offered by Hofer and Schendel (1978), Steiner (1979), Aaker (1984), Nutt (1984), and Peters (1985).

Mission of the Organization

The first step in development of strategies is to identify the general direction in which the organization is headed. It is the responsibility of senior management to provide the institution with goals and an operating philosophy to serve as a guide to direct the future of the firm. The communication of corporate purpose, scope of operations, self-concept, and image to important stakeholders occurs through a mission statement. As noted by Pearce (1982):

> The principal value of a mission statement as a tool of strategic management is derived from its specification of the ultimate aims of the firm. It thus provides managers with a unity of direction that transcends individual, parochial, and transitory needs. It promotes a sense of shared expectations among all levels and generations of employees. It consolidates values over time and across individuals and interest groups. It projects a sense of worth and intent that can be identified and assimilated by company outsiders (i.e., customers, suppliers, competitors, local committees, and the general public). (P. 24)

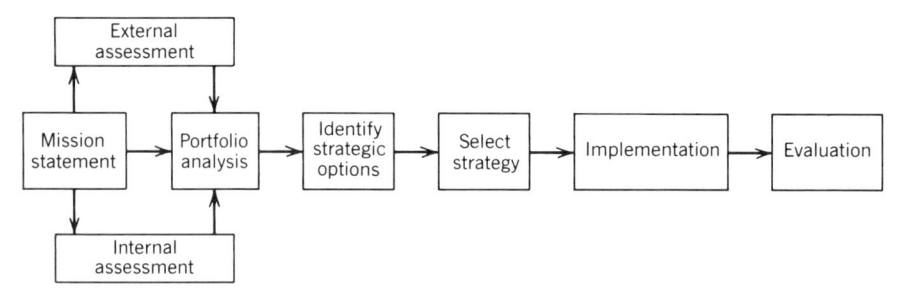

Figure 2.3. The strategy development process in health services organizations.

Some authors suggest that an organization requires two distinct types of mission statement (Hax and Majluf, 1984). One is concerned with articulating the mission of the entire firm and is developed at the *corporate* level simultaneously with the grouping of products/services into SBUs and the enunciation of a corporate philosophy. These elements (corporate mission statement, philosophy, and selection of SBUs) are known as the *vision* of the firm. A second more detailed type of mission statement is required at the *business* level for each SBU. Whether developed at the corporate or business unit level, the mission statement usually specifies the products or services that will be provided, the markets that will be served, and the manner in which competitive leadership will be attained.

Product Definition

In identifying the product line of services, management defines the *scope* of services that the organization will offer by determining the breadth of the line that will be provided. Will a complete range of patient needs be met, or will the organization target a specific service? A multispecialty group may decide to change its product definition by broadening its services with the addition of a specialty such as neurology. This decision allows the group to capture revenues that may have been lost to the group when patient referrals were made to other neurologists who were not members of the group.

Defining how an organization wishes its services to be perceived by consumers in relation to competitors providing the service is known as *positioning*. One example is positioning on the basis of a price-quality spectrum. An organization may attempt to be perceived as providing comparable services at a lower cost than its competitors. Emergicenters attempt to convince consumers that the centers can provide minor emergency service comparable to a hospital emergency room more rapidly and at a lower cost.

Products and services benefit from a positioning strategy that is clearly articulated to the target market for that service. Six potential approaches for a positioning strategy have been identified by Wind (1982, pp. 79–81):

- Positioning on specific product features
- Positioning on benefits, problem solutions, or needs

- Positioning for specific usage occasions
- Positioning for user category
- Positioning against another product
- Product class dissociation

Market Definition

Defining the market for a business requires identifying the consumer groups to be served by the organization. The most logical delimiter for a health services organization is the *geographic boundary* of its service area. A hospital may identify its primary service area as a region that has a high relevance index (Griffith, 1972, p. 76), or a generalist hospital consulting firm may decide not to offer its services beyond the midwestern states in which it is located.

A market also may be defined by the *consumers that desire the product*. The consulting firm that is restricting its services to the Midwest also may identify its clients as senior managers of hospitals that are operating between 100 and 300 beds. This group may value the general management consulting package that is being provided more than executives in more complex facilities who purchase functional consulting packages.

The process of defining markets allows management to decide the *market segments* that will be served. That is, when the total market is broken into elements that share common properties, organizations are able to develop appropriate services, and to provide those services more efficiently, for groups that have common needs as identified through segmentation.

Distinct Competence

The "distinct competence" of an organization refers to an advantage the institution holds over its competitors. The advantage may emanate from the competitive advantages of cost leadership, differentiation, or focus, as discussed previously. Alternatively, an organization may possess an asset that provides it with an advantage over competitors. A hospital may have a reputation for a nursing staff that is outstanding and well-respected within the community. A nursing home may have evolved into an organization offering a complete array of vertically integrated services for the elderly. A hospital system may have facility sites that are conveniently located near transportation arteries. Articulating the distinct competence for the organization will help management to design strategies that reinforce strengths.

External Assessment

An external analysis focuses on elements that are relevant for organizational performance but are outside the institution's boundary. The purpose of this assessment is to determine the major opportunities that might be available to the firm and any potential threats that might prevent the organization from achieving desired outcomes. This analysis often consists of identifying and analyzing patient groups or consumers, competitors, industry conditions, and general environmental factors.

Consumer Groups

The first steps in external analysis are to identify the markets currently being served by the institution, to select markets that potentially could be served, and to determine how to segment these markets. As just noted, the process of segmentation groups consumers into clusters that share common characteristics. The demand for services by consumers should be relatively homogeneous within the group and heterogeneous with respect to demand by other groups. Analyzing consumer groups assists management in identifying increases or decreases in demand that might be associated with changing needs or requirements by the groups that are studied. This knowledge will assist in making investment decisions across alternative product/market opportunities.

Competitor Analysis

The result of competitor analysis is identification of the threats or opportunities that will occur from the probable moves and reactions of competitors (Sammon, 1984, p. 97). This information is obtained by building a competitor's response profile (Porter, 1980), which includes for each competing firm the future goals, current strategy, assumptions about market conditions, and current strengths and weaknesses.

This profile should include an estimation of the financial goals of the competitor as well as a determination of its interest in long-term versus short-term performance of the institution. The extent to which the competitor will stress quality versus cost in the provision of services is important to note. Recording the values and beliefs of senior managers of the competitor will provide insight into the reasons for initiating specific services and programs. Knowledge of the competitor's organizational structure tells who the key decision makers are and exposes the methods they use to handle the medical staff and their ability to respond rapidly to your initiatives.

A competitor's assumptions about its own operation and the health services industry also should be a component of the profile. How have its managers assessed their organization? What do they believe are its strengths and weaknesses? How will these perceptions influence likely corporate thrusts? What do they view as future utilization patterns for the health services industry? What are the *actual* strengths and weaknesses of the competition?

The foregoing information can be used to determine how the competitor will respond to new services that you may decide to offer or to your decision to terminate some of your current services. It also can be used to anticipate strategic thrusts the competitor may be planning.

Analyzing competitors also improves understanding of consumer responses to product offerings. Cognition of the strategies, operating practices, and successes and failures of competing organizations helps management to know the attributes that appeal to various market segments. For example, close monitoring of an innovative service provided by a major competitor will allow management to decide which benefits and features will be included in their own version of the service.

Additionally, this analysis validates consumer analysis and provides clues to market opportunities that may become available (Abell and Hammond, 1979). It is possible that identification of the products/markets served by competitors will result in recognition of an area unserved by other institutions.

Industry and Environmental Analysis

A third major component of external assessment involves trends in the industry as well as overall environmental conditions. *Industry analysis* is intended to identify the competitive factors that lead to success in a given product/market and to determine the relative attractiveness of a industry/market for the firm. The forces that drive competition (Porter, 1980) and the manner in which firms are organized for delivery of services are areas to examine. Moreover, securing projections of demand and number of competitors currently providing services or anticipated to begin service delivery for the market allows you to pinpoint the stage in the life cycle for the industry.

The environment provides the context in which industry operating rules and practices exist. *Environmental analysis* is concerned with identifying trends and major events that potentially have an effect on an industry and, ultimately, the strategy of an organization. Just as the environment is one step removed from and above an industry, environmental analysis is related to, but broader than, industry analysis. The environmental analysis may be divided into five components (Aaker, 1984), and all these areas should be monitored.

1. *Technology.* New technologies, the life cycle of current technology
2. *Government.* Legislative and regulatory actions, tax policies, values concerning the health industry
3. *Economics.* Interest rates, economic health of local firms, general economic conditions
4. *Culture.* Life-style trends that affect consumption of health and related services
5. *Demographics.* Trends in age, income, education, geographic location

A major component of industry and environmental analysis for health services organizations must include projections of the future availability of selected types of health professionals. In many cases state licensure boards provide counts of individuals practicing in an area. These figures may be misleading, however, because some will retain their licenses without intending to practice. Historical labor force participation rates are helpful in determining the size of the current manpower pool.

To determine the potential size of the pool from which new employees may be recruited, it is necessary to ascertain the numbers of individuals in professional training. Historical data on the success rate of the organization in attracting new recruits to work in the institution will provide at least a crude measure of the anticipated inflow of future workers.

Internal Assessment

An internal assessment is intended to provide a detailed understanding of the strategically important attributes of the organization. The outcome of this assessment should be a listing of the strengths of the firm that might be used for competitive advantage. In addition, problems or weaknesses that might hinder performance must be surfaced.

An inventory of current internal capabilities, resources, operating characteristics, and actions provides the basis for this assessment. Webber and Peters (1983, p. 59) suggest that the following areas be examined:

- Management and governance
- Functional programs and services
- Human resources
- Medical staff
- Financial resources and results
- Physical facilities
- Basic values and culture of the organization
- Interrelationships of the areas above

Although a description of each of these areas is beyond the scope of this chapter, some discussion of the human resources and medical staff inventory is required. Management must maintain separate data bases on personnel and medical staff that can be used both for the day-to-day operation of the institution and for strategic planning.

For strategic purposes, these data bases include demographic information, career information, a skills profile, productivity measures, and utilization statistics. These data offer management the basis for understanding the characteristics of its personnel and medical staff. This information can be used for determining how well the organization is currently functioning as well as for identifying the organization's future internal capabilities for the implementation of new programs and services. Information systems for human resource management are discussed in more detail in Chapter 6.

Once current capabilities, resources, operating characteristics, and actions of the health care organization have been inventoried, these attributes can be compared to performance levels that are standards in the industry or are desired by management. This assessment will yield a list of attributes viewed as institutional strengths and a list of organizational weaknesses. These attributes, when combined with the previously developed external opportunities and threats, provide data to be used in assessing the institution's portfolio of goods and services.

Portfolio Assessment

The analysis of the portfolio of SBUs operated by the firm consists of an examination of the service mix of the organization to determine whether the overall array of business units is appropriately balanced. This analysis typically requires that the SBUs of the organization be placed on a matrix that contains two parameters. One parameter describes characteristics (e.g., desirability) of the market or industry in which the SBU is competing, while the other illustrates the strength of the business unit in that unique market. The techniques used to conduct portfolio analysis include the market growth rate–relative market share analysis known as the Boston Consulting Group (BCG) business grid, the industry attractiveness–competitive strength analysis based on a General Electric (GE)-business screen, and the industry maturity–competitive approach analysis known as the product life cycle portfolio matrix.

BCG Business Grid

One method of examining the holdings of an organization is based on pioneering work conducted by the Boston Consulting Group in the early 1960s. The easily understood analysis focuses on the growth rate for the market of a business and the relative market share that the business commands. The annual market growth rate for a business is plotted vertically, versus the relative market share of the SBU to its largest competitor (Fig. 2.4).

The market growth rate illustrated in Figure 2.4 ranges from 0 to 22 percent, although values above or below these numbers are possible. Since a growth rate above 10 percent is considered high, the vertical axis is divided into high versus low growth markets at 10 percent.° The relative market share, shown in log scale with equal distances representing the same percentage increase, has the midpoint of the horizontal axis as 1. This occurs when the organization's share exactly equals its largest competitor. A value of 2 on the scale occurs when the firm's SBU has twice the sales as the next strongest firm, and a market share that is half that of the strongest competitor is represented by the 0.5 value. The matrix is divided into four quadrants by these midpoints. The relative share and growth rate of these markets suggests management decisions for the SBUs operating in these four markets.

The lower left quadrant contains the *cash cows*. These SBUs operate in low-growth markets with relatively large market share. The strong market position suggests that

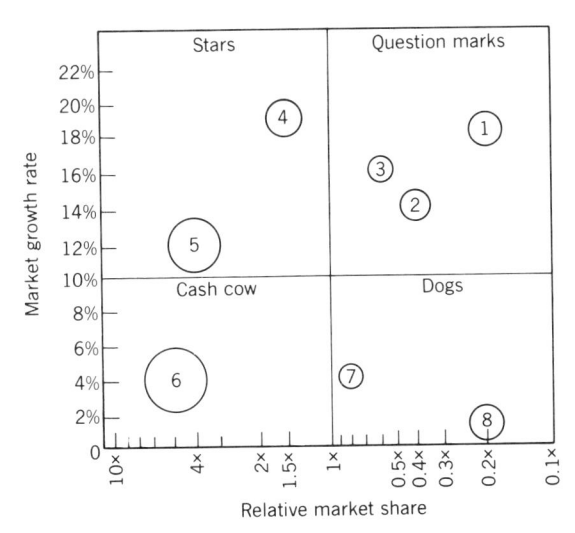

Figure 2.4. The Boston Consulting Group's growth-share matrix. Reprinted with permission from B. Heldey, "Strategy and the business portfolio," *Long Range Planning*, February 1977, p. 12. Reprinted with permission from *Long Range Planning*, Copyright 1977, Pergamon Books, Ltd.

°The cutoff for high versus low growth rates is normally selected as the average growth for that industry, assuming that all SBUs are competing in the same industry (Hax and Majluf, 1984, p. 129).

such units should be able to generate positive cash flow, given their potential for economies of scale and higher profit margins. Investment needs should not be great, since typically cash cows are situated in mature markets with low growth rates. Funds from these businesses are used to support SBUs in other, high-growth markets.

The upper left quadrant contains the *stars,* which are high-growth, high-share SBUs. They require funds to support growth in demand for services, but they also are in a position to perform well financially because of their strong competitive position as market leaders. Stars usually do not generate surplus cash because this high-growth market attracts new firms and the organization must reinvest to maintain share. These SBUs may be cash users if competitors begin major efforts to gain relative market share that must be countered.

The upper right quadrant contains *question marks,* also known as wildcats. These SBUs, which have relatively low market share in rapidly growing markets, need to fund growth but are not able to obtain funding internally because they are not far enough down the experience curve. These businesses are in the worst position because they combine growth needs with low market share; they should be grown to become stars or divested.

The lower right quadrant contains the *dogs*—units that are not profitable because of a weak competitive position in a low-growth market. Since increases in market share are costly in low-growth markets, these businesses need to be phased out when possible.

Once the SBUs have been plotted on the matrix, the organization must determine how balanced the portfolio appears. Too many question marks or dogs, or an inappropriate mixture of exclusively stars or cash cows, causes management to attempt to change its investment mix.

Although the BCG approach has merits, it should not be used in the health care industry without certain modifications. For example, health providers have multiple stakeholders with numerous, often conflicting goals and objectives for the organization, and the allocation of capital based on market growth alone is not a responsible approach to this complex situation. Additionally, the success of market share leadership associated with cost reductions is meaningful only in volume businesses. Health providers concentrate on small segments of the market, however, can focus on a defensible niche that is not easily penetrated by market leaders. Thus, market share leadership is not always an adequate criterion for determining the strength of a business.

GE Business Screen

Recognizing that market growth rate is only a proxy for market attractiveness and relative market share is one measure of the competitive strength of a firm, an approach was developed by GE in the early 1970s to provide composite measures of market attractiveness and competitive strength.

Industry attractiveness is represented on the vertical axis by the factors that are considered relevant. This is accomplished by first identifying the criteria that are important to the firm. These may be based on institutional objectives as well as on determination of the desires of the organizational stakeholders. Second, the criteria are weighted, since all are not anticipated to be equally important. These weights should add to 1. Third, individual services or business units are rated. Rating may be on a 5-

point scale with 1 being very unattractive and 5 being very attractive. Finally, each weight is multiplied by each rank, and the results are summed to determine the overall score for a service. An example of this rating process is shown in Table 2.1.

In a similar manner, the competitive strength of the organization in each market should be assessed. The key success factors for competitiveness in each service are identified. The relative importance of these is then determined. Third, the strength of the organization on each success factor is noted. Finally, the overall strength is identified by multiplying each weight times each rank, and the results are summed to obtain the overall score for a service. An example is shown in Table 2.2.

Once the value for each service has been identified, the services can be plotted on a matrix (see Fig. 2.5). By examining the matrix, analysts can determine how well the services of the firm fit the overall values of the firm and whether the services are strong enough to succeed in the market.

Product Life Cycle Matrix

The life cycle approach is analogous to the biological cycle of embryonic development, growth, maturity, and decline experienced by living matter. Products and services (Kotler, 1984), organizations (Kimberly et al., 1980), and industries (Porter, 1980) are

TABLE 2.1. Health Service Market Attractiveness

	Weight	Score	Total
Profitability	0.20	5	1.00
Payer mix	0.10	4	0.40
Market size	0.20	5	1.00
Growth potential	0.15	3	0.45
Referral network	0.20	4	0.80
Ease of entry	0.05	2	0.10
Level of investment required	0.10	4	0.40
	1.00		4.15

TABLE 2.2. Hospital Strengths

	Weight	Score	Total
Competitive advantage	0.05	5	0.25
Market share	0.10	4	0.40
Public perception	0.15	4	0.60
Medical staff support	0.35	2	0.70
Location	0.15	5	0.75
Plant and equipment	0.15	4	0.60
Support staff	0.05	4	0.20
	1.00		3.50

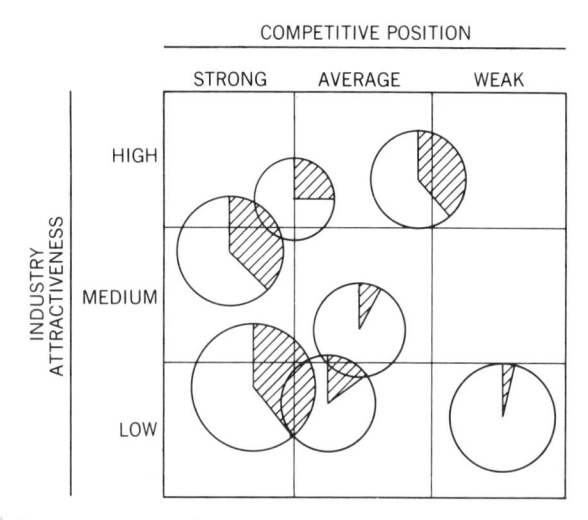

Figure 2.5. Market attractiveness/business position matrix. Reprinted by permission from *Strategy Formulation: Analytical Concepts* by Charles W. Hofer and Dan Schendel. Copyright © 1978 by West Publishing Company. All rights reserved.

believed to follow a similar pattern from inception to decline. Each stage in the life of a service, business unit, or industry is characterized by changes in demand for the service, number of competitors, profitability and cash flow, and other features of market conditions.

The life cycle model in Figure 2.6 represents an expected or "normal" life cycle; the curve of any given product or industry will not always have the same shape illustrated here. Some services may have an introductory phase followed by rapid growth and sharp, immediate decline. Other services may have long periods of maturity and never enter decline, while still others may enter decline and then observe a revival in demand.

The products and services in the health services industry may be at any stage of

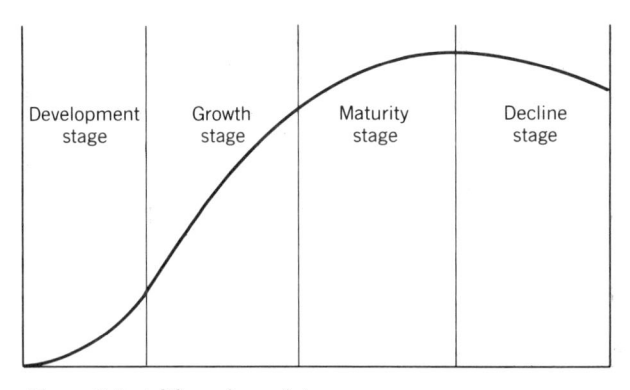

Figure 2.6. A life cycle model.

Formulating Organizational Strategy 39

development on the life cycle. Health Central (1983) developed Figure 2.7 to illustrate their view of the location of selected services on the curve.

The general market conditions associated with each phase of the life cycle have been identified by Porter (1980) for industry evolution and by Kotler (1984) for product evolution. In the first phase, or *introductory* stage, the service has just been introduced and consumer knowledge is limited, with resulting low utilization and revenue levels. Managerial time must be devoted to developing markets for the services because organizational capacity is underused. Providers are uncertain as to the technology that will be most effective because no common standards have evolved. Since product quality may be erratic, control should be exerted over service delivery, possibly through narrowing the scope of service to ensure that operations are better managed. Image and credibility with the financial community may be critical because of the funding required to support growth and the uncertainty associated with new services.

During the second phase, the *growth* stage, consumer knowledge of the service increases and demand rises. Technical quality improves considerably. Reliability and differentiation become more critical, since consumer expectations increase. Organizations experience increased use of existing capacity and profits grow as costs are spread over larger volume. More providers enter the market and offer a broader scope of services, which further expands the market. The potential exists for segmented services because of diverse consumer demand. Since the opportunities for shifts in relative market share are greatest during this stage, major efforts may be directed toward market penetration.

At *maturity,* Kotler (1984) identifies three substages as follows:

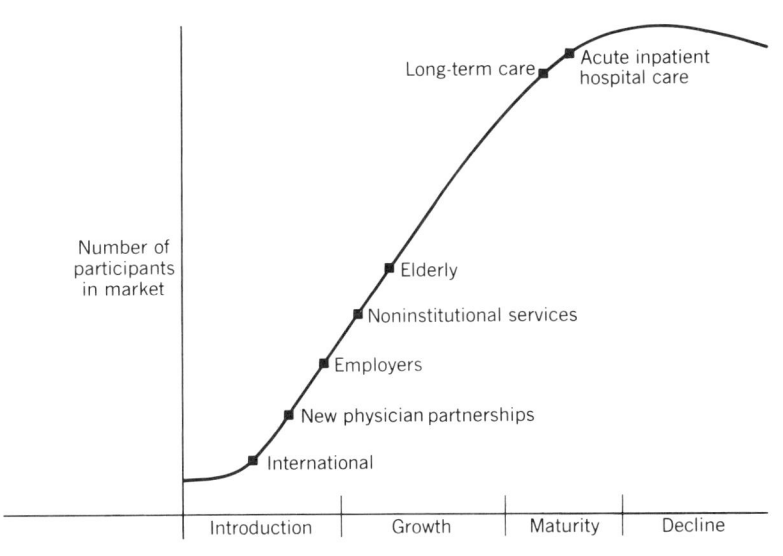

Figure 2.7. Positions of seven health care services on a life cycle model. From HealthOne Corporation. *Environmental Assessment–Health Care: New Dynamics, New Markets.* Minneapolis, MN: HealthOne Corporation, Copyright 1983.

In the first phase, growth maturity, the sales-growth rate starts to decline because of distribution saturation. There are no new distribution channels to fill, although some laggard buyers still enter the market. In the second phase, stable maturity, sales become level on a per capita basis because of market saturation. Most potential consumers have tried the product, and future sales are governed by population growth and replacement demand. In the third phase, decaying maturity, the absolute level of sales now starts to decline, and customers start moving toward other products and substitutes. (p. 367)

The slow rate of demand growth and overcapacity leads to intense competition for market share. Consumers are more cost conscious and technically knowledgeable. Only sophisticated cost analysis can provide the data required to identify and prune unprofitable services from the broad service line initiated during the growth stage. Also, correct pricing is essential, since cross-subsidization through average cost pricing may not be possible. Service delivery innovation is directed toward identifying lower cost delivery methods.

Defending remaining market share is important. Existing consumers are encouraged to increase their use of the scope of services offered by the institution, an approach that is less costly than winning new consumers. Providers become more selective in terms of the groups to whom they will offer services.

The final phase, *decline,* is evidenced by a significant drop in demand. This reduction may occur rapidly or gradually. It is not certain whether the downturn is permanent or is a symptom of a short-term condition that will self-correct. If the downturn represents a sharp movement toward market extinction, management must withdraw as soon as possible to prevent significant losses. A gradual decline, on the other hand, suggests that management might be able to harvest some profits from the market by controlling service delivery costs and efficiently providing services. Alternatively, pockets of demand may remain that allow highly selective marketing to be initiated toward those niches. Finally, a decline may be temporary, and aggressive marketing tactics could renew demand.

Strategy Identification and Selection

The foregoing methods for analyzing portfolios contribute to management understanding of the organization-environment interface. These conceptual tools, combined with important data on market conditions and internal capabilities, are valuable aids in determining future directions for health care organizations. Decision algorithms have been developed to augment the previously described portfolios.

Strategic decisions for business areas of organizations are facilitated by the BCG business grid. Since market growth rate is beyond the control of management (Abell and Hammond, 1979), focus shifts to market share and allocation of funds. One successful long-term strategy is financing efforts to increase market share of question marks from cash generated by cash cows. This produces stars that eventually become cash cows. This "success sequence" is shown in Figure 2.8. Question marks not in an adequate competitive position to become stars should not receive infusions of cash and should be allowed to descend to become dogs with the decline in market growth rate.

The GE-type business screen also provides insights into appropriate portfolio investment decisions. Figure 2.9 illustrates these strategies in the nine blocks of the matrix. Kottler (1984, p. 55) suggests that harvesting or divesting should occur when

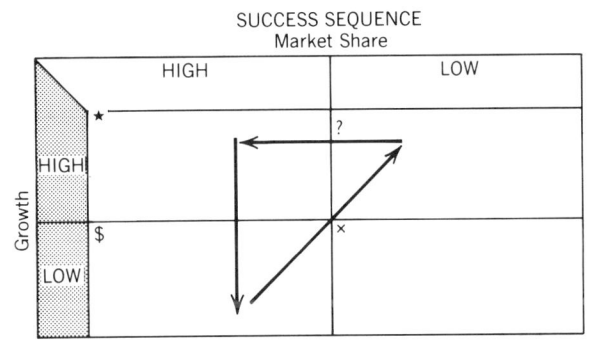

Figure 2.8. Success sequence. From "The Product Portfolio." Boston: The Boston Consulting Group, 1970. Perspectives No. 66.

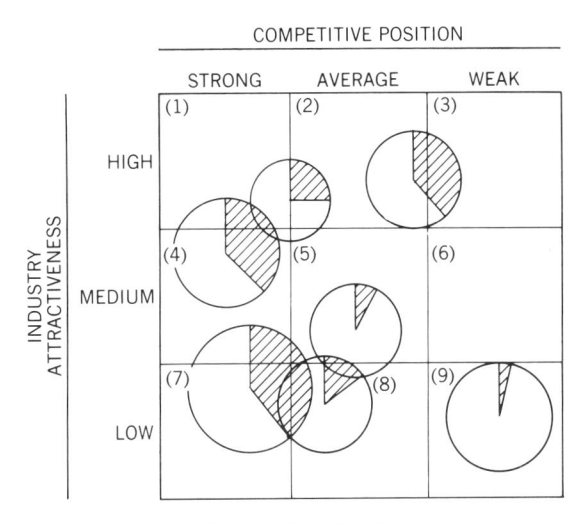

Figure 2.9. Strategies for investment decisions based on the GE-type business screen. Reprinted by permission from *Strategy Formulation: Analytical Concepts* by Charles W. Hofer and Dan Schendel. Copyright © 1978 by West Publishing Company. All rights reserved.

competitive position and industry attractiveness are low (blocks 6,8,9); investment and growth should occur when they are high (blocks 1,2,4).

The product life cycle matrix provides direction for managers interested in strategy formulation. Hillestad and Berkowitz (1984) suggest a *strategy-action match* to synchronize the organization's life cycle stage with the present phase in the marketplace life cycle to determine appropriate managerial action. This match is illustrated in Figure 2.10. During the service introduction, an organization should "go for it" and strive for overall market leadership, limit service variations, concentrate on quality, establish high prices, and make related moves.

Differentiation is attempted during the growth stage. At maturity, a "necessity"

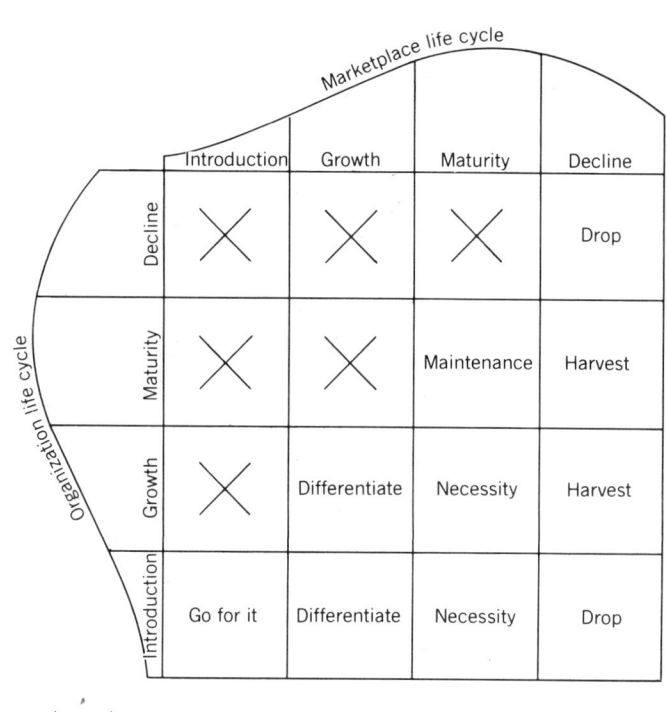

	Marketplace life cycle			
Organization life cycle	Introduction	Growth	Maturity	Decline
Decline	X	X	X	Drop
Maturity	X	X	Maintenance	Harvest
Growth	X	Differentiate	Necessity	Harvest
Introduction	Go for it	Differentiate	Necessity	Drop

Legend
X = Position cannot occur

Figure 2.10. Strategy action match matrix. From S. Hillestad and E. Berkowitz, *Health Care Marketing Plans: From Strategy to Action.* Homewood, IL: Dow Jones-Irwin, 1984.

approach is used if for competitive reasons, the organization must initiate a service it does not currently provide even though it will not obtain market leadership. "Maintenance" is an attempt to retain market share without undue investment of funds. Finally, the decline stage requires that an organization either harvest or divest itself of the service. Interestingly, Porter (1980) suggests that a firm may not only harvest or divest during decline, but also may attempt to find a defensible niche or maintain a market share leadership position, if conditions warrant.

ORGANIZATIONAL STRATEGY AND HUMAN RESOURCES MANAGEMENT

Human resources functions should have a direct supportive relationship to the formulation and implementation of organization strategy. *The strategic management of human resources is concerned with ensuring that qualified personnel are available to staff the portfolio of business units that will be operated by the organization.* This assurance means that qualifications for operation of the units must be determined, appropriate personnel must be recruited and selected, manpower to meet future needs must be developed, and adequate rewards must be provided to attract and retain productive employees.

Unfortunately, only a few, progressive organizations seek to include human resources management activities in the strategy planning process, even though data are available to support decision-making (Tichy et al., 1984). The general role of human resources functions in the strategy of health care organizations is discussed in the next section. Then, the relationship of these functions to the portfolio of services managed by the organization is described.

Functions of Human Resources Management

Recruitment and Selection

When hiring individuals into the organization and moving personnel internally to positions elsewhere in the firm, matching qualified individuals to appropriate jobs is crucial for organizational performance. A health care organization must design a selection system that supports its chosen strategy. For example, if an organization plans to diversify, management must have careful analysis of types of persons needed to staff and manage the new enterprise. Diversification by large health systems may be hindered if new products or services are developed but not managed well. Early development of selection criteria and hiring of individuals for the diversification will ensure that staff is in place when the new venture is initiated.

The organization also must monitor the internal flow of individuals and identify persons whose attributes match those required for the emerging business strategies. Skills that mark employees as upwardly mobile may change if shifts occur in the strategic thrust. Inpatient acute care managers need the ability to work well with the medical staff, numerous community groups, and others. To determine who is eligible for career advancement in this situation, interpersonal abilities (along with other skills) must be assessed. However, new ventures in health insurance, real estate, or other entrepreneurial activities may require greater emphasis on bargaining skills. Thus, the types of data used to identify candidates for advancement and the priority weights attached to the measures may change with different market thrusts.

Comparably, health care organizations must match their key executives to business units pursuing strategies that require unique skills. With the advent of portfolio management in health services, organizations have begun to offer an array of services situated at different stages of the life cycle. As will be discussed later, services at different stages of development require distinct managerial capabilities. Individuals should be placed with services at development stages that match their competence.

Development

Identification of needed skills and active management of employee learning for the future in relation to explicit corporate and business strategies will contribute to organizational stability. The process of enhancing an individual's present and future skills most frequently occurs through training on the job. Health care organizations must develop job rotation sequences and other early career developmental tracks aimed at producing people capable of handling key positions.

Health systems that have competed exclusively in one market segment, such as inpatient acute care, have specific types of experiences planned for developing executives. With the emergence of alternative service delivery systems and diversification into services only tangentially related to health care, human resources managers must devote considerable attention to planning early career experiences for management personnel assigned to operate these new ventures.

Appraisal

Obviously, the entire human resources planning system depends on valid, reliable appraisals of positions and employees. The recruitment, selection, and placement functions require accurate descriptions of the skills needed in all positions. Then, the capabilities of candidates for those positions must be assessed systematically to ascertain that an appropriate match has been achieved.

Compensation systems require support from valid appraisal methods. Rewards can be allocated on the basis of performance only if performance standards for a job have been established and the performance of the incumbent can be measured.

Finally, management development depends on accurate data from the performance appraisal system. Summary information on the strengths and weaknesses of managerial personnel can be analyzed to bring out training and development needs.

Compensation

To be effective in securing behaviors that are desired from managers, compensation must be tied to the performance needed by the organization. Compensation systems appear to influence factors that in turn affect organizational outcomes. Those who are attracted to work for an organization and remain there are swayed by the types of compensation system used with high performers; these frequently include merit-based compensation systems (Mobley, 1982).

Persons responsible for designing compensation systems need to understand that the systems should drive managers toward long-term goals. If rewards are tied to short-term results, it will be difficult to get managers to focus on long-term goals that may not be reached for years. The difficulty of designing compensation systems under these circumstances is discussed in the next section.

Portfolio Analysis and Human Resources Management

The portfolio analysis methods described previously have positive and negative attributes in terms of their ability to guide the human resources management process of health care organizations. The GE-type business screen is a thoughtful method for drawing out often implicit views of outcomes desired by organizations and converting these to explicit measures. However, this method gives few clues as to the relationship of the portfolio of services being provided to human resources management issues that must be addressed.

In a comparable manner, the BCG business grid does an excellent job of guiding the suggested flow of funds among business unit choices. However, it too fails to lend itself to clarification of planning requirements in the area of human resources.

The product life cycle method has more intuitive appeal. The stages of the cycle of development for services suggests market conditions that lead to critical issues that must be faced by senior management. These issues, in turn, require select elements of the human resources system be used to address the issues that are raised. In addition, the skills required of managers vary according to the stages of the life cycle of the services being managed. Identification of the competencies needed assists in the formulation of criteria for selection, appraisal, compensation, and development of managerial personnel across the business units of the organization.

The sections that follow examine each of the four stages of the life cycle to identify the management issue that must be faced. The human resources function or tactic that is critical during the stage is noted. Finally, the managerial skills most important for handling business units during that stage of the life cycle are listed.

Introduction

As noted previously, during the introductory stage the service is not well-known and considerable time must be devoted to identification and development of markets. Because the best methods for producing the service are relatively unknown, the scope of services offered must be limited, and managers must concentrate on maintaining the quality of services offered. The dominant competitive issue to be faced by the organization is associated with identifying methods for *developing the market.*

Since this is a new venture for the institution, the major concern for the human resources management system is *recruiting and selecting* competent managerial personnel to staff the enterprise. Additionally, others must be recruited to staff all the positions required for the new unit to be able to deliver services. Internal or external sources of personnel must be considered. Criteria for selection of appointments must be developed.

Because of the uncertainty associated with the technology required to deliver the service, the managers responsible for these new ventures must be able to handle ambiguity well. Rapid changes in the methods used to provide the service or in the expectations made of employees will require good interpersonal skills to handle the disturbances that are likely to occur.

Managers and other employees of the health services organization must be entrepreneurial, ready to exploit new markets or variations in service delivery techniques that improve operations. Knowledge of new services marketing tactics is critical. Skills in financing the new services or in budgeting and projecting the economic results of business unit operations are required.

Growth

The growth stage is characterized by increased consumer knowledge of the service and a rise in consumer demand. The largest shifts in relative market share occur during this stage. Improvements in quality and in the availability of differentiated services provide a competitive edge for health care organizations that are able to institute these enhancements. Thus, the significant managerial issue that must be handled during the growth stage is related to the ability of the organization to *meet market demand* for the service.

The abilities to respond to rapid changes and to maintain or increase market share

are improved by the organization whose managers and other health professionals are exceptionally knowledgeable about opportunities occurring in the market and have technical knowledge of the production capabilities of their units. The human resources function that takes on increased importance during this stage is concern for *management development* activities.

Improving the current and future effectiveness of managers during the growth stage is critical for several reasons. First, the managers selected to operate these business units may have general management training but little knowledge of health services, or they may be health professionals without formal training in business and marketing principles. During this stage managers must have both technical product knowledge and marketing-management competence. Required marketing skills include methods for increasing market share or tactics for beginning market penetration. Thus, human resources managers must ensure the planning of experiential or technical training activities that develop the expertise necessary in a growth market.

As demand for the service increases, more organizations will enter the market and current providers will expand operations. Thus, demand will grow for managers knowledgeable about operation of these units, and this is a second reason that management development is critical now. For example, health maintenance organization (HMO) development has mushroomed during the past 5 years, and projections are for more HMO sites to begin operations in the near future. However, large HMO systems have devoted inadequate attention to identifying and developing managers for these new sites. Thus, a seller's market exists for managers with HMO experience. Corporations interested in expanding their HMO operations will be unable to implement such plans successfully unless they can develop human resources for projected growth.

Maturity

During maturity, the rate of increase in growth of demand drops sharply. Consumers are more cost conscious and knowledgeable about service features. The scope of services that was broadened during the growth stage may be a liability now, and pruning of services occurs. *Competition* is the dominant competitive issue facing the organization. Cost analysis and cost-cutting methods are required to keep the services of the organization competitively priced. Defending market share in a mature to declining market is essential for continued survival.

Appraisal and *compensation* are crucial functions performed by human resources during this stage. The character of compensation systems changes from loose, informal methods appropriate for a rapidly changing environment to a more structured, formal approach as industry maturity takes place. Human resources personnel must spend more time and effort developing and fine-tuning this formal system to ensure equity. Compensation also may take on increased significance to managers of mature service units, since less intrinsic satisfaction may be derived from managing these units during periods of increased competition.

Valid appraisal of managerial performance is important: the reduction in demand for an organization's services may be the result of overall market conditions, but it may arise from judgmental errors by management. Additionally, managers of business units may be eligible for promotion into senior positions. The appraisal system must be able to identify successful managers at the business unit level, so that compensation will be equitable, and simultaneously identify those who have talents needed for higher levels of management or for corporate positions.

Decline

The final stage of market conditions is associated with significant drops in consumer demand. An organization may decide to divest itself of the business, to remain in the market for an extended period and harvest profits from the service, or to identify a suitable market niche for the organization; only if it is believed that demand for the service will be relatively enduring will an attempt be made to maintain a market share leadership role. The dominant issue or concern facing management will depend on the strategic decision that is made. Divestment simply means finding a suitable buyer for the service or terminating the business unit. Attempting a harvesting position suggests that *cost control* must occupy managerial attention. Conversely, finding a market niche or attempting to maintain leadership suggests that *redevelopment* is to occur.

When there is a decline in demand for services, fewer business units may be operated and cost reduction strategies are in order. Thus, the major activities for human resources management involve *outplacement* services for employees who will not be retained and *selective retention* of personnel and their reassignment in other operating units of the organization.

Human Resources Management

The foregoing discussion described the management issue during each stage of the life cycle of a product, identified the human resources function or tactic critical during each stage, and suggested managerial skills required for handling business units at various stages of growth. This information is illustrated in Figure 2.11. The association of a human resources function or tactic with each stage of the life cycle does not mean that other functions are not necessary during any given stage. This illustration is used to identify the issues, tactics, and skills that are *most critical* during each stage.

This material serves as a guide to suggested interrelations that should exist between human resources management functions and strategic planning and management activities within health services organizations. Other chapters of the book provide greater detail on the methods for accomplishing these activities.

SUMMARY

Several planning approaches have been used by health services organizations, but emphasis must be placed on strategic planning and strategic management in today's competitive marketplace. Organizational strategies that might be used by health services institutions require determining the "business" of the organization and the methods that will be used to compete in those businesses.

The formulation of the organization's strategy begins with a well-articulated mission statement. Analysis of external opportunities and threats as well as internal strengths and weaknesses provides input in determining the alternative services that a health services organization should include in its portfolio. Portfolio analysis also suggests

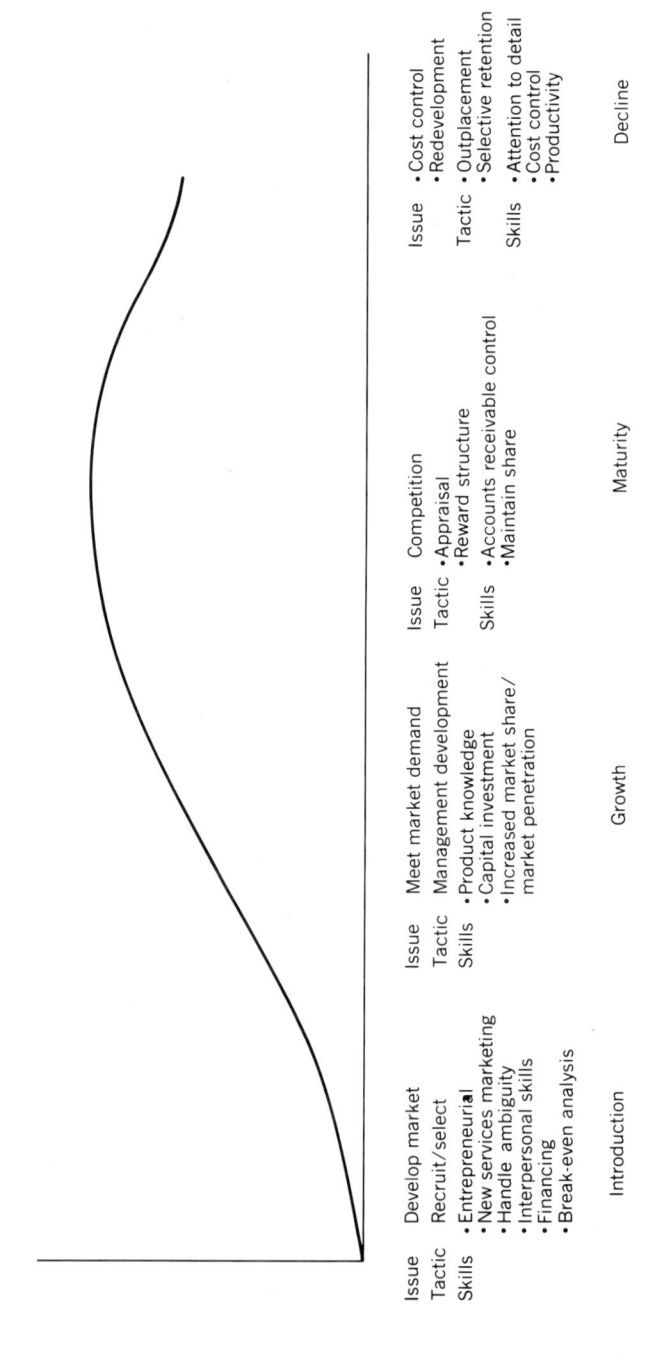

Issue	Develop market	Issue	Meet market demand	Issue	Competition	Issue	Cost control
Tactic	Recruit/select	Tactic	Management development	Tactic	Appraisal	Tactic	Redevelopment
Skills	•Entrepreneurial	Skills	•Product knowledge	Skills	•Reward structure		Outplacement
	•New services marketing		•Capital investment		•Accounts receivable control		Selective retention
	•Handle ambiguity		•Increased market share/		•Maintain share	Skills	•Attention to detail
	•Interpersonal skills		market penetration				•Cost control
	•Financing						•Productivity
	•Break-even analysis						

Introduction Growth Maturity Decline

Figure 2.11. Integration of the human resources management process into a life cycle model.

strategies that should be followed for the services the organization decides to offer.

Relationships exist among business strategies, stages of product life cycle, and human resources functions. Life cycle stages suggest critical issues that must be handled under different market conditions. These issues require the deployment of varying elements of the human resources system to address the issues; the skills needed by managers must vary according to the stages of the life cycle of the service. Identification of the competencies needed assists in development of criteria for selection, appraisal, compensation, and development of managerial personnel across the business units of the organization.

REFERENCES

Aaker, D. A. *Developing Business Strategies.* New York: Wiley, 1984.

Abell, D. F., and J. S. Hammond. *Strategic Market Planning.* Englewood Cliffs, NJ: Prentice-Hall, 1979.

Alfirevic, J., J. Nackel, and D. M. Shade. "Prospective payment and case mix management." *Osteopathic Hospitals,* pp. 11–15 (August 1983).

Andrews, K. *The Concept of Corporate Strategy.* Homewood, IL: Dow Jones-Irwin, 1971.

Ansoff, H. I. "The state of practice in planning systems." *Sloan Management Review,* 18:2, 1–24 (Winter 1977).

Ansoff, H. I. *Implanting Strategic Management.* Englewood Cliffs, NJ: Prentice-Hall, 1984.

Berman, H. J., and L. E. Weeks. *The Financial Management of Hospitals,* 4th ed. Ann Arbor, MI: Health Administration Press, 1979.

Dillon, R. *Zero-Base Budgeting for Health Care Institutions.* Rockville, MD: Aspen Systems, 1979.

Drucker, P. F. *Management: Tasks, Responsibilities, Practices.* New York: Harper & Row, 1974.

Griffith, J. R. *Quantitative Techniques for Hospital Planning and Control.* Lexington, MA: Lexington Books, 1972.

Hall, W. K. "Survival strategies in a hostile environment." *Harvard Business Review,* pp. 75–85 (September–October 1980).

Hax, A. C., and N. S. Majluf. *Strategic Management: An Integrative Perspective.* Englewood Cliffs, NJ: Prentice-Hall, 1984.

Health Central. *Environmental Assessment: Health Care, New Dynamics, New Markets.* Minneapolis: Health Central System, 1983.

Hillestad, S. G., and E. N. Berkowitz. *Health Care Marketing Plans: From Strategy to Action.* Homewood, IL: Dow Jones-Irwin, 1984.

Hofer, C. W., and D. Schendel. *Strategy Formulation: Analytical Concepts.* St. Paul, MN: West Publishing, 1978.

Kimberly, J. R., R. H. Miles, and Associates. *The Organizational Life Cycle.* San Francisco: Jossey-Bass, 1980.

Kotler, P. *Marketing Management: Analysis, Planning, and Control,* 5th ed. Englewood Cliffs, NJ: Prentice-Hall, 1984.

Marrus, S. K. *Building the Strategic Plan.* New York: Wiley, 1984.

McNerney, W. J., et al. *Hospital and Medical Economics: A Study of Population, Services, Costs, Methods of Payment, and Controls,* 2 vols. Chicago: Hospital Research and Educational Trust, 1962.

Mobley, W. H. *Employee Turnover: Causes, Consequences, and Control.* Reading, MA: Addison-Wesley, 1982.

Moore, R. F., Ed. *AMA Management Handbook.* New York: American Management Association, 1970.

Nutt, P. C. *Planning Methods: For Health and Related Organizations.* New York: Wiley, 1984.

Pearce, J. A. "The company mission as a strategic tool." *Sloan Management Review,* pp. 15–24 (Spring 1982).

Perlin, M. S. *Managing Institutional Planning.* Rockville, MD: Aspen Systems, 1976.

Peters, J. P. *A Strategic Planning Process for Hospitals.* Chicago: American Hospital Publishing, 1985.

Porter, M. E. *Competitive Strategy.* New York: Free Press, 1980.

Porter, M. E. *Competitive Advantage.* New York: Free Press, 1985.

Pyhrr, P. A. *Zero Base Budgeting.* New York: Wiley, 1973.

Sammon, W. L. "Competitor intelligence: An analytical approach." In W. L. Sammon, M. A. Kurland, and R. Spitalnic, Eds., *Business Competitor Intelligence: Methods for Collecting, Organizing, and Using Information.* New York: Wiley, 1984, pp. 91–143.

Steiner, G. A. *Strategic Planning.* New York: Free Press, 1979.

Tichy, N. M., C. J. Fombrun, and M. A. Devanna. "The organizational context of strategic human resource management." In C. Fombrun, N. M. Tichy, and M. A. Devanna. *Strategic Human Resource Management.* New York: Wiley, 1984, pp. 19–31.

Webber, J. B. and J. T. Peters. *Strategic Thinking: New Frontier for Hospital Management.* Chicago: American Hospital Publishing, 1983.

Wind, Y. J. *Product Policy: Concepts, Methods, and Strategy.* Reading, MA: Addison-Wesley, 1982.

CHAPTER 3

The Organizational Life Cycle and Strategic Human Resources Management

Myron D. Fottler
Howard L. Smith

In recent years, particular emphasis has been placed on the need to study organizations as they evolve over their life cycle. As mentioned in Chapter 2, each stage in the organizational life cycle may be associated with particular issues and problems as well as particular human resources issues and challenges.

The present chapter discusses the nature of the organizational life cycle, strategies for managing organizational decline, and appropriate human resources management strategies and tactics at each stage of the organizational life cycle. Examples from the health care industry are also provided.

ORGANIZATIONAL LIFE CYCLES

People, products, markets, and even societies have life cycles, which include birth, growth, maturity, old age, and death. At every life cycle stage, a typical pattern of behavior emerges that poses particular challenges to management. As the organization passes from one phase of its life cycle to the next, different roles are emphasized and the different role combinations that result produce different individual and organizational behaviors.

However, it should also be remembered that at any point in the organizational life cycle, various organizational subunits and the organization as a whole may be in different phases of the cycle. Such a situation may necessitate some strategies and tactics different from those required elsewhere in the organization. For example, if a hospital suffering declining inpatient census and revenue decides to start an HMO, it may need to recruit a director who is more entrepreneurial than most of the present hospital administrators. The compensation package would probably involve some incentives for the director based on enrollment and revenue, even though such incentives may not exist within the hospital itself.

Several authors have discussed organizational life cycles and the nature of problems at each stage. We use a four-stage model of start-up, growth, maturity, and decline based on the work of Adizes (1979) and Quinn and Cameron (1983). By observing the development of organizations through the life cycle stages, it may be possible to predict the major problems, decisions, and opportunities to be faced by organizations and to provide some suggestions for appropriate organizational responses.

Start-up Phase

The start-up phase for most organizations is usually typified by creativity and entrepreneurship. Marshalling resources, creating an ideology, and finding an ecological niche are the emphasized activities. The dominant competitive issue is to identify methods to develop a market for the particular good or service. Since the best methods of producing the service are not well-known, the scope of services offered must be limited and managers must concentrate on maintaining the quality of services offered.

Managers must be entrepreneurial and develop or exploit new markets or new methods of delivering existing services. At the same time, they must be adept at raising needed capital and recruiting new personnel with appropriate skills. Building a team from the newly recruited personnel is a challenge that requires either development of an ideology (a concept of what the organization is about) or charismatic leadership.

Producing results is a key value. What counts at this stage of organizational life is not what one thinks, but what one does. Most people in such an organization, including the CEO, are out selling and doing. There are hardly any staff meetings.

The start-up organization is highly centralized and best described as a one-man show. There are few policies, systems, or procedures, and sometimes only rudimentary budgets. The administrative system is very primitive. There is no long-range planning.

The organization is so busy doing that it is usually inundated by short-term tactical pressures. Management often misses some long-term opportunities. When it does perceive opportunities, it tends to move fast and to make decisions intuitively. Since it lacks experience, almost every opportunity becomes a priority.

The dangers to the organization are twofold. First, the organization is spread so thin that it may run out of capital. Second, the personification of its managerial process (i.e., the founder's trap) may preclude development of appropriate administrative structures and processes.

The Growth Phase

The growth phase is characterized by an emphasis on meeting market demands from consumers as well as development of an internal management structure. Increases in consumer knowledge and consumer demand for the service occur at this stage. The organization finds and exploits one or more market niches on the basis of quality or price differentials. There is an emphasis on responding to changes in market demand and exploitations of new markets to continue the pattern of growth.

To handle the growth in revenues and employment while complying with various external requirements, administrative structures are developed. At this stage, there is still a high level of cohesion among employees and commitment to the organization. Face-to-face communication and informal structures are common. There is a sense of mission and dedicated service to the organization. However, the administrative role is increasing in importance, and more time is spent on planning and coordinating meetings. Training programs and various standardized personnel procedures are established. A computer is installed.

These activities may take time away from the entrepreneurial role (which enhances long-run survival and growth) or from producing results in the short term. A "healthy" growth phase is one in which the production of short-term results is temporarily sacrificed for the sake of integrating the required administrative systems.

There is often a tension between those seeking more stability and predictability (i.e., the administrative orientation) and those seeking continued growth and entrepreneurship. The first group feels that policies and organizational systems are necessary if the organization is going to survive and grow in the long run, whereas the second group feels that the first group is stagnating the organization.

At the latter stages of the growth phase, the organization has a healthy balance between a short-term results orientation, a long-term entrepreneurial role, and appropriate administrative policies and procedures. Rates of growth in revenue and profits (or "surplus") have become more stable and predictable.

However, at the latter stages of the growth phase some early signs of aging begin to appear. There may no longer be a great gap between the aspiration levels of top management and expected achievements. As the group becomes more satisfied with the status quo, energy and pressure for change are reduced. The mental aging of top management, the infeasibility of increasing market share, and an ambiguous organizational structure with unclear lines of authority and responsibility can also reduce aspiration levels in the latter stages of the growth phase. When these effects are present, momentum based on past actions becomes the prime determinant of present growth. Slowdown is inevitable.

The Mature Organization

In the mature organization, the rate of increase in revenue and profit drops sharply. There are no serious declines or losses, but the organization has reached a rough equilibrium with its environment. The range of services that expanded during the growth phase may be narrowed as unprofitable or low-volume services are pruned. Competition increases in many of the service lines, and there is an emphasis on efficiency as a means of achieving or maintaining competitive prices.

The maturity phase is also characterized by *formalization* and *control*: Procedures and policies become institutionalized, goals are formalized, flexibility is reduced, and conservatism predominates. The administrative system of policies and procedures is enhanced and institutionalized. Employees spend more time with each other than working on producing results. A climate of personal friendships develops, meetings proliferate, the climate becomes more formal, and a sense of urgency is lost. New ideas are received without enthusiasm, and there is reluctance to rock the boat. An eagerness to excel and a "results orientation" declines. The climate is relatively stable.

What counts in the mature organization is not what you did or why you did it, but *how* you did it. There is an emphasis, for example, on "dressing for success." The managerial dress style would be appropriate for a funeral or a wedding. Form is more important than function.

Individual managers are rewarded for loyalty and for not making waves. Criticisms and threatening questions are not expected or allowed. The organization climate rewards conformity. Those unable or unwilling to adapt to such a climate typically leave. The result is an organization that has become even more insular.

To maintain growth in revenues, the mature organization is inclined to raise prices rather than to generate new products or penetrate new markets. Eventually prices are raised to the inelastic portion of the products' demand curves. Total revenues decline.

Sometimes the mature organization acquires a growth organization to compensate for its own inability to grow or the lack of internal entrepreneurial activity. In other cases, the cash-rich, mature organization is acquired by a smaller, growing organization. In either case, the marriage can be very rocky. The two cultures usually are not compatible. The mature organization's policies and procedures may stifle the growth organization, or perhaps the management of the growing organization is unable to cope with the problems of the mature organization.

In the case of some mature organizations, decline is averted when the organization is able to decentralize, expand its domain (or market), renew its adaptability to the environment, and develop new subsystems. Health care organizations that develop into horizontally or vertically integrated multi-institutional systems represent one example of the phenomenon.

The Decline Stage

In the decline stage, the organization can no longer pretend that the problems identified above are temporary. Significant decreases in consumer demand for services are occurring. The strategic decision to be faced by top management is whether to sell the organization to potential buyers, identify a suitable market niche and try to maintain market share in that submarket, or remain in the present markets and attempt to harvest profits. Which strategy is most appropriate depends on internal and external factors such as whether a suitable buyer exists or whether the organization has the potential to undercut the competition in terms of price.

Within the declining organization, a fight for personal survival begins. Managerial paranoia is rampant. Managerial behavior accelerates the organizational decline as managers fight each other rather the competition. Creative capabilities are directed into ensuring personal survival by eliminating or discrediting fellow employees rather than creating better products and services. The more productive employees are feared because of their productivity and either leave or are fired.

This process eventually results in a stifling bureaucracy, which allows no innovation or change. Very little, if anything, gets done. Systems, rules, procedures, and forms predominate. Conflict is minimal or nonexistent. The organization buffers itself from the external environment by allowing few linkages and ignoring relevant information from (or about) that external environment.

In the decline portion of the life cycle, the administrative role predominates. Employees assume that unless something is explicitly assigned and permitted, it is probably not allowed or expected. Employees do not initiate or take chances. They wait to be told what to do. Finance and accounting managers have much greater power than production or marketing managers. Budgetary controls are emphasized, and new expenditures are rarely approved.

Summary

The start-up and growth phases of the organizational life cycle tend to emphasize strategies Miles and Snow (1978) describe as "prospector." The prospector organization innovates early in the face of product/market opportunities. It also pursues a particular type of "differentiation" strategy (Porter, 1980). This involves creating value that is perceived as unique. It may take many forms, including unique product or service attributes, quality, or service. Costs are not a key ingredient.

At the operational level, there is an emphasis on open communications, informality, and productivity. While the administrative structure is developing during the growth phase, the emphasis is still on entrepreneurial activity and producing results. Aspirations exceed achievements.

During the maturity and decline phases the strategic emphasis is on either cost leadership or focus (Porter, 1980). Cost leaders are often market share leaders and pay great attention to asset use, employee productivity, and discretionary expenses. Consumers purchase the products or services primarily because they cost less than equivalent offerings by competitors. Focus involves competing in a narrow segment based on consumer type, service type, geography, or other factors. In Miles and Snow's (1978) terms, organizations involved with cost leadership or focus tend to be "defenders" or "reactors." A defender maintains a stable offering and may choose to exploit its stability in the form of low costs or other competitive weapons. A "reactor" follows the lead of other organizations and consequently fails to achieve the benefit of early exploitation.

At the operational level, the maturity and decline phases are characterized by bureaucracy and declining internal response to the external environment. The administrative structure becomes an end in itself and innovation is discouraged.

In a later section of this chapter we examine in more detail the nature and causes of organizational decline, appropriate strategies and tactics for managing decline, and organizational constraints in managing decline. From the latter half of the 1980s into the 1990s most health organizations will experience organization decline in whole or in part. The management of declining organizations requires skills, strategies, and tactics different from those used in the management of growing organizations. In addition, the management of human resources in the decline phase of the organizational life cycle requires strategies and tactics different from those used during the growth phase. These differences are discussed in the final section of the chapter.

REINSTITUTING GROWTH

Although the preceding sections have described the typical stages through which organizations pass if they are not actively managed, decline and death are not inevitable. Proactive management can strategically direct an organization to seek out and exploit new markets, reduce unprofitable operations, and avoid or reduce bureaucracy. Even organizations that are in the decline phase may be reinvigorated by "new blood," new ideas, or both. In the health care industry, two major approaches to reversing stagnation or decline are transformational leadership and corporate restructuring.

Transformational Leadership

Transformational leadership typically occurs when there is a significant "performance gap" between what the key stakeholders desire and what they receive from the organization. It typically involves a new CEO or leadership team, which challenges the existing norms, values, and culture. There is an attempt to transform a stagnant or declining organization by adding new people with new ideas and goals, developing a new culture, reducing the influence of certain key individuals not attuned to the new goals, and creating a new vision of the organization and its mission. New leadership is usually required in this process because the old leadership is either unable or unwilling to make such fundamental changes.

According to Kaluzny et al. (1987), the major collective task of any organization is to negotiate an acceptable accommodation with its environment. As the environment changes, so too must the organization, or it risks decline and death. A stagnant or declining organization has obviously failed to respond to environmental change. Transformational leadership, one response to this failure, triggers a search for solutions by the organization.

Organizations initiate actions that have the least invasive effects on themselves and yet achieve acceptable performance (Shortell and Wickizer, 1984). Time and other resources for solving problems faced by various individuals and departments are limited. As a consequence, the search for solutions is also restricted and begins with existing routines and standard operating procedures. Thus, searches are usually executed in a hierarchical fashion, beginning with those that have the least invasive implications for the organization (Smith and Kaluzny, 1986).

Kaluzny et al. (1987) propose the following hierarchical order for transformations required when there is a "performance gap:"

1. *Environmental manipulation.* Activities or programs that affect the environment, to make it more compatible with the organization and its ongoing operations
2. *Process modifications.* Activities or programs that affect internal operating processes without affecting the basic structural configuration or basic mission of the organization
3. *Strategic modifications.* Activities or programs that affect the basic goal or mission of the organization
4. *Reconfiguration.* The development of a new organizational form

Transformational leadership is usually necessary to accomplish the latter two stages and probably not necessary to accomplish the first two stages since the transformations are not as significant and fundamental. However, even with transformational leadership, the process of making strategic modifications or structural reconfigurations is problematic. Attempts to make an organization more compatible with its environment through strategic or structural transformations so that it can continue or resume growth are not always successful.

Since the culture of an organization tends to select and retain certain types of people and to reinforce certain types of behavior, change does not come easily or quickly. Individuals are very resistant to any change they perceive as adverse. Some combination of "new blood," deletion of "old blood," communication of new values and goals, reinforcement of the new values and goals, board support, and the passage of time is required for successful implementation of strategic or structural change.

Such attempted transformations require reinforcement at the operational level, including human resources management activities. Recruitment and selection of all new employees, but particularly managerial employees, must reinforce the new values, goals, and missions. Those with experience or knowledge in the newer growth areas are preferred. Likewise, those with experience or knowledge related to projected areas of decline or deletion should be avoided. Performance appraisal and compensation practices should also reinforce the new values, goals, and missions.

Corporate Restructuring

A second major method by which some health care organizations have been able to avoid stagnation and decline is corporate restructuring. This means adding or deleting divisions, adding new services, and/or becoming part of a multiunit system through merger. Clearly, it is not necessary for an organization to choose *either* transformational leadership *or* corporate restructuring. Both may be implemented simultaneously and often are.

As a solution to their failure to grow or out of concern for simple survival, many hospitals have turned to various collaborative forms referred to as multihospital systems, or simply as systems (Fottler et al., 1982). Such systems are defined as two or more hospitals owned, leased, or contract-managed by a central organization.

The alleged advantages of systems include increased access to capital, reduction in duplication of services, economies of scale, improved productivity and operating efficiencies, access to management expertise, enhanced employee career mobility, easier employee recruitment, improved access through geographical integration of various levels of care, improvement in quality through increased volume of services for specialized personnel, and increased political power to deal with planning, regulation, and reimbursement issues (Fottler and Vaughan, 1987).

The rapid growth in systems between 1960 and 1980 initiated speculation that the majority of all hospitals would join systems by 1990. However, the actual growth has been more moderate than predicted by some. In 1980 systems accounted for 31 percent of all hospitals and 35 percent of all beds. By 1985 these percentages had grown to 38 and 39 percent, respectively (Fottler and Vaughan, 1987). During the same period, the number of systems shrunk from 267 to 250, indicating that consolidation was occurring. A smaller number of larger systems is emerging.

Horizontal integration (i.e., buying additional hospitals) accounted for the bulk of growth in systems during the 1970s. However, from 1980 to 1985 systems have emphasized vertical integration (i.e., purchase of ambulatory facilities) while de-emphasizing horizontal integration. The net impact of these trends is to make it more difficult for unaffiliated, free-standing hospitals to survive in the future by joining systems (Fottler and Vaughan, 1987).

A recent study of acquisition strategies of multihospital systems found that all systems studied were expected to become vertically and horizontally integrated health care services organizations (Alexander et al., 1985). No system saw itself solely as a provider of hospital services. Instead the strategy called for provision of both insurance and health care delivery. Insurance services were usually linked to participation in preferred-provider or health maintenance organizations. Health care services were viewed on a continuum of ambulatory-primary care, acute inpatient care, and postdischarge recuperative or chronic care services.

Systems have become more particular about the hospitals or other organizations they are willing to acquire. They look primarily to favorable market and management fundamentals in determining which hospitals are likely candidates for merger or acquisition. Limits to the future growth of systems, in addition to their reluctance to purchase marginally profitable facilities, include limits on access to capital, antitrust constraints, and a lack of leadership vision (Alexander et al., 1985).

What are the human resources implications of acquisition of free-standing facilities by large systems? From the viewpoint of the acquired institution the worst-case alternative to not being taken over is that all employees will lose their jobs. In the less urgent event of a slow decline, employment will decline by attrition, salary increases will be small or nonexistent, and opportunities for advancement few. So the benefits to the employees of the acquired institution may include continued employment, merit salary increases, and career mobility. In addition, research indicates that systems attempt to stabilize the work force and to improve personnel recruitment and retention policies (Alexander et al., 1985).

However, the chief executive officer in the acquired hospital is seldom kept on. Department directors and other administrative staff may also lose their jobs after acquisition. If the hospital is overstaffed, nonadministrative personnel may be laid off.

Management of the system faces the challenge of modifying the culture of the newly acquired institution, integrating it into the existing system, and returning it to profitability. In addition to implementing new accounting and management information systems, the system must recruit, retain, appraise, and compensate employees in the newly acquired organization in a way that will reinforce the values, missions, and goals it wishes to pursue. Interim conflicts are to be expected. Complete integration of the newly acquired organization is usually a long-term process.

Thus, the search for continued growth is eventually problematical in all organizations. Only proactive management policies that continually examine the organization's structure, strategy, and processes can hope to perpetuate such growth. Even in these cases, however, the challenges are significant and short-term setbacks almost inevitable. Human resources management strategies and tactics must reinforce the strategic and structural changes management is attempting to make.

ORGANIZATIONAL DECLINE

The term "decline" has two principal meanings in the organizational literature (Durham and Smith, 1982). First, it is used to denote a cutback in the size of an organization's work force, profits, budget, and clients. It has also been used to describe the general climate or orientation of an organization that shows evidence of stagnation, bureaucracy, and passivity. This latter condition of decline may or may not result in a loss of revenue. Decline-as-stagnation does not necessarily imply an absolute decreased revenue, whereas decline-as-cutback does. Stagnation is more often reflected as a decrease in the rate of increase than as an actual decrease. Stagnation is more likely to occur during periods of abundance, whereas cutbacks are more likely during periods of scarcity. This chapter focuses primarily on decline-as-cutback.

The most frequently cited causes of organizational decline reflect administrative inadequacies, administrative motivation, and external factors. In the for-profit sector, the first two causes are considered primary, whereas in the nonprofit and public sectors the third appears to be the most important. Problems of decline in hospitals are typically due to changes in their external environments, particularly regulatory and reimbursement changes (i.e., prospective pricing systems). Lack of support in the external environment may also result from the inability of administrators to adapt to changes in that environment. (Levine, 1978; Pfeffer and Salancik, 1978).

Table 3.1 indicates the major causes (both external and internal) of decline in health care organizations in the 1980s. The reasons for organizational decline are not these changes alone or mismanagement per se. Rather, decline is due to the inability of health care executives to adapt to those changes by making necessary modifications in strategy or tactics.

Levine (1979) has defined four basic causes for loss of external environmental support. *Organizational atrophy* or *inertia* results when an organization habitually maintains programs based on their previous (rather than present) utility. Previous success may desensitize managers to environmental changes, and they therefore become more vulnerable to future failure. *Political vulnerability* is reflected in a public agency's inability to resist budget cuts. Factors contributing to such vulnerability (according to Levine) include small size, internal conflict, changes in leadership, lack of a base of expertise, and the absence of a positive self-image and history of excellence.

TABLE 3.1. Major Causes of Organizational Decline in Health Care in the 1980s

I. External factors
   A. Declining support in the external environment
      1. Shift from "rapid-growth" to "slow-growth" industry
      2. Shift from "open-ended" to "close-ended" funding
   B. Increased emphasis on competition and cost containment
      1. Oversupply of hospital beds and physicians
      2. Prospective reimbursement for Medicare patients based on average regional costs for 467 diagnosis-related groups (DRG)
      3. State laws (Massachusetts and New Jersey) limiting reimbursement
      4. Competitive bidding in California for both public and private insurance contracts
      5. Unwillingness of private corporations and private insurers to allow significant "cost shifting" to private patients
      6. Growth of employer coalitions to reduce costs
   C. Capital shortage

II. Internal factors
   A. Lack of management experience in managing retrenchment
   B. Inability of management and other members of the dominant coalition to correctly diagnose the causes of decline
   C. Unwillingness or inability of management to implement necessary changes due to inertia, lack of understanding, or a weak power position

Problem depletion is a major cause of loss of legitimacy. Benson (1975) has stressed the importance of legitimacy as an organizational resource and has noted the tendency of some public agencies to overemphasize the acquisition of resources and to overlook the value of cultivating political acceptance. Public funding support for particular organizations, for example, tends to rise and fall depending on the perceived legitimacy of the values such agencies embody. *Environmental entropy* results from the reduced capacity of the environment to support a particular organization. Whereas the literature on business organizatios emphasizes finding another market niche, the literature in the public sector emphasizes scaling down operations.

Changes in Ecological Niches

A major external cause of organizational decline is a significant change in the ecological niche in which the organization exists. A niche is the environmental habitat of a population of organizations. It can be defined as a set of physical, biological, and social conditions that provide resources for or place constraints on the performance of an organizational population (Zammuto, 1982). Moreover, niches in a social exchange system are partially determined by the willingness of people to purchase the products or services produced by a population or organizations (Boulding, 1981).

The organizational domain refers to that part of the population's niche in which each organization operates. An organization's domain is defined by the clientele served, the technology employed, and the services or products produced (Meyer, 1975). It is identified by examining the major activities pursued by an organization (Cameron, 1981). It is that part of the niche that each member of the organizational population creates and inhabits.

Changes in ecological niches may produce conditions of decline for a population of organizations. Adaptation by individual organizations in an evolving niche occurs through domain responses. Organizations modify their domains in response to changes in the population's niche (Zammuto and Cameron, 1985). For example, a niche that remains constant in shape might be shrinking in size. A domain response would be to begin shifting activities to other niches that are not shrinking.

The two dimensions of niche change (shrinkage or shifting) can be used to construct a typology of decline. The cells in Figure 3.1 represent four different types of environmental decline that vary by type of change in niche configuration and by the continuity of change (Zammuto and Cameron, 1985). Organizations that encounter a continuous decrease in the size of their niche experience *erosion.* This form of environmental decline reflects a gradual reduction in the amount of organizational activities the niche will support. An example of erosion is provided by the experience of many elementary school systems during the 1970s. A decreasing birthrate during the 1960s produced a gradual and continuous reduction in enrollments during the 1970s, requiring many school systems to reduce their level of operations.

Organizations that encounter a discontinuous decrease in the size of their niche experience *contraction*. A sudden decrease in available resources or in the demand for the goods and services produced is often the cause of contraction. The sudden contraction of the market for nonfilter cigarettes in 1954 after widespread publicity linking smoking to various health hazards (Miles and Cameron, 1982) provides an example of this type of decline.

Dissolution is the form of decline that occurs when one niche evolves into another. For example, the graduate shift in the demand for educational programs during the

	Continuity of Environmental Change	
	Continuous Change	Discontinuous Change
Change in Niche Size	Erosion	Contraction
Change in Niche Shape	Dissolution	Collapse

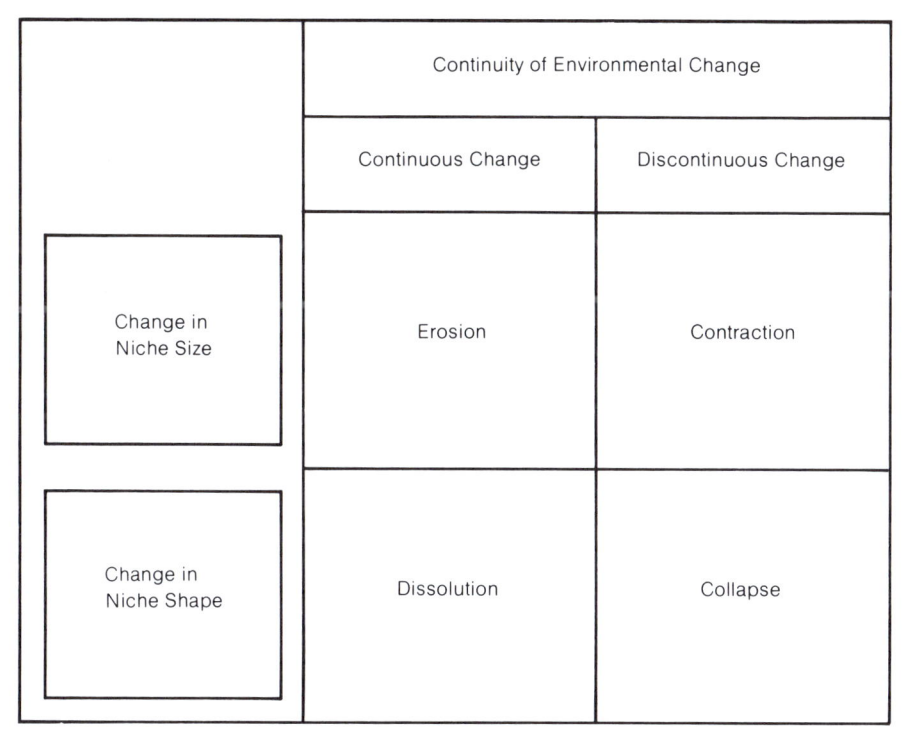

Figure 3.1. A typology of environmental decline. From Zammuto, R. F., and K. S. Cameron. Environmental decline and organizational response. In B. M. Staw and L. L. Cummings (Eds.), *Research in Organizational Behavior*, Vol. 7. Greenwich, Connecticut: JAI Press, 1985. Reprinted by permission.

1970s, from the humanities and social sciences to the professions and applied life and physical sciences, created conditions of dissolution for the college and university population (Zammuto, 1984).

Collapse refers to a rapid, dramatic condition of decline in which the existing niche of a population is more or less abolished and replaced by a new niche demanding different forms of performance. For example, fluorocarbon producers in the United States experienced collapse during the 1970s after scientists predicted that fluorocarbon propellants and coolants would deplete the ozone layer in the upper atmosphere and result in globally higher rates of skin cancer. The federal government quickly banned the use of fluorocarbons in most products (Post, 1978), and the niche for producers of other propellants, such as carbon dioxide, rapidly expanded.

The cell that best describes the health care industry in general during the late 1980s and 1990s is dissolution. The health care niche is experiencing *continuous* change in *shape*. Of course some segments, such as inpatient care, may erode or contract. This typology is discussed later in the chapter when we examine potential managerial strategies and tactics for preventing or reversing decline.

Table 3.2 outlines the major common symptoms of health care organization decline, including individual symptoms, organizational symptoms, and organizational outcomes.

TABLE 3.2. Common Symptoms of Organizational Decline in Health Care Organizations

I. Individual symptoms ("crisis syndrome")
    A. Increased withdrawal behavior
    B. Increased stress and cognitive rigidity
    C. Increased interpersonal conflict
    D. Poorer decision quality
    E. Reduced productivity, job satisfaction, and commitment
    F. Increased resistance to change
    G. Poorer physical and mental health
    H. Shorter time horizons and fantasy

II. Organizational symptoms
    A. Planning
        1. Management failure to admit that deterioration exists
        2. Inability and/or unwillingness to make necessary changes
        3. Leadership vacuum and lack of a clearly defined strategy
        4. Inability of the organization to adapt to changes in the external environment
        5. More unproductive meetings
        6. Lack of innovation and emphasis on routine functions
        7. Short-term orientation with little long-range planning
        8. Persistence of ineffective strategies
    B. Organizing
        1. More centralization and formalization of structure
        2. An increasing proportion of administrators
        3. Lack of a system for generating organizational renewal
    C. Directing and motivating
        1. Less encouragement of subordinate involvement and participation
        2. Increased interunit conflict
        3. Increased employee resistance to change
        4. Failure to define objectives and reward employees based on contribution
        5. Unwillingness of competent managers to remain with deteriorating organization
    D. Decision making
        1. Decision-maker responses that exacerbate the problem
        2. Emphasis on "group think"
    E. Controlling
        1. Lack of quantitative performance measures
        2. Too many controls
        3. Control by complaint
        4. Emphasis on efficiency rather than effectiveness

III. Organizational outcomes
    A. Revenues increasing more slowly than expenses
    B. Competitive weakness
    C. Low occupancy rates

Early Symptoms

Whetten (1980a, 1980b) has observed that our society inculcates values of self-determination and self-confidence. We are taught that success can be willed and that no obstacle should deter us from reaching an objective. As a result, administrators may fail to admit that decline exists until it has become apparent to all (Scott, 1976; Starbuck and Hedberg, 1977). Budget cutbacks may be viewed as *temporary* due to transient economic conditions. As a consequence, it is not unusual to observe administrators who are unwilling and/or unable to plan for necessary changes. Other common planning symptoms include a leadership vacuum, lack of a clearly defined strategy, various administrative extremes (e.g., too much innovation in the development of services, or too little), and failure to adapt to rudimentary changes in the external environment (Ford, 1980a).

Early organizing symptoms include the lack of a system for generating organizational renewal and extremes in formalization and centralization (Levine, 1979; Miller, 1977). High degrees both of formalization and centralization, as well as low degrees of each, are associated with decline (Thompson, 1967). Highly structured organizations tend to filter information, while "loose" structures tend to provide too many gaps in their communications systems.

The process of motivation usually suffers from a failure to define objectives and reward employees based on contribution, as well as an unwillingness of competent administrators to remain with the deteriorating organization (Scott, 1976). Perhaps as a result, decisions made by administrators at this point are often inappropriate and tend to exacerbate the problems associated with decline (Hedberg et al., 1976). The area of control is often deficient in the early stages of decline because a negative feedback control system is lacking, as well as quantitative performance measures (Miller, 1977; Summers, 1977). It should be noted that studies discussing these early symptoms are based on casual observation and case studies.

Later Symptoms

While the earlier symptoms tend to persist, new and potentially more damaging symptoms develop as the decline process continues. Much of our present understanding of these symptoms is a result of inferences from studies on the effects on *individuals* of crisislike events such as disasters and sudden bereavement or personal loss (Fink et al., 1971; Janis and Mann, 1977). Responses to these events tend to follow a pattern of initial shock (numbed inaction), then "defensive retreat" (denial, refusal to change), and then a period of acceptance of the new situation and eventual adjustment to it. When this pattern is applied to organizations, it is the first two stages, which have been labeled "the crisis syndrome," that have received the most attention (Hermann, 1963). Empirical evidence based on studies of a wide variety of organizations lends support to this conceptual framework (Dunbar and Goldberg, 1978; Hall and Mansfield, 1971; Starbuck et al., 1978).

One of the most interesting case studies of the crisis atmosphere that usually accompanies the onset of organizational deterioration examined two state hospitals forced to merge and to reduce their work force in the face of budgetary cutbacks (Jick, 1979a, 1979b). The characteristics of the crisis syndrome emerged in full view as job insecurity and general uncertainty swept the organization. Each dimension of the early crisis phases was clearly observed, as interpersonal relations, communication, and leadership deteriorated.

This research creates a picture of considerable stress experienced by the majority of employees as a result of budget cutbacks (Holsti, 1978). The morale of those who remained in the organization was severely damaged, and their productivity declined. Many other skilled employees left for what they saw as better opportunities. The net result was understaffing due to a combination of employee attrition and loss of motivation among those who remained. Studies of the crisis syndrome have empirically demonstrated these symptoms over and over again in nonprofit public sector organizations.

The crisis syndrome typically is accompanied by a series of organizational effects that are much more serious from the viewpoint of the long-term viability of the organization. First, administrators often begin to participate in an increasing number of meetings to plan short-run strategy or resolve conflicts related to the obvious organizational decline (Jick, 1979a). A system for administering change may be almost totally lacking (Summers, 1977). Also, there tends to be less long-range *planning* (sometimes none) and more orientation to the short run (Summers, 1977; Whetten, 1980b). Long-range planning is difficult due to increased interpersonal and interunit conflict (Levine, 1979), which makes it harder to get good information for problem identification and problem solution purposes (Hall, 1976; Smart and Vertinsky, 1977; Starbuck et al., 1978). The tendency to centralize decision making may be attributable to the environmental hostility, which demands quick responses and rapid coordination of organizational activities (Mintzberg, 1979). To survive in the short run, the organization chooses centralization. This tendency does not contradict the earlier discussion indicating that some declining organizations exhibit extreme decentralization and low degrees of formalization. Ford explains this phenomenon by proposing that a structural hysteresis occurs, wherein the impact of growth and decline on administrative ratios is asymmetrical (Ford, 1980b).

Third, the *motivation* of employees in the later stages of organizational deterioration tends to be extremely problematical. There is less encouragement of subordinate involvement and participation in all areas. The organization exhibits a complicated tangle of rules and regulations, extensive paperwork and forms, and the need of involve several departments to achieve even minor goals. Many individual symptoms (noted above), such as employee resistance to change and conflict, occur at the departmental and organizational level as a result of these motivational problems.

Fourth, the crisis environment affects *decision-making* by discouraging innovative solutions to problems and encouraging "group think" (Janis, 1972). This tendency toward conformity is reinforced by the high turnover rates, and by the increased proportion of long-term employees that results when those who leave are not replaced. Employees who stay in a declining organization are more likely to share common norms and values, to reach consensus more easily, to have similar perceptions of the organization, and to be more committed to it (Starbuck, 1976). As a result, habitual or routine ways of perceiving the environment tend to predominate. Research has shown that high degrees of identification with the organization and long employee tenure are negatively related to certain measures of effectiveness and creativity (Rondine, 1975).

Finally, the *control* system at the later stages tends to operate by complaint (Summers, 1977). The assumption is that if a particular function is not performed up to the standards expected, someone will complain. A reduction in complaints, or their absence, is viewed as an indication that all is well. There might also be too many or too few controls (Miller, 1977). The control system also tends to emphasize efficiency measures rather than effectiveness measures, which might have greater long-term potential for reversing the deterioration process (Miller and Friesen, 1980; Whetten, 1981).

It is emphasized that both the crisis syndrome and the tendency of administrators to respond to severe decline through increasing centralization and formalization have been well documented through empirical work. The other later symptoms of decline discussed in this section have much less empirical support and should be viewed as hypotheses for further testing.

STRATEGIC AND TACTICAL RESPONSES

Strategic responses refer not only to reactions or adaptations of the organization, but also to practice and anticipatory moves as well. Organizations are not completely at the mercy of an immutable environment, with no choice but to adapt. As noted earlier, they can implement strategies that influence and even alter the population niche in which they exist (Miles and Cameron, 1982; Pfeffer and Salancik, 1978). Strategic responses, therefore, refer to actions taken by organizations to alter their domains and influence the niche in which they operate. Our model contains strategies of five types: domain defense, domain offense, domain creation, domain consolidation, and domain substitution (Zammuto and Cameron, 1985).

Domain defense strategies are oriented toward preserving the legitimacy of the existing domain of activities and buffering the organization from encroachment of the environment. These strategies are externally oriented and frequently involve attempts to manipulate or change the nature of the organization's environment (e.g., generating coalitions of support among key resource providers). Domain offense strategies are designed to do more of what the organization already does well. That is, these strategies help to generate organizational slack by expanding current activity levels (e.g., expanding the market for current products or introducing new forms of a product). Domain creation strategies supplement current domains of activity with completely new domains. Diversification and innovation are often a part of domain creation strategies (e.g., forming a subsidiary in another domain, where current organizational expertise can still be utilized).

Domain consolidation strategies involve reducing the size of the domain occupied by the organization. Activities peripheral to the core domain of the organization are eliminated, and the organization becomes more specialized (e.g., divesting unprofitable divisions or products). Domain substitution strategies are designed to replace one domain with another. In the extreme case, the carrying capacity of the original niche completely disappears; no evidence of the earlier domain is left after domain substitution occurs. For example, when the cure for polio was found, the March of Dimes became an organization oriented toward birth defects. Table 3.3 provides some examples of specific approaches some health care organizations are currently implementing under each of the five generic strategies.

The strategic responses presented in the model are expected to be the most effective strategies for coping with the particular type of decline identified. However, as mentioned before, they are not expected to be the only strategies selected by organizations under these conditions, because many types of strategy can be pursued

TABLE 3.3. Some Examples of the Five Strategic Alternatives for Managing Decline in Health Care Organizations

Domain Defense
- Developing or enhancing political coalitions to reduce threat
- Becoming a preferred provider for employees of particular organizations
- Activating boards of trustees for political action

Domain Offense
- Providing enhanced services to a target segment of the service community (e.g., no-delay admitting procedures for certain employee groups)
- Increased advertising of services through various media
- Sponsoring and/or participating in community health care events

Domain Creation
- Opening new clinics in areas of wellness, stress management, and chemical dependency
- Offering contract management services to other hospitals
- Opening satellite facilities with ambulatory services

Domain Consolidation
- Divesting unprofitable or low-volume inpatient services
- Divesting unprofitable units (e.g., HMO)

Domain Substitution
- Substituting free-standing emergency clinic services for in-hospital emergency services
- Substituting commercially insured or self-pay patients for Medicare or Medicaid patients

concurrently or in sequence. That is, organizations may perform more than one strategy at a time, or they may pursue one strategy for a time and then change to another. The strategies identified in this model, however, are more likely to foster successful adaptation to the different conditions of decline than are domain strategies of other types.

Figure 3.2 summarizes the four conditions of decline together with their impact on competition, the types of organization likely to be successful, and the appropriate strategic response when the conditions of decline are predicted in advance so that there is adequate lead time (e.g., continuous change) and when they are not (discontinuous change).

Erosion

Under the condition of erosion, competition within the population will gradually increase as the niche shrinks. Specialist organizations will be better able to adapt because of their greater efficiency. Generalist or strategist organizations have fewer economies of scale in resource acquisition and utilization.

When erosion is predicted, domain offense is likely to be the best strategic response. The goal of domain offense is to expand the resource base and the market share, and implementing this strategy helps the organization to ensure that it will maintain or increase its share of the existing niche, even though that niche is getting smaller. This

| | Contingency of Environmental Change ||
	Continuous Change	Discontinuous Change
Change in Niche Size	**Erosion** Competition: Slow increase Successes: Specialists Strategies: Domain offense predicted domain consolidation not predicted Tactics: Small, incremental change (fine-tuning)	**Contraction** Competition: Rapid increase Successes: Specialists Strategies: Domain defense if predicted; domain consolidation if not predicted Tactics: Change by deletion; substantial selective or across-the-board cutbacks
Change in Niche Shape	**Dissolution** Competition: Moderate increase Successes: Generalists Strategies: Domain defense then domain creation if predicted; domain substitution if not predicted Tactics: Search for new alternatives; change by addition	**Collapse** Competition: Overall decrease Successes: Strategists Strategies: Domain creation if predicted; domain substitution if not predicted Tactics: Change by substitution; trial-and-error use of past solutions

Figure 3.2. Strategic and tactical responses to four types of organizational decline. From Zammuto, R. F., and K. S. Cameron. Environmental decline and organizational response. In B. M. Staw and L. L. Cummings (Eds.), *Research in Organizational Behavior*, Vol. 7. Greenwich, Connecticut: JAI Press, 1985. Reprinted by permission.

strategy prevents the organization from being squeezed out of the shrinking niche by competition from stronger or more specialized organizations.

When erosion is not foreseen, or when the organization is faced with a short lead time, domain consolidation is predicted to be the more appropriate initial strategy. Because implementing domain offense often requires substantial lead time, an organization is expected to focus on strategies of consolidation. This has been the response of many colleges and universities to declining revenues during the 1980s. At the University of Wisconsin, for example, three different incremental cuts in the budget were received throughout the 1980–1981 academic year. These cuts had not been unanticipated because the budget had already been appropriated by the legislature (Kauffman, 1982). The response of the institution was to eliminate all nonessential expenditures and activities. Photocopying was restricted, as was faculty travel. Contingency funds were eliminated, and the availability of noncore courses was limited. Resources were consolidated and focused solely on core areas of acknowledged high priority.

Since erosion involves continuous shrinkage in the size of the niche, it is not likely to present an immediate threat to organizational survival. Management tactics will therefore emphasize small incremental adjustments, fine-tuning, and a redistribution of resources to improve efficiency. A conservative stance of "weathering out the storm" is typical (Hall and Mansfield, 1971). Examples of what health care organizations have done to fine-tune their organizations to improve efficiency include hiring freezes to reduce labor costs, reeducation of physicians concerning costs, and general institutional cost containment programs.

Contraction

Under conditions of contraction, there is a discontinuous reduction in niche size. Competition increases dramatically within the industry as each organization tries to protect its share of services in the face of declining demand. Again specialist organizations have a competitive advantage because of their greater efficiency. Given the lack of adequate lead time, organizational failures are much greater among nonspecialist organizations compared to the erosion cell.

Strategically, when contraction (e.g., sudden shrinkage in niche size) is predicted, domain defense strategies are the most likely to be successful. The organization must preserve its legitimacy and ensure that its domain remains viable, even though the niche is smaller. Domain defense strategies are implemented to help buffer the organization from environmental encroachment. When contraction is not predicted, domain defense strategies, which require lead time to implement, are not likely to occur. Therefore, domain consolidation to preserve the organization's core domain is expected. The organization has little time and few resources with which to respond when an unexpected decline in the resource base occurs, and consolidation, or marshalling resources around a core area of expertise, is the most likely strategy.

The most common tactic in response to contraction is a "threat-rigidity response" (Staw et al., 1981), which leads to cutback and retrenchment (e.g., change by deletion). Minor adjustments are unlikely to be effective because of the suddenness of the niche shrinkage. Thus, the magnitude of the tactical response is usually larger than in the erosion cell. Elimination of weaker services and significant layoffs of employees are examples of tactics used by health care organizations to respond to contraction.

Dissolution

Under conditions of continuous change in niche shape, there will be a moderate increase in competition as the carrying capacity of the original niche is reduced by movement toward a new niche. Generalist organizations will exhibit a greater adaptive potential than specialist organizations because their broader domains offer more options for responding to changes in niche shape. Increased failures will be experienced by specialist organizations with the "wrong" specializations.

The dissolution cell describes continuous environmental change: Here a decrease in the acceptability of organizational outcomes creates a change in the shape of a niche. This type of decline is likely to produce incremental, but proactive, adjustments, and the search for new alternatives will be a central activity. Because organizations must change the nature of their activities to survive, change by addition is most likely to be effective. Organizations facing decline by dissolution, therefore, may actually choose to *expand* for the sake of adapting.

When dissolution is foreseen, effective organizations are predicted to implement domain defense strategies in an effort to build political slack and to preserve their legitimacy and the acceptability of their outputs. The orientation is toward exploiting the environment so that the impact of niche change on the organization may be inhibited. Domain defense is expected to be followed, however, by domain creation strategies so that the organizational domain can be located in a more viable part of the niche. Domain creation diversifies organizational domains and decreases the risk of having a continued state of dissolution, where the outputs of the organizations finally become completely unacceptable.

When dissolution is not predicted in advance, domain substitution is the strategy likely to be most effective. Caught by surprise, the organization is required to change domains to survive. Confronted by qualitative niche changes, the organization is expected to engage in different activities to facilitate its entry into the new niche. With no advance prediction of decline, there is little possibility of altering the environment itself (as in domain defense), so the strategic emphasis must point toward guaranteeing organizational survival. Substituting a supported domain in the evolving niche for the threatened domain is therefore a likely response of successful organizations.

Examples of the expected tactics to deal with dissolution in the health care field are long-range planning and market research to diagnose outlook for various market segments. Open discussion of the possibility of offering new and innovative ambulatory and long-term care services is also important.

Collapse

In the fourth cell, decline by collapse, the shape of the niche changes suddenly. Competition is likely to decrease because the rapid collapse of the original niche drives many competitors out of business. Strategists—organizations that can move quickly to take advantage of rapidly evolving portion of the niche—are most likely to be successful.

Experimentation with past successful activities is the most likely tactical response, since threat produces a constriction of the new alternatives considered in search procedures, although it does not constrict search per se (Staw et al., 1981). Structural adjustment is expected to occur by substituting more acceptable activities for old activities, but the substituted activity is expected to be within the institution's area of expertise. The first satisfactory alternative is likely to be accepted because there is neither the time nor the inclination to search widely (Hall and Mansfield, 1971; Hermann, 1963).

When collapse is predicted, domain creation strategies can be implemented. These strategies allow the organization to form and nurture new domains of activity in addition to current domains. The organization becomes more generalized or diversified in an attempt to adapt to the collapsing niche. The goal of domain creation is to maintain support for the activities conducted in what remains of the original niche while expanding into the evolving portion of the niche or into other less threatened niches.

Nonpredicted collapse leaves the organization with little choice but to substitute an already existing domain that has some support for the domain that is now threatened.

Lead time is not available to nurture a diversification effort (i.e., domain creation). Instead, the organization concentrates on maintaining viability. Domain substitution, therefore, is the most likely strategic choice.

CONSTRAINTS IN THE MANAGEMENT OF DECLINE

Health care executives are typically not prepared to cope with, or to manage effectively, conditions associated with organizational decline. Even if they are prepared, there are severe constraints on their freedom of action.

First, the experience of most health care executives has largely been in responding to conditions of growth. Abundant financial resources and steadily increasing demand for services made conditions of growth almost universal during the 1950s, 1960s, and 1970s. Expansions of physical plant facilities and services were typical.

Second, the values and ideology of our culture emphasize growth and expansion as indicative of effectiveness. Whetten (1980a) has noted that large size is widely lauded as a desirable organizational characteristic because it enhances economies of scale and the organization's ability to absorb the shocks accompanying environmental change. Managers are typically evaluated positively if they produce more, obtain a larger budget, or expand their organizations. When the reverse occurs, negative evaluations normally result.

Third, factors operating in the institutionalized environments of health care organizations often severely constrain the available strategies and tactics. For example, the ability of many public sector health care organizations to modify their domains is limited by the corporate charter. Domain creation or consolidation strategies may not be legal. Moreover, the civil service systems under which they manage their human resources may constrain tactical options. Minimum staffing, personnel licensure, promotion policies, and indigent care requirements are all examples of such constraints.

Finally, administrators in both health care and non-health care organizations tend to define conditions of decline exclusively as a resource allocation problem or a problem of efficiency (Cameron, 1983). They respond conservatively rather than innovatively. Whetten (1981) has noted several examples of administrators in higher education who have emphasized internal resource allocation aimed at operating more efficiently at the expense of longer term strategies for ensuring effectiveness.

The same phenomenon has been observed in health care organizations and is attributed to the conservative nature of the dominant coalition and its unwillingness to propose solutions not compatible with the existing organizational culture (Fottler et al., 1986). Not only are these conservative approaches contrary to prescriptions from management theory, but empirical evidence indicates that the exclusive emphasis on conservative, efficiency-oriented coping mechanisms leads to ineffective performance and even organizational death (Hedberg et al., 1976; Starbuck et al., 1978; Whetten, 1980a).

Whetten (1981) has argued that managers tend to focus on efficiency at the expense of effectiveness when facing decline and to respond conservatively rather than innovatively for at least six reasons. First, organizational effectiveness has been extremely difficult to define and measure, particularly in service industries. Since efficiency is more easily measured than effectiveness, it is given more attention by administrators. Most institutions have budget (efficiency) monitoring devices in place, but few have any mechanisms to measure effectiveness.

Second, the stress resulting from having to face conditions of decline compels individuals to engage in conservative and self-protective behaviors (Whetten, 1980a). A common side effect of decline is personal stress among managers. Research has shown that the consequences of such stress are (a) engaging in anxiety-reducing behaviors at the expense of problem-solving behaviors, (b) reducing the risk of mistakes (which are more visible under conditions of decline) by becoming conservative, (c) restricting the communication network, (d) reducing the number of participants in decision making, (e) enforcing rules more closely, (f) rejecting contrary evidence more readily, (g) perceiving the tasks and decisions to be more difficult, and (h) being prone to "group think" (Hall and Mansfield, 1971; Hermann, 1963; Janis, 1972; and Staw et al., 1981).

Third, there is a tendency to pursue strategies that were successful in the past (e.g., during conditions of abundance and growth) even though they are inappropriate under current conditions of decline (March and Simon, 1958). In times of abundance, the major concerns were related to resource allocation rather than resource acquisition. The same pattern often carries over to the decline phase of the organizational life cycle, where it is clearly inappropriate except as a stopgap to buy time for implementation of a long-term resource acquisition strategy.

Fourth, the tendency toward passive leadership by administrators is also consistent with attribution theory, according to which individuals tend to believe that success (e.g., growth) is due to personal factors, and failures (e.g., decline) to environmental factors beyond their control. Since the causes of decline are outside the administrator's control, no proactive responses are forthcoming.

Fifth, health care institutions are frequently structured as loosely coupled systems, governed by committees and semiautonomous units. They are similar to political organizations in having multiple stakeholders to satisfy, each with vested interests, some of which are conflicting. Multiple interests groups and semiautonomous units make consensus on any decision unlikely, and resistance from some group to almost any strategy is virtually guaranteed. A consistent, innovative strategy for coping with decline is difficult to develop.

Miles and Snow (1978) refer to organizations without consistent strategies as "reactors." Research has found them to be the least effective of all organizational types (Snow and Hrebiniak, 1980). Since stakeholder conflict is heightened under conditions of decline, administrators are even more likely to "satisfice" by adopting conservative (e.g., efficiency-oriented) strategies to ameliorate this conflict.

Sixth, many creative and innovative managers are the first to leave an organization when decline occurs (Hirschman, 1970). These individuals are often the most marketable to other organizations. Declining organizations often do not attempt to retain these individuals since their entrepreneurial talents may not be perceived as valuable under decline conditions. The managers who are left as decline accelerates are often the least innovative ones, who tend to play it safe.

Finally, innovation itself may be viewed as one of the causes of decline. When rapid growth is the norm, organizations often experiment and expand to the point that services and facilities are difficult to maintain with stable or declining resources. Previous innovations are now viewed as financial burdens. One response is to eliminate innovative services and to avoid instituting alternatives.

HUMAN RESOURCES ISSUES AND THE ORGANIZATIONAL LIFE CYCLE

Matching Human Resources to the Organizational Life Cycle

There are a number of human resources issues that are more and less important at different points in an organization's life cycle. The growth phase poses problems and challenges different from those of the decline phase and thus requires both different business strategies and different human resources strategies.

Assume that a hospital in the maturity phase of its organizational life cycle has diversified by acquiring an exercise/wellness center. Should the compensation package of the parent be extended to the new addition, which competes in an industry where benefits are not as generous? Or again, assume that a hospital whose inpatient business has been stable for many years finds that its health maintenance organization is growing very rapidly and holds a strong position in an expanding market. Are the hospital's traditional criteria for selecting, promoting, and compensating health care executives appropriate to the management of this growing sector? Do personnel policies in recruitment, selection, training, and compensation developed during a hospital's growth cycle still fit during the maturity phase and the decline phase?

These examples indicate not only that different human resources policies are called for during different phases of the life cycle, but that a given organization might have different subunits at different points in their life cycles at the same time. In the latter case the application of uniform personnel policies to all subunits would be inappropriate.

The conditions of decline facing some segments of the health care industry (e.g., inpatient hospital care) stem from both federal policy (e.g., prospective payment systems) and increased competition. The duration of the decline promises to be long term for these segments. The conditions of decline are serious enough to threaten the survival of many health care organizations, whose choice of domain is generally limited by their stakeholders. Multiple stakeholders exist both inside and outside these organizations, demanding different types of performance and threatening to withdraw support. The demand for traditional services is decreasing, and public support is tenuous.

For all these reasons, the management of organizational decline in health care organizations is particularly problematic. Once a strategy for preventing or reversing organizational decline has been developed, its successful implementation depends on how well human resources are managed.

Matching Human Resources to Strategic Objectives

The examples above also illustrate the problem of matching human resources to strategic management of the organization. Effective organization and management of human resources is both a precondition for successful implementation of strategy and a reflection of that strategy. Attracting and retaining the right people, motivating and rewarding them for good performance, designing appropriate training programs, and planning for the replacement of key people are fundamental to the implementation of the organization's strategy.

Although these principles are not in dispute, health care executives have been slow to adapt them to the portfolio concept of managing diverse organizational activities. Whereas the theme of personnel management has typically been to develop uniform policies and procedures, the portfolio concept of strategy formulation maintains that both human and capital requirements vary according to the strategic posture of the business unit. Human resources systems must be congruent with the strategy of the business unit they are meant to support, even if this leads to considerable variation in programs across the organization.

A business unit positioned in a growth industry with a strategy calling for investment, risk taking, and aggressive pursuit of market share will need different kinds of people and will manage them differently from a business unit positioned in a mature or declining industry. The challenge, therefore, is to support a diversity of business unit strategies while maintaining equity and consistency with overall organizational objectives. All too often, however, those charged with the management of human resources are excluded from both the process of strategy formulation and implementation.

Life Cycle, Strategy, and Human Resources Management

Table 3.4 integrates both the preceding discussion and the human resources management functions covered in upcoming chapters. Five stages in the organizational and/or service unit life cycle are delineated, including two conditions of decline: terminal and reversible. Generic strategies most appropriate at each stage are noted, along with specific human resources strategies and tactics required for successful implementation of these strategies.

Start-Up Phase

In the start-up stage, the dominant value is entrepreneurship and the major generic strategy is domain creation. The human resources area is typically underdeveloped, informal, and designed to support the dominant entrepreneurial values and strategies. There is little or no human resources planning, and career planning is informal because career opportunities develop naturally as the organization grows.

Because of the high mortality rates of new organizations, there is an understandable emphasis on the short run, "the here and now." Consequently, the recruitment and selection of risk takers (i.e., entrepreneurs) to meet present needs for domain creation is critical. The emphasis is on external recruitment because the organization does not have a large enough base of experienced personnel to rely on internal recruitment for most positions.

There is little formal training, and most of what exists occurs on the job rather than in classroom settings. The emphasis is on creating a cadre of professional, technical, and managerial personnel to meet the requirements for the emerging activities. Performance appraisals tend to be informal and geared to evaluating the individual employee's future contributions. Compensation tends to reflect a policy of incentives based on market share, successful risk taking, and service improvements. Fringe benefits are limited. The key issue in labor relations is to prevent a union election victory by meeting employee needs in other ways (e.g., good communications, informality, charismatic leadership, competitive wages, unique growth opportunities).

TABLE 3.4. The Relationship of Organizational/Service Unit Life Cycle Stage, Generic Strategies, and Human Resources Management Strategies and Tactics

	Organizational or Service Unit Life Cycle Stage				
	Start-up	Growth	Maturity	Reversible Decline	Terminal Decline
Dominant Values	Entrepreneurship	Revenue growth Market share	Competitiveness	Entrepreneurship	Cost control
Generic Strategies	Domain creation	Domain offense Domain creation	Domain defense Domain offense	Domain defense Domain offense Domain creation	Domain consolidation Domain substitution
Strategies and Tactics					
1. Major challenge	1. Recruit/select for present activities	1. Recruit/select for future activities	1. Develop appraisal/reward systems	1. Recruit/select for new activities	1. Manage reduction in force/redeployment
2. Human resources planning	2. Little or none	2. Meet critical manpower needs through job information system and flexible retirement	2. Greater effort to motivate and retain best employees; cross-functional job rotation	2. Planning for selective retention; more closely integrated with strategy; broader job tasks	2. Planning for phased disengagement and outplacement
3. Career planning	3. Informal; natural career growth	3. Rewards for developing subordinates; career counseling	3. Provide opportunities for job enrichment, lateral transfers, formalized career ladders	3. Avoid early retirement; match new needs with skills of those retained	3. Little or none

74

4. Recruitment/ selection	4. Recruit risktakers for present activities; fill existing positions mainly from external sources	4. Recruit for present and future activities; long-range perspective; emphasis on selection of "best" candidates internally and externally	4. Recruit executives skilled in reversing decline; recruit entrepreneurs	4. Little or none; recruit executives skilled in service termination
5. Training/ management development	5. Little formal training; on-the-job training to meet skill requirements of new activities	5. Generic formal training to meet requirements in growth areas; upgrade key personnel	5. Training to meet organization needs	5. Little or none
6. Performance appraisal	6. Informal; emphasize potential for future contributions	6. Measure formal subordinate development as well as future potential; assessment centers	6. Appraise to determine retention based on present and future contribution	6. Appraisal based on managers' ability to manage disengagement
7. Compensation	7. Based on gaining market share, willingness to take calculated risk, and service improvements; more incentives; fewer benefits	7. Based on building effective organization and subordinate development; more benefits	7. Based on generating new resources; incentives for entrepreneurs	7. Based on risk avoidance, cost control, and annual profits; flexible separation packages
8. Labor relations	8. Preventive labor relations	8. Preventive labor relations or efforts to prevent greater union penetration	8. Preventive labor relations, or minimize impact of union by "giveback" bargaining	8. Preventive labor relations or negotiations to terminate services

Integrate training with career ladders; produce skills necessary to implement strategy

Emphasize employee productivity

Based on contributions to government/consumer relations or bottom-line performance; fewer incentives; more benefits

Preventive labor relations or productivity bargaining

Less external recruitment; more internal recruitment; fewer positions to be filled; marketing specialists recruited

A health care organization in the start-up stage will emphasize recruiting marketing managers to develop new markets (domain creation) and physicians to provide the necessary services once these markets have been established. When these two groups are in place, other professional and administrative staff are recruited as needed.

Growth Phase

The growth phase is usually accompanied by increasing formalization of policies and procedures to minimize inequities in different subunits of the organization. Part of this formalization process is increased long-range planning and recruitment/selection for staffing *future* as well as present needs. Job-posting and job information systems become institutionalized to enhance internal recruiting. Retirement systems tend to be flexible enough to allow the organization to retain its most productive experienced employees irrespective of age.

During the growth phase managers should be rewarded for developing subordinates and helping them plan their careers within the organization. Recruitment and selection activities are not only more formalized, but they are based an objective data generated through job analysis so that the "best" candidates are chosen: this usually means those who can provide or expand existing services (domain offense) or provide/create new services in growth areas (domain creation).

Training also becomes more formalized during the growth phase and tends to focus on developing the individual employee through generic approaches (i.e., training that is not specific to the employee's present position). This type of training facilitates upgrading of key personnel to growth areas offering greater responsibility. Performance appraisal also tends to be both more formal and more developmental than in the start-up phase. There is an emphasis on subordinate development as well as the individual's future potential. Assessment centers to predict the future potential of administrators are common in the growth phase.

Compensation for managers is usually based on their contribution to the building of an effective organization to cope with the experienced growth as well as subordinate development. Benefits for all employees increase and become institutionalized during the growth phase because resources are available to fund such benefits and the desire exists to recruit top talent to meet growing demands.

Labor relations continues to be preventive during the growth phase unless some portion of the organization's employees become unionized. In this event, a policy of containment is usually adopted wherein the organization attempts to constrain further union gains among the work force and to prevent significant union contract inroads into what are viewed as management prerogatives.

There are presently fewer health care organizations in the growth phase than there were in the 1950s, 1960s, and 1970s. Growth organizations either provide ambulatory services (e.g., wellness) or services to the elderly (e.g., nursing homes). In the past, the existence of a cost-plus reimbursement environment allowed growing organizations to become bureaucratized and also overspecialized in terms of work force skills. Growing health care organizations today tend to be leaner and to employ personnel with broader skills. As in the case of non-health care organizations, the growing phase is characterized by human resources management activities aimed at employee development (through training, job posting, and career counseling) to fill future openings.

Maturity Phase

In the maturity phase of development, an organization's human resources system becomes institutionalized and standardized. The goal is to respond to both legal and employee pressures for equity, while allowing the organization to continue doing what it is already doing with perhaps some expansion of existing activities. This requires that the organization become more competitive in its existing markets in terms of price, quality, or service. Achieving such a competitive edge requires sophisticated appraisal and reward systems to measure and reward employee productivity.

Human resources planning emphasizes motivating and retaining the best employees through job enrichment, cross-functional job rotation, lateral transfers, and an emphasis on internal recruitment when higher level positions are open. Promotion opportunities, however, are more limited than during the growth phase. This situation is partly offset by the more formal nature of career planning, which allows employees to follow well-defined career paths and to be aware of the qualifications required for higher level positions in their own career paths. The organization's emphasis is on internal recruitment; little external recruitment occurs for positions other than entry level. Since competitiveness requires consumer awareness of the organization and its services, more marketing specialists are recruited.

Training is also geared toward preparing employees for higher level positions as well as improving their productivity in their present positions. Training needs are identified by analyzing employee deficiencies noted in the performance appraisal forms. The compensation package often rewards those who are effective in dealing with government or consumers as well as bottom-line performance. There are fewer incentive packages at this stage and more employee benefits.

Preventive labor relations continues for nonunion organizations in the maturity phase, while those that have been organized attempt to remain or become competitive by engaging in productivity bargaining. The latter approach is an attempt to control unit labor costs by making compensation increases contingent on productivity increases (Fottler and Maloney, 1979).

Health care organizations in the maturity phase often attempt to compete by increasing their marketing efforts aimed at both patients and physicians. This means that the marketing role is enhanced, and marketing specialists, often from other industries, are recruited to coordinate the effort. At the same time, physicians are evaluated more critically in terms of their contribution to revenue (Smith and Fottler, 1985). Pressure is exerted on those who are far out of line with their colleagues with respect to resource use for particular diagnoses. Upgrading of personnel appraisal systems for other employees is also common during the maturity phase.

Reversible Decline Phase

When an organization is experiencing decline that is viewed as reversible, domain defense followed by domain offense and domain creation are the appropriate generic strategies. The organization is likely to have administrative personnel capable of implementing the first two strategies, since these are implemented during the previous maturity phase. Consequently, a key human resources challenge is to revitalize the organization and to recruit entrepreneurs who are skilled in domain creation.

Human resources planning becomes more closely aligned with the strategy of reversing decline. The trend toward job specialization is reversed as job task definition becomes broader. Selective retention of the employees capable of contributing to the new areas of growth is emphasized. Many organizations encourage early retirement when trying to reverse decline, but this often leads to loss of the experienced, skilled employees most needed for future growth. A better alternative is to provide flexible work alternatives for key employees (e.g., phased retirement, part-time employment).

The reversal of decline requires the recruitment of executives who have had success in reversing similar declines. Executives who have experienced only growth or stability usually will not manage the situation well, as noted earlier in the chapter. The recruitment of entrepreneurs to develop new growth areas is also a critical challenge.

The training programs at this phase are geared to meet immediate organizational needs for domain defense, offense, or creation. Individual employee development for the future is not emphasized. The role of the performance appraisal system is to assess both present contribution and future potential in the areas of future growth. Compensation is often based on potential contribution to new areas of growth, and entrepreneurs are able to negotiate incentive packages. Labor relations remains preventive unless unionization has already occurred. In this event, the emphasis is on "giveback" bargaining, which calls for management aggressiveness in winning compensation and work rule concessions from the union.

Hospitals across the country are presently experiencing some degree of decline in their inpatient areas. While they are lobbying and marketing to protect their revenue base and to expand current services (domain defense and domain offense), they are also planning and implementing domain creation strategies. Among the new activities has been development of HMOs, substance abuse clinics, exercise facilities, free-standing emergency clinics, and consulting services of various types.

These attempts to reverse decline in core areas have required hospitals to recruit personnel experienced in the new areas, together with marketing personnel. These individuals are often recruited from other industries and are offered incentive compensation packages that reward the successful development of the opportunity. Failure is penalized by loss of the position. At the same time, present employees are being appraised with greater sophistication in an attempt to determine their present productivity and future potential contributions. Obviously, scrutiny of physician practice patterns to determine contributions to both revenue and costs is a major part of this effort.

Terminal Decline Phase

Once the organization has determined that a service unit or the whole organization is facing terminal decline with no possible reversal, the emphasis is on the systematic management of reduction in force and redeployment to other areas in the organization (in the case of a service unit). Control of costs (efficiency) is important, since the goal is to avoid losing money during the terminal phase.

There is little or no career planning, training, or management development because future opportunities in the service unit or the organization are nonexistent. Human resources planning emphasizes phased disengagement and outplacement of current employees (Fottler and Schuler, 1984). The only recruitment that occurs involves hiring executives who are skilled at phasing out organizations or subunits of organizations.

Performance appraisal is based on the manager's ability to manage disengagement with a minimum of conflict and avoidance of further economic losses if possible. There is no future orientation. Rewards for executives are based on risk avoidance, cost control, and annual profits. Separation packages for executives are flexible and individually negotiated. If there is a labor union, negotiations may be held to agree on the process of termination and on the benefits/services terminated employees may expect to receive.

When a health care organization faces terminal decline, it often attempts to find a buyer (e.g., a large chain) to which it can sell out. However, the chains are becoming more selective, and this option may be shrinking for many declining organizations. If all other avenues for reversal of the decline have proved fruitless, the organization begins to plan for termination. This means communicating the decision to terminate and the reasons for this decision. Plans for phased reduction in force and outplacement then need to be implemented.

SUMMARY

All organizations and subunits of organizations evolve and go through predictable phases. At each stage in the organizational life cycle, different strategies for growth, revitalization, and survival are appropriate. Successful formulation and implementation of these strategies are required if the organization is to avoid premature decline and death. The implementation of such strategies requires appropriate human resources strategies and tactics as discussed above. Ignorance of the appropriate linkages between business strategies and human resources strategy/tactics may mean organizational failure even if the strategy itself is appropriate.

Health care organizations need to be aware of where they are in terms of their organizational and service subunit life cycles. This knowledge will help them to develop and implement appropriate organizational strategies supported by reinforcing human resources strategies and tactics. The increasingly competitive health care market will penalize organizations that fail to formulate and implement appropriate strategies.

Management of the health organization also needs to be aware of differences in life stage and strategies in different subunits. These differences require different human resources strategies and tactics to successfully implement the subunit strategy. Bureaucratic uniformity in the human resources system across the entire organization may stifle some units and provide too much flexibility to others. Health organizations will have to determine what policies should be uniform (and therefore centralized) and which ones need to be flexible and nonuniform (and therefore decentralized).

REFERENCES

Adizes, I. "Organizational passages: Diagnosing and treating life cycle problems in organizations." *Organizational Dynamics,* 8:3–25 (Summer 1979).

Alexander, J. A., B. L. Lewis, and M. A. Morrisey. "Acquisition strategies of multihospital systems." *Health Affairs*, 4:49–66 (Fall 1985).

Benson, J. K. "The interorganizational network as a political economy." *Administrative Science Quarterly*, 20:229–249 (1975).

Boulding, K. E. *Evolutionary Economics*. Beverly Hills, CA: Sage, 1981.

Cameron, Kim S. "Domains of organizational effectiveness in colleges and universities." *Academy of Management Journal*, 24:25–47 (1981).

Cameron, Kim. "Strategic responses to conditions of decline: Higher education and the private sector." *Journal of Higher Education*, 54:359–380 (July–August, 1983).

Dunbar, R. L. M., and W. H. Goldberg. "Crisis development and strategic response in European corporations." In C. F. Smart and W. T. Stanbury, Eds., *Studies in Crisis Management*. Toronto: Butterworth, 1978.

Durham, John W., and Howard L. Smith. "Toward a general theory of organizational deterioration." *Administration and Society*, 14:373–400 (1982).

Fink, S. L., J. Beak, and K. Taddeo. "Organizational crisis and change." *Journal of Applied Behavior Science*, 7:15–37 (1971).

Ford, J. D. "The administrative component in growing and declining organizations: A longitudinal analysis." *Academy of Management Journal*, 23:615–630 (1980a).

Ford, J. D. "The occurrence of structural hysteresis in declining organizations." *Academy of Management Review*, 5:561–575 (1980b).

Fottler, M. D., and W. F. Maloney. "Guidelines to productivity bargaining in the health care industry." *Health Care Management Review*, 4:375–388 (1979).

Fottler, M. D. and D. W. Schuler. "Reducing the economic and human costs of layoffs." *Business Horizons*, pp. 9–15 (July–August 1984).

Fottler, M. D. and D. G. Vaughan. "Multihospital systems." In L. F. Wolper and J. J. Pena, Eds. *Health Care Administration: Principles and Practice*. Rockville, MD: Aspen Systems, 1987.

Fottler, M. D., J. R. Schermerhorn, J. Wong, and W. H. Money. "Multi-institutional arrangements in health care." *Academy of Management Review*, 7:67–79 (1982).

Fottler, M. D., H. L. Smith, and H. J. Muller. "Retrenchment in health care organizations: Theory and practice." *Hospital and Health Services Administration*, 31:29–43 (October 1986).

Hall, Douglas T., and Roger Mansfield. "Organizational and individual response to external stress." *Administrative Science Quarterly*, 16:533–547 (1971).

Hall, Roger I. "A system pathology of an organization: The rise and fall of the old *Saturday Evening Post*." *Administrative Science Quarterly*, 21:185–211 (June 1976).

Hedberg, Bo L. T., Paul C. Nystrom, and William H. Starbuck. "Camping on seesaws: Prescriptions for a self-designing organization." *Administrative Science Quarterly*, 21:41–65 (1976).

Hermann, Charles F. "Some consequences of crisis which limit the viability of organizations." *Administrative Science Quarterly*, 8:61–82 (1963).

Hirschman, Albert O. *Exit, Voice and Loyalty*. Cambridge, MA: Harvard University Press, 1970.

Holsti, O. R. "Limitations on cognitive abilities in the face of crisis." In C. F. Smart and W. T. Stanbury, Eds., *Studies in Crisis Management*. Toronto: Butterworths, 1978.

Janis, Irving L. *Victims of Group Think.* Boston: Houghton Mifflin, 1972.

Janis, Irving L., and Leon Mann. "Emergency decision-making: A theoretical analysis of responses to disaster warnings." *Journal of Human Stress,* pp. 35–48 (June 1977).

Jick, T. D. "Process and impacts of a merger: Individual and organizational perspective", unpublished doctoral dissertation. Cornell University, Ithaca, NY, 1979a.

Jick, T. D. "Mixing qualitative and quantitative methods: Triangulation in action." *Administrative Science Quarterly,* 24:602–611 (1979b).

Kaluzny, A. D., B. J. Jaeger, and K. M. Habib. *Multi-Institutional Systems Management.* Owings Mills, MD: National Health Publishing, 1987.

Kauffman, Joseph F. "Some perspective on hard times." *Review of Higher Education,* 6(1):69–78 (Fall 1982).

Levine, C. H. "Organizational decline and cutback management." *Public Administration Review,* 38:316–325 (1978).

Levine, C. H. "More on cutback management: Hard questions for hard times." *Public Administration Review,* 38:316–325 (1978).

March, J. G., and H. Simon. *Organizations.* New York: Wiley, 1958.

Meyer, M. W. "Organizational domains." *American Sociological Review,* 40:599–615 (1975).

Miles, R. E., and C. C. Snow. *Organizational Strategy, Structure, and Processes.* New York: McGraw-Hill, 1978.

Miles, R. H., and K. S. Cameron. *Coffin Nails and Corporate Strategies.* Englewood Cliffs, NJ: Prentice-Hall, 1982.

Miller, D. "Common syndromes of business failure." *Business Horizons,* pp. 43–53 (November 1977).

Miller, D., and P. H. Friesen. "Momentum and revolution in organizational adaptation." *Academy of Management Journal,* 23:591–614 (1980).

Mintzberg, H. *The Structuring of Organizations.* Englewood Cliffs, NJ: Prentice-Hall, 1979.

Pfeffer, J., and G. R. Salancik, *The External Control of Organizations: A Resource Dependence Perspective.* New York: Harper & Row, 1978.

Porter, M. *Competitive Strategy.* New York: Free Press, 1980.

Post, J. E. *Corporate Behavior and Social Change.* Reston, VA: Reston, 1978.

Quinn, R., and K. S. Cameron. "Life cycles and shifting criteria of effectiveness." *Management Science,* 29:33–51 (January 1983).

Rondine, T. "Organizational identification: Issues and implications." *Organizational Behavior and Human Performance,* 13:95–109 (1975).

Scott, William G. "The management of decline." *Conference Board Record,* pp. 56–59 (June 1976).

Shortell, S. M., and T. M. Wickizer. "New program development: Issues in managing vertical integration." In J. R. Kimberly and R. E. Quinn, Eds., *Managing Organizational Transitions.* Homewood, IL: Irwin, 1984.

Smart, C., and I. Vertinsky. "Designs for crisis decision units." *Administrative Science Quarterly,* 22:640–657 (1977).

Smith, D. B., and A. D. Kaluzny. *The White Labyrinth: A Guide to the Health Care System*, 2nd ed. Ann Arbor, MI: Health Administration Press, 1986.

Smith, H. L., and M. D. Fottler. *Prospective Payment: Managing for Operational Effectiveness*. Rockville, MD: Aspen Systems, 1985.

Snow, C., and L. Hrebiniak. "Strategy, distinctive competence, and organizational performance." *Administrative Science Quarterly*, 25:317–336 (1980).

Starbuck, William H. "Organizations and their environments." In M. Dunnette, Ed., *Handbook of Industrial and Organizational Psychology*. Skokie, IL: Rand McNally, 1976.

Starbuck, William H., and Bo L. T. Hedberg. "Saving an organization from a stagnating environment." In H. Thorelli, Ed., *Strategy + structure = Performance*. Bloomington: Indiana University Press, 1977.

Starbuck, William H., Arent Greve, and Bo. L. T. Hedberg. "Responding to crisis." *Journal of Business Administration*, 9:111–137 (1978).

Staw, B. M., L. E. Sandelands, and J. E. Dutton. "Threat-rigidity effects in organizational behavior: A multilevel analysis." *Administrative Science Quarterly*, 26:501–524 (1981).

Summers, J. "Management by crisis." *Public Personnel Management*, 4:194–200 (1977).

Thompson, J. D. *Organizations in Action*. New York: McGraw-Hill, 1967.

Whetten, D. A. "Sources, responses, and effects of organizational decline." In J. Kimberly and R. Miles, Eds., *The Organizational Life Cycle*. San Francisco: Jossey-Bass, 1980a, pp. 342–374.

Whetten, D. A. "Organizational decline: A neglected topic in organizational science." *Academy of Management Review*, 5:577–588 (October 1980b).

Whetten, D. A. "Organizational responses to scarcity—Exploring the obstacles to innovative approaches to retrenchment in education." *Educational Administration Quarterly*, 17(3):80–97 (1981).

Zammuto, Raymond F. *Assessing Organizational Effectiveness: Systems Change, Adaptation, and Strategy*. Albany, NY: SUNY Press, 1982.

Zammuto, Raymond F. "Are liberal arts colleges an endangered species?" *Journal of Higher Education*, 55:184–211 (March–April 1984).

Zammuto, Raymond F., and Kim S. Cameron. "Environmental decline and organizational response." In B. M. Staw and L. L. Cummings, Eds., *Research in Organizational Behavior*, Vol. 7. Greenwich, CT: JAI Press, 1985, pp. 223–262.

PART TWO
Structural Systems

CHAPTER 4

Technology and Human Resources Management

Peggy Leatt
Bruce Fried

There are many different technologies in health services organizations, all of which have an impact on how health services managers are able to do their work. One of the major sources of technological uncertainty evolves from the overwhelming complexity and diversity of *medical* technologies. The practice of medicine has become increasingly specialized, with each group of specialists developing a set of technologies. How are we to keep track? Most new medical technologies are tested through clinical research before they are made readily available; however, findings from clinical research are not always clear-cut, so there is often much unpredictability about whether a new technology should be adopted. Decisions about the adoption of a new technique for patient treatment are clearly in the domain of physicians. When human and financial resources are involved, however, the decisions to use new technologies have broader implications and are by necessity the responsibility of health services managers.

As an illustration, let us explore the case of a community general hospital where the physicians on staff are interested in acquiring a CT scanner, mainly because they would like their patients to have the best services possible. Also, however, the physicians have heard that some patients are transferring to a neighboring hospital that already has computerized tomography equipment. How does the hospital decide whether to buy a CT scanner? The decision cannot be made by the physicians alone because the new equipment will most likely require extra space and additional human resources. At this point it becomes the responsibility of the hospital manager to be involved in assessing the potential impact this new technology would have on the hospital. Some of the most important questions to be discussed by the senior management team include: How does this technology fit the mission of the hospital? What other resources, including human resources, are necessary? and How will the new equipment affect the services currently being provided in other departments?

Although medical technology tends to be the predominant one that influences the operations of health services organizations, other *paraprofessional* groups bring their own technologies or sets of techniques to the organization. Nurses and social workers, for example, may not require special equipment to carry out their work, but the social-psychological support to patients that is usually provided by nurses and social workers is an important component of their technological repertoire.

Consider a 100-bed nursing home which for the past 20 years has served primarily elderly persons. The average patient is over 80 years of age, and two-thirds of the patients are women. The board of the facility has decided to close half the beds and remodel the physical space to provide services to teenage schizophrenics, who are mostly male. This decision was made on the basis of a grant that was available from the federal government. However, the type of social-psychological support the nurses and the social workers have been providing to elderly women is not appropriate for the new clientele. Accordingly, this change will necessitate the retraining of nurses and social workers to impart the knowledge, skills, and attitudes required for providing support to teenagers with schizophrenia.

One of the biggest changes currently taking place in health services organizations is the introduction of computers to manage both clinical and management information systems. Data processing, communications, and office automation, as branches of information technology, are becoming increasingly important components of health services, especially in hospitals. The whirlwind of activities in *information* technology is enough to catch even the most experienced health services managers off guard. Many hospitals have extensive in-house capabilities with respect to the decisions that must be made about choices of hardware, software, networks, and so on. There are obvious human resources implications. For example, new occupational groups are emerging to handle the complexities of information technology; these include highly specialized technicians who not only understand computers and their programming but also have some insights into the information needs of health services managers and other occupational groups.

Another example of the impact of computers on human resources might be found in a large teaching hospital. Suppose it has been decided to computerize the patient information system to allow all patient data, including orders for tests, drugs, and treatments, to be inputted into the computer terminal at each patient's bedside by the physician. These orders will be automatically communicated to the relevant service departments. The introduction of this system, however, will make the 40 unit clerks in the hospital redundant because they will no longer be required to transcribe the orders for the physicians. The union local is very interested in the unit clerks' case because it wants to ensure that alternative employment is made available to them in the hospital.

One type of technology often underplayed in its importance for human resources is the set of techniques used by managers. *Management* technology includes the tools used by managers in relation to strategy development, goal setting, organizing, motivating, and evaluating. Although there are numerous examples of management technologies, one that has considerable impact on human resources is the type of performance appraisal system the organization adopts. Performance appraisal systems have important potential for motivating employees and providing appropriate rewards;

however, the systems tend to become routinized so that filling out the form often becomes more important than providing feedback to individuals workers. Most managers find providing negative feedback difficult, so this task tends to be either avoided or depersonalized through the use of evaluation forms.

THE THEORY OF TECHNOLOGY IN ORGANIZATIONS

Although the classical approaches to management such as the scientific management and human relations schools recognized the importance of work in organizations, during this early period the concept of technology was not explicitly recognized. Dubin in 1958 and Woodward in 1965 pioneered research on the importance of technology in organizations. Technology was defined as the tools, instruments, machines, and technical formulas basic to the performance of work. The definition also included the functional importance of the tasks, and the rationale for the methods employed. Woodward (1965) developed a continuum of technological *complexity* (unit, mass, and process modes), which ranged from the simplest form of production known (i.e., production of units to customers' requirements) to the most complex (i.e., continuous flow production). Harvey (1968) criticized the ordering of Woodward's complexity continuum because he saw the continuum in reverse. To Harvey, the unit, mass, and process modes of production were not arranged in order of technological complexity; rather they move toward technical simplicity. He described technology in terms of its degree of *diffuseness* as an indicator of the frequency of production change.

Thompson (1967) was concerned with the technical rationality of organizations: "Perfection in technical reationality requires complete knowledge of all cause and effect relations so that the techniques can be predicted to produce the desired outcomes" (p. 14). Thompson described three types of technology that focused on the conversion process to transform inputs into outputs. A *long-linked* technology involves serial interdependence of actions; for example, A must be completed before B, and B must be completed before C. A *mediating* technology links clients to customers wishing to be interdependent, such as banks. An *intensive* technology has a variety of techniques depending on feedback from the object being transformed. As an example of the latter technology, Thompson used the general hospital, where the type of service provided varies according to a patient's needs at a particular time.

Technology was conceptualized in terms of work processes or transformation processes by Hickson et al. (1968); however, technology was approached at the work-flow level rather than at the level of the total organization. Hickson et al. described operations technology in terms of production continuity (as did Woodward) and work-flow integration. This latter dimension included the degree of mechanization and automation of equipment, the degree of work-flow rigidity, and the means employed for assessing performance.

Using transformation processes as the focus, Whisler (1970) developed a concept of information technology because he was interested in identifying the impact of computers on organizations. Information technology was defined as the sensing, coding, transmitting, translation, and transformation of information. Whisler saw information as a technology of control; he believed older technologies were an extension of man's muscle (tools), yet the newer technologies were an extension of man's brain with the potential of being a partner or even taking control.

Perrow (1965, 1967) was interested in technology in human service organizations as well as in industrial organizations. Perrow (1965) described technology as the process whereby basic material is taken into an organization, certain acts are performed on it in series, and the material is altered in a desired fashion. Perrow (1970) described the work performed in organizations according to its degree of *routineness*. Two conditions were considered necessary for routineness: there must be well-established techniques that are sure to work, and these must be applied to similar raw materials. In these circumstances, there is little uncertainty about what is to take place and little variety in the tasks to be performed. Where work is nonroutine there are few well-established techniques and the raw materials are varied.

Raw materials were considered to be the most important factor in human service organizations because actual human beings, not inanimate objects, are the focus. Three characteristics of raw materials were described: their variability, their instability, and the extent which they are understood. Although polar situations of routine versus nonroutine were described, technology was seen by Perrow as involving two dimensions: the nature of the raw materials and the nature of the techniques for transforming the raw materials. These two dimensions form a stimulus-response set for workers. The stimuli are the raw materials on which workers must operate, and the responses are the actions performed by the workers to produce the desired changes. Perrow maintained that it is workers' *perceptions* of the raw material that are important. If they perceive the material as stable and uniform (regardless of whether it is objectively so), they will act on it in a standardized way. Accordingly, organizations may attempt to standardize raw materials to reduce variability and minimize exceptional cases (Perrow, 1970).

The critical characteristic of the techniques used in the transformation process is the nature of the search process undertaken to find an appropriate technique, especially when exceptions occur. The search process may be logical and analytic, for example, where techniques can be applied with predictable outcomes. Search behavior may be unanalyzable when outcomes are unpredictable and the search relies on intuition, inspiration, guesswork, or some other unstandardized procedure (Perrow, 1967). The dimensions of Perrow's technology concept are summarized in Figure 4.1.

Four different technological situations may occur, shown as cells 1, 2, 3, and 4 in Figure 4.1, and different types of health services organizations may be illustrative of the four sets of technological circumstances. For example, in a large, complex teaching hospital (cell 2) the raw materials (patients) are typically unstable, variable, and not well understood; and often the search processes to find appropriate treatment methods are difficult to identify or analyze. In contrast, in a nursing home (cell 3) patients are relatively stable, uniform, and well understood; and the techniques for caring for the elderly are relatively well defined. Situations 1 and 4 are less clear-cut and less easy to exemplify. However, organizations providing psychiatric services might illustrate cell 4, insofar as patients' problems are not well understood but do tend to be treated in standardized ways such as with drug therapy. Alternatively, cell 1 may be illustrated by organizations providing services to oncology patients. The characteristics of certain cancers may be well understood, but the search for appropriate techniques is unanalyzable; that is, a variety of treatments may be used with unpredictable outcomes.

Building on the work of Perrow, Hasenfeld (1983) wrote:

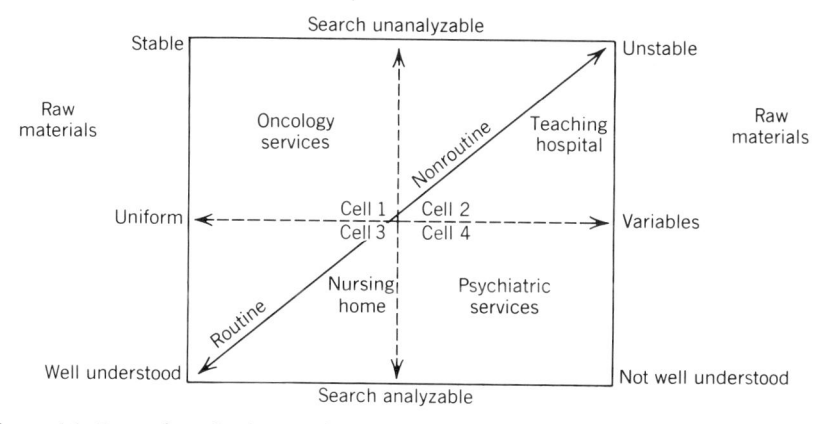

Figure 4.1. Perrow's technology techniques. Adapted from Perrow, C. A framework for the comparative analysis of organizations. *American Sociological Review*, 1967, 32, 194–208, Figure 1 on p. 196.

[A] human service technology [is] a set of institutionalized procedures aimed at changing the physical, psychological, social, or cultural attributes of people in order to transform them from a given status to a new prescribed status. The term "institutionalized" denotes that the procedures are legitimated and sanctioned by the organization. (p. 111)

According to Hasenfeld (1983), most human service technologies have several components, which tend to coincide with the way in which clients are processed through organizations. These are as follows.

1. *Recruitment and selection.* There is a set of procedures to recruit and select clients that confers eligibility on the clients to be served. These procedures may take place, for example, in the emergency room or admitting department of a hospital.
2. *Assessment and classification.* This phase comprises the period of assessment and diagnosis (classification) of the patients' or clients' problems.
3. *Status transformation.* During this stage a set of techniques is administered to bring about the desired outcomes. In health services organizations direct treatment and care are provided to patients in stage 3.
4. *Termination and certification.* In this stage judgments are made and procedures followed to confirm that the client's status has indeed been transformed and the relationship between the client and organization can be terminated. In hospitals this type of certification is most likely provided by a physician's written order that the patient may be discharged.

One of the most important human resources factors associated with Hasenfeld's approach to technology is the nature of client/staff relationships. Clearly, human beings entering health services organizations have ideas of their own about how they should be treated. Not everyone, especially among today's better informed consumers, is willing to follow exactly the instructions of physicians or other health professionals; thus the extent to which status transformation can take place may be severely limited.

In addition, health services organizations are staffed by human beings, with their own ideas about what should be taking place, with the result that the validity of the application of various techniques or treatment procedures may be open to a variety of interpretations. These factors are particularly crucial in health services organizations where life-or-death decisions are made.

Empirical Research on Technology and Structure

The study of technology in organizations has been considered important because of the potential relationship between technology and organizational design. Contingency theory (e.g., Perrow, 1970; Woodward, 1965), for example, suggests that to maximize efficiency and effectiveness, an organization's structures should be designed to match the type of technology used by the organization. To optimize efficiency and effectiveness where the technology is nonroutine or unpredictable, it has been suggested that the organizational structure should be flat and decentralized; where the technology is routine or certain, the structure may be hierarchical and more bureaucratic (Ford and Slocum, 1977). Health services organizations, for the most part, tend to have nonroutine technologies and, therefore, are more likely to have flatter organizational structures and greater decentralization in decision making, and to employ more professional staff.

There has been little research examining relationships between technology and structure in human service organizations. Hage and Aikers (1969), in their study of health and welfare agencies, found that routine work was associated with a more centralized and formalized structure and that agencies with nonroutine work employed more professional staff. However, in a study of 144 work groups in 13 health departments Mohr (1971) did not find a relation between the uniformity, complexity, and analyzability of technology and a measure of participatory style of management.

In nursing units in hospitals, Kovner (1966) found relationships between variability and predictability in technology and unit structure although only 8 units were included in this investigation. Comstock and Scott (1977), in their study of 142 nursing units, concluded that task predictability, when measured by nurses' perceptions, was found in units with more centralized and standardized structures.

Leatt and Schneck (1982), using Perrow's concept of technology, examined the variations in technology in nursing units ($n = 139$) of nine types in 25 hospitals. Differences in technology were found; for example, units such as intensive care, pediatrics, and psychiatry were more complex technologically than units such as long-term care and obstetrics. The units with the more complex technologies were generally found to have more dencentralized organizational designs, fewer formalized procedures, and a larger proportion of professional staff. When work stress was examined as an outcome of the match between technology and structure, the complexity of the technology was found to be a more important predictor of stress than other contingency variables such as environment and size. In some situations, structural design factors seemed to influence or modify levels of stress, but these findings were not clear-cut (Leatt and Schneck, 1985).

Alexander and Randolph (1985) studied 27 nursing units to explore the relationships between technology, structure, and quality of nursing care. The findings of this research suggested that the fit between the structure and the technology was more important than either of these factors alone in predicting quality of care.

In summary, organizational theorists have recognized the importance of technology in organizations and its potential implications for organizational design in both industrial and human services environments. There has been relatively little research in health services organizations, especially in relation to the impact of medical technologies, information technologies, and management technologies on the human resources function. The research that has been done has frequently used varying levels of analysis (total organization, unit, or individual), limiting the comparability of findings.

Environment, Strategy, and Technology

As discussed in Chapter 2, the development of a corporate strategy for an organization depends partly on an assessment of the organization's external environment; such an assessment is seen as a systematic attempt to articulate opportunities that might be presented to the organization, as well as threats that must be dealt with. The central premise underlying these activities is that organizations have the capability of adapting to varying environmental conditions (Hickson, et al., 1971). These adaptations help the organization to survive over the long run (Aldrich, 1979) and short run (Pfeffer and Salancik, 1978).

A number of authors have argued that the environment only partially dictates the behavior of organizations and that strategic choices made by top managers are also critical determinants of structure and processes (see, e.g., example: Chandler, 1962; Child, 1972; Weick, 1969). Top managers are able to assess the environments associated with their organizations' product lines and then choose appropriate technologies, structures, and processes. The correct selection of these factors influences the extent to which the organization can meet its goals (Randolph and Dess, 1984). This perspective assumes that organizations through top managers are purposive, rational, and proactive.

Figure 4.2 illustrates the potential relationships between an organization's environment, strategy, and technology. The complexity of the relationships among these factors may be demonstrated through an example of a community general hospital located in a small resort town on the southern California coast. There are 200 acute care beds, and a range of ambulatory care programs is offered. The most important external environmental factor facing the board of the hospital is the changing demography of the community the hospital serves—increasing numbers of elderly persons are finding this area to be an ideal retirement center. Currently 25 percent of the population are over 65 years of age, and this proportion is increasing. The elderly, who often have multiple health problems, are being admitted to the hospital for both minor and major procedures; once they are in hospital they often stay for extended periods. The senior management team has clarified that services for the elderly are an important part of the hospital mission; however, they are concerned about the effect of the increasing lengths of stay on the hospital's overall performance. Senior management's range of strategic choices is as follows: add more acute care beds, cut back on the services to the elderly, or, provide alternative services for the elderly. Close examination of the health problems of the hospital's elderly patients revealed that the majority of admissions were for ophthalmological procedures, especially cataract surgery. The hospital, therefore, made a strategic decision to set up a new program in gerontological ophthalmology, which would provide cataract extractions on a day surgery basis.

92   Structural Systems

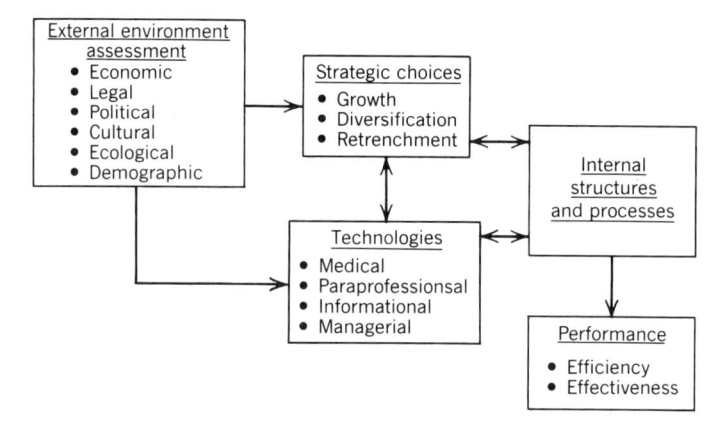

```
┌─────────────────────┐
│ External environment│
│   assessment        │      ┌──────────────────┐
│ • Economic          │      │ Strategic choices│
│ • Legal             │─────▶│ • Growth         │
│ • Political         │      │ • Diversification│
│ • Cultural          │      │ • Retrenchment   │◀────┐
│ • Ecological        │      └──────────────────┘     │   ┌──────────────┐
│ • Demographic       │               ▲ │             └──▶│   Internal   │
└─────────────────────┘               │ ▼                 │  structures  │
           │                 ┌──────────────────┐        │ and processes│
           │                 │   Technologies   │◀──────▶└──────────────┘
           │                 │ • Medical        │                │
           └────────────────▶│ • Paraprofessionsal│              │
                             │ • Informational  │                ▼
                             │ • Managerial     │        ┌──────────────┐
                             └──────────────────┘        │ Performance  │
                                                         │ • Efficiency │
                                                         │ • Effectiveness│
                                                         └──────────────┘
```

Figure 4.2. Environment, strategy and technology.

This mini-case provides an example of a particular strategic choice influencing the form of technology to be used. The technology (in keeping with Perrow's definition) included the space, equipment, operating room, and day beds required for the new program. It also included the knowledge that this type of surgery could be safely performed on a day patient basis with predictable success. In terms of human resources needs, it was necessary to employ a gerontological ophthalmologist who understood the medical technology, as well as other key personnel to organize the program. Information technology was necessary to monitor the clinical and financial outcomes of the program.

Although this example serves to illustrate how certain strategic choices may lead an organization to select one or more particular technologies, it is conceivable that the relation between strategy and technology could be reversed. In other words, there is the potential for a new technology, developed internally, to influence senior management's strategic choices. For instance, a rehabilitation center may have an exceptionally strong prosthetics service, with biomedical researchers who are at the cutting edge of their field. Strategically, the senior management may decide that the development of prostheses should become an important product line and that other health services organizations in the community will be offered the opportunity to purchase prosthetic services from the center. In this case the technology of the organization could influence the strategy selected by the senior management team.

Levels of Analysis in Organizations

So far in this chapter we have focused on the organization itself as the level of analysis; that is, the strategy of the total organization has the potential to influence the choices of technologies, which, in turn, can determine the type of organizational design that is appropriate. The relationship between technology and design may be examined at several different levels of analysis, as shown in Figure 4.3.

All levels are important in designing organizations. For example, in designing home health services, it is necessary to examine internally the *cluster of work groups* in the organization to ensure they are grouped and coordinated appropriately, the design of

```
Individual positions
      ↓
  Work group
      ↓
Cluster of work groups
      ↓
 Total organization
      ↓
Network of organizations
      ↓
    System
```
Figure 4.3. Levels of analysis.

each individual *work group* (Fry and Slocum, 1984), and the design of each *individual position*. It would also be essential to examine the *network of organizations* in which the home health services was located, as well as the larger health *system* (Provan, 1984). As Mintzberg (1983) has pointed out, the individual positions are the building blocks for the whole organization design.

Technology may impact on all these levels of analysis. Clearly, the complexity of the tasks to be performed will influence the nature of the responsibilities that can be assigned to individual positions. The design of a work group in terms of composition and structure may be influenced by what is known about the techniques to be used to achieve the goal of the work group. Different work groups in organizations are likely to experience different levels of complexity of technology. For example, the technology of a hospital's housekeeping or laundry department will be more routine technology than that of the emergency room or operating theater. The larger the unit of analysis (e.g., total organizations and networks), the more likely it is that there will be considerable variations among the units in the complexity of the technologies.

As pointed out in Chapter 2 and elsewhere (see, e.g., Kimberly, 1980; Kimberly and Quinn, 1984), organizations may pass through several different stages in a life cycle and may have varying design needs at different times. For example, positions, work groups, clusters of work groups, and total organizations may require different designs when they are newly formed and when they are experiencing a transition (Hoare, 1983; Kimberly and Quinn, 1984). Such design changes may be brought about by changes in technology or may in fact represent a need for a change in technology at a particular time.

THE INFLUENCE OF TECHNOLOGY ON THE HUMAN RESOURCES FUNCTIONS

Technology thus has a pervasive effect on health services organizations and on their employees. In managing human resources, it is necessary to take into consideration a number of factors related to the technology in the organization as well as within particular subunits. In this section, we describe the critical technological factors that should be considered in job design, personnel selection, performance appraisal systems, reward system development, and occupational development activities.

Technology and Job Design

Designing individual positions in organizations involves specifying how many tasks a particular position should contain and how specialized these tasks should be. It also requires decisions about the extent to which the tasks may be standardized and what knowledge and skills are necessary (Mintzberg, 1983). All these decisions require detailed understanding of the technology to be used. Various approaches have been used to determine the design of individual positions, and these are reviewed below.

Job design, which is concerned with the ways in which organizations assign tasks and responsibilities to individual workers, has been recognized as important since the early nineteenth century. For example, the scientific management school had a pervasive influence on job design because it clearly pointed out to managers that this function was essential and that specialization and routinization of tasks may increase efficiency (Porter et al., 1975). The main goal in job design is to increase the motivation, satisfaction, and productivity of the persons occupying the positions.

Historically, a number of different approaches have been used to facilitate the process of matching individuals to particular jobs. One attempt to minimize boredom from being in a routine job too long was *job rotation;* that is, individuals would be assigned for a period of time through several related jobs. This approach was used in nursing departments in hospitals in the 1950s and 1960s. Nurses would be asked to rotate through different nursing units to relieve the monotony of working with the same kind of patients. This approach was not for the most part successful. The nurses were unable to develop their specialized technical expertise in a particular area and, consequently, did not find the work rewarding. In this instance, the specialization of the technology did not permit job rotation.

An approach that emerged in the 1940s and 1950s in response to overspecialization of tasks advocated by the scientific management school was *job enlargement.* Instead of breaking down work into smaller components, which tended to become repetitive and monotonous, job enlargement combined several tasks together to form a more complete whole. Often the enlargement of the tasks and responsibilities was accompanied by greater freedom for individuals to choose the procedures for completing the work and also their own pace of work. Suppose, for example, that medical technologists in a clinical laboratory are responsible for carrying out a variety of routinized and standardized clinical tests. With a specialization approach each technologist is assigned to complete one single type of test on all patients who require it. With a job enlargement program, each technologist is assigned patients and is able to perform the full range of tests that has been prescribed for his or her patients. One of the main criticisms of job enlargement programs is that they often leave the basic nature of the jobs unchanged and may simply result in more work for individual employees.

The *job enrichment* approach to the design of work is based on Herzberg's two-factor theory of motivation and satisfaction. Herzberg (1968) suggested that characteristics of jobs may be categorized as motivators or as hygiene factors. Motivators are factors intrinsic to the job and include responsibility, recognition, achievement, advancement, and growth and development. Hygiene factors include aspects of the job content, such as working conditions, salary, security, supervision, status, and organizational policies and procedures. Herzberg indicated that motivators can influence levels of satisfaction of individual workers, whereas hygiene factors can influence only levels of dissatisfaction. Job enrichment involves a basic increase in the amount of control individuals have over their own work. This control in turn acts as a motivator by increasing responsibility, recognition, and opportunities for growth. An example of job enrichment in health services organizations is found in the development of primary care nursing. Historically, nurses were assigned work on a daily basis or shift in terms of tasks or procedures or in terms of patients. In contrast, a primary care nurse is assigned a case load of patients and assumes responsibility for the nursing services to be provided to those patients throughout their stay. Primary care nursing was introduced in psychiatry, where the technology (patients and techniques) was such that continuity of patient-nurse interaction was considered critical.

Job enrichment programs, however, have not always been successful because they often fail to take into consideration the needs of individuals. Not everyone is ready for job enrichment, and not everyone wants it (Ferris and Gilmore, 1984).

Recently, the design of work has focused on the specific characteristics of the job itself that appear to have an impact on how people respond to their work. Hackman and Oldham (1976, 1980), based on work of Hackman and Lawler (1971), developed a *job characteristics model* that attempts to identify the conditions under which jobs generate high levels of internal work motivation. The model, shown in Figure 4.4 assumes that when internal work motivation is high, good performance on the job generates positive feelings and provides self-reward. Poor performance leads to negative feelings and, therefore, no personal award. Three conditions must be met to increase internal work motivation: The worker must (a) experience personal meaningfulness of the work, (b) experience responsibility for outcomes of the work, and (c) have knowledge of the actual results of the work activities. Three job characteristics—skill variety, task identity, and task significance—contribute to the person's experiences of meaningfulness of the work. Experienced responsibility for the outcomes of the work is derived from the degree of autonomy the person has. Knowledge of the results of the work is achieved by way of feedback. Three moderators are described as having the potential to influence the extent to which individuals may be able to achieve internal work motivation. These moderators include the person's knowledge and skills, his or her needs for growth, and the context of the job, including working conditions, pay, and supervision.

Hackman and Oldham (1980) have developed a job diagnostic survey to measure all the components in the job characteristics model. Once the diagnosis has been made, changes can be implemented according to certain principles, which are summarized in Figure 4.5.

The relationships between technology in health services organizations and the main approaches to job design are illustrated in Figure 4.6. Relatively routine tasks can be broken down into highly specialized activities and detailed standards can be developed for their performance. Workers can then be asked to carry out the tasks on a repetitive basis. For example, consider a hospital employee who has the monotonous job of wrapping instruments ready for sterilization in a central supply room. Two design approaches are relevant. First, job rotation might be used (i.e., having different people

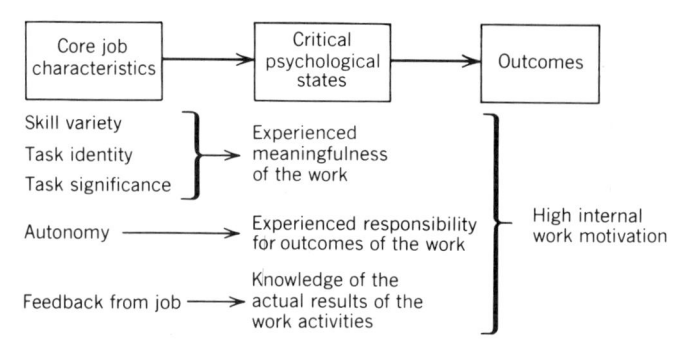

Figure 4.4. Job characteristics that foster three psychological states. From Hackman & Oldham, *Work Redesign*, © 1980, Addison-Wesley Publishing Company, Inc., Reading, Massachusetts. Figures 4.2 and 4.3. Reprinted with permission.

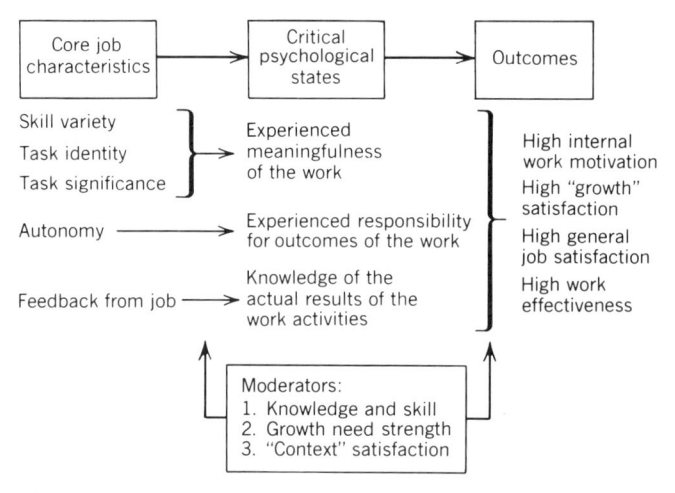

Figure 4.5. The complete job characteristics model. From Hackman & Oldham, *Work Redesign*, © 1980, Addison-Wesley Publishing Company, Inc., Reading, Massachusetts. Figures 4.2 and 4.3. Reprinted with permission.

rotate into the position). The cost to the organization would be low in this case because there are few specialized procedures to learn. Second, job enlargement might be used to broaden the scope of the job to include such activities as packaging other supplies and delivery instruments to the users.

At the polar end of the continuum, nonroutine tasks, it is most likely that the job characteristics model is applicable. As is discussed in Chapter 7, health services organizations employ highly professionalized staff. Physicians, nurses, social workers, and pharmacists represent highly specialized occupational groups, and professionals expect to be able to work relatively autonomously and to be making decisions under conditions of uncertainty. The higher the professional status of an occupational group, the more likely its members are to be expected to respond to complex technologies and

```
Routine tasks  <──────────────────────>  Nonroutine tasks
Job rotation          Job enrichment         Job characteristics model
Job enlargement
                  ╲_____╱
                Degree of professionalization
```

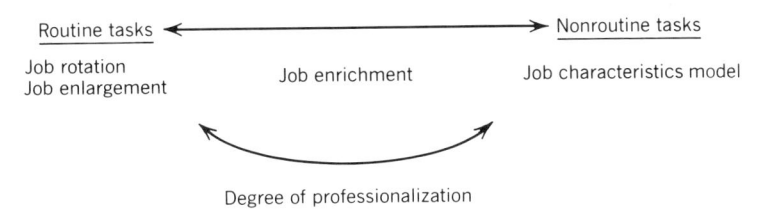

Figure 4.6. Technology and job design methods in health services organizations.

to make decisions about exceptional cases. Professionals are also more likely to be responsive to internal motivating factors. The service ideal underlying professional commitment is entirely compatible with Hackman and Oldman's model.

The technological conditions under which the job design method of job enrichment is applicable are not clear. As noted earlier, not everyone wants job enrichment, and some workers suspect that employers see job enrichment as a means of increasing employee workloads and responsibility with no corresponding raise in pay. Job enrichment may have greater potential with an occupational group such as nursing, which is seeking to achieve higher professional status and whose individual members desire further growth and development. Where the technology is extremely nonroutine, such as that of physicians in a hospital emergency ward, the job may be already overenriched because of the high degree of technological uncertainty.

Technology and Personnel Selection

Selection is the process by which an organization chooses the person or persons who best meet the criteria for a particular position under the organization's current environmental conditions (Glueck, 1978). In this section, we discuss the relation of the type of technology involved in a particular job to the development of selection criteria and to the process of selection. The relationship between the technology of an entire organization and the choice of a system of selection is also outlined.

A prerequisite to the selection process is, of course, an analysis of the particular job under consideration. Job analysis involves "defining the job and discovering what the job calls for in employee behaviors" (Dunnette, 1966, p. 69). This process includes a description of the requisite tasks and activities, as well as the necessary employee behaviors and characteristics associated with successful job performance (Levine et al., 1983). The methods of job analysis used in health services organizations include observation, questionnaires, interviews, and diaries kept by individual employees. The important point is that the job analysis should lead directly to a job description and the specification of required qualifications to do the job. For example, an analysis of the job of medical social worker would include specification of duties and responsibilities, relationships to other departments and activities, and the skills, training, and experience required of job incumbents. Subsequently, a job description can be written, which summarizes the job. Finally, the job specification—a statement of the qualifications required to do the job—is developed. For a medical social worker, these may include the following:

- A master's degree from an accredited school of social work
- Three or more years of postgraduate experience in a medical setting
- Ability to work effectively on a multidisciplinary team
- Knowledge of the diseases of the aged
- Ability to conduct individual and family therapy
- Knowledge of community support agencies and services

It is evident that some of these job specifications are easier to evaluate than others. Determining an applicant's education and experience is relatively uncomplicated, while evaluating a person's ability to work with other professionals and conduct therapy does not lend itself to simple methods of assessment.

The most critical factor is to develop criteria that adequately predict successful job performance. The degree to which a set of criteria predicts successful job performance depends on the reliability and validity of the criteria. In the context of selection, reliability is the extent to which a set of criteria accurately measures a particular quality. For example, a strong cognitive understanding of diseases of the aged may be a criterion for the job of home health care coordinator. Job applicants may be evaluated on the basis of this criterion through an interview, written tests, examination of transcripts of prior educational performance, and checking of references—these procedures may provide a consistent and reliable measure of an individual's knowledge in this area. However, a person's knowledge of the diseases of the aged may be an inconsistent predictor of success on the job because this knowledge, by itself, may not represent the quality of the work that may be performed on the job. Thus, the content validity of this criterion is questionable.

Alternatively, it may be determined that content knowledge is not as important to the work of home care coordinator as judgment, leadership qualities, and interpersonal skills. Such procedures as interviews, observation of persons in action, and simulations may then be used to evaluate applicants. Construct validity is established to the extent that these procedures in fact measure an individual against the defined criteria. In summary, criteria that are highly reliable may have low validity if they do not reflect the work on the job or do not measure the established criteria.

The development of valid selection criteria is a recurrent problem in health services organizations. We suggest that there is an important relationship between the nature of the job-related technology and the development of selection criteria. Because of the rapidity of technological change in health services organizations, for certain jobs, selection criteria are in a state of flux because of the volatility of the technological requirements themselves. Technological changes are ongoing and essentially irreversible, whether we are referring to the use of automation and computers, new inventions, or innovative techniques and procedures. Technological change, therefore, necessitates the selection of individuals capable of effectively employing new technologies. For example, the growing sophistication and complexity of financial management technologies requires people skilled in these procedures. New positions are being designed to match the necessary technology. Because many of these technologies are new, requirements for successful performance in these jobs are not always clear.

New developments in technology may also change the relative importance of a position to a health services organization and thereby alter the criteria for successful performance and the processes of selection. For example, prospective payment systems have dramatically changed the role of the medical records department; the position of medical records administrator has expanded in importance to the point that the performance of this person may be critical to the very viability of the organization. New skills are required of individuals assuming this role. Because this expanded role is new, however, earlier job specifications may be obsolete, and the characteristics and performance of former job incumbents are likely to be irrelevant in predicting success. The selection process itself may be affected by the expanded role. Because of the increased importance placed on medical records administration, the selection process for this position may have to be upgraded. For example, additional personnel, representing groups with which the medical records department must now interact (such as physician and nurses), may need to be involved in the selection process.

While technological change requires the acquisition of some new skills, other skills may become obsolete. For example, the accounting clerk, trained and skilled in the management of debits and credits, may not be needed in hospitals that have developed automated data processing systems. There is thus the possibility of technological advances turning "workhorses" and "stars" into "deadwood" (Odiorne, 1984). Technological change may also require individuals to adopt new social skills. For example, star performers who previously operated in privileged or isolated circumstances may find themselves working in teams or project groups requiring a level of interpersonal skills not heretofore needed. Thus, the establishment of valid criteria to predict success is increasingly difficult in an environment of rapidly changing technology in which the associated job requirements are altered accordingly.

The next stage of the selection process is the decision itself. Selection decisions may be made in a variety of ways, but usually, criteria are established and applicants are assessed according to these criteria. Objective and subjective data may be used in assessing job applicants. Objective data include information on education and experience, references, and the results of performance tests. Subjective data, which enter into almost every selection decision, are generally of a judgmental and intuitive nature. Subjective data are usually obtained from interviews, references, and other indirect sources of information about the applicant.

Once criteria have been established and information about applicants has been collected, decisions must be made about the manner in which criteria will be combined and weighed. Two methods have been suggested as valuable ways to combine information: intuitive judgment and systematic combination (Feldman and Arnold, 1983; Meehl, 1954). The intuitive judgment method requires an individual in the organization to review all the information on applicants and to base a decision on an overall impression. Systematic combination, on the other hand, involves quantification of all information on applicants, with weights assigned to different criteria. Here, selection is not an intuitive process, although judgment may be involved in assessing applicants according to the criteria. The process is essentially objective, and the applicant receiving the highest score is selected for the position. It has been found that the best selection decisions are made when both objective and subjective information about applicants is used, and that information should be combined systematically to arrive at a selection decision.

What is the relationship between the technology required of a particular position and the selection process, especially in terms of the choice of selection criteria? Figure 4.7 shows that jobs may be classified along two dimensions. The first dimension, which is related to the degree of routineness of the tasks, is the level of understanding of the job-related technologies, or means-ends relationships. For some jobs, such as a laboratory technician or housekeeper, the technology is clearly routine. For others, such as a home care coordinator and clinical nurse specialist, the work is more nonroutine (i.e., it is less clear how tasks are to be accomplished). The second dimension reflects the degree to which job specifications and corresponding valid and reliable selection criteria are available. Although a well-planned and well-executed job analysis for certain jobs will yield job specifications and selection criteria, the reliability and validity of selection criteria for some jobs is difficult. For example, there may be relatively strong confidence in criteria for hiring a person for a hospital pharmacist position. The applicant's education, experience, and references probably will provide most of the information required for a selection decision. The job of intensive care unit (ICU) nurse, however, may be quite different. Appropriate training and experience are important, but the job of an ICU nurse requires considerable teamwork and the ability to handle stress. The criteria for assessing these qualities are generally less valid and reliable than criteria for other areas.

Figure 4.7 indicates that the choice of an appropriate selection strategy depends on the level of understanding of job-related technologies, the confidence one has in the job specifications, and the reliability and validity of selection criteria. Cell 1 includes the jobs for which there is a high level of understanding of job-related technologies, as well as high confidence in job specifications and selection criteria. Such jobs as food

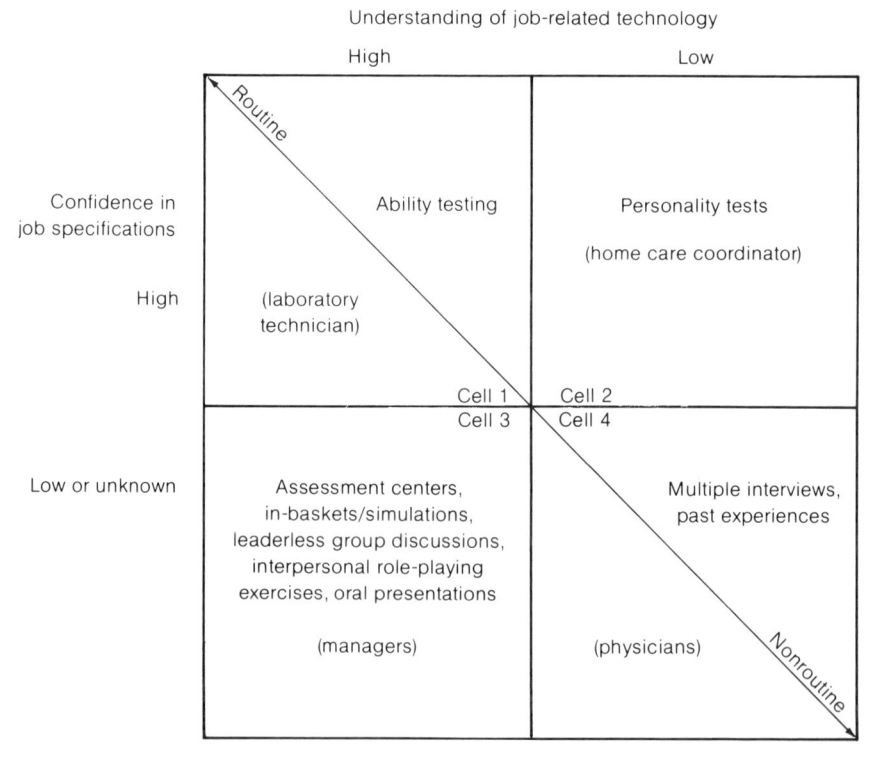

	Understanding of job-related technology	
	High	Low
Confidence in job specifications — High	Ability testing (laboratory technician) — *Routine*	Personality tests (home care coordinator)
	Cell 1	Cell 2
	Cell 3	Cell 4
Low or unknown	Assessment centers, in-baskets/simulations, leaderless group discussions, interpersonal role-playing exercises, oral presentations (managers)	Multiple interviews, past experiences (physicians) — *Nonroutine*

Figure 4.7. Type of selection process as determined by job-related technological certainty and confidence in job specifications.

tray assembler, X-ray developing machine operator, laboratory technician, and cashier fall into this category. For each of these jobs, the required tasks are clear and unambiguous and there is high confidence in the validity of job specifications and selection criteria. Objective criteria may be established for these positions, including credentialing and ability testing. The important factor in criteria development is ensuring content validity, that is, that the skills or abilities evaluated are truly representative of the work involved in the job.

Cell 2 includes jobs for the which there is low understanding of the job-related technologies but high confidence in the job specifications and the reliability and validity of selection criteria. The positions of home care coordinator and psychiatric aide are examples. The home care coordinator, for example, must possess a community orientation, have knowledge about community resources, and understand the medical and social dimensions of disease. The job specifications are, therefore, clear. However, the job of home care coordinator has considerable uncertainties. Despite similarities among cases (e.g., formal admitting and discharge criteria to be followed), each case in fact represents a new situation in search of a solution. This search, which is different for each case, requires intuitive judgment and unstandardized analytic approaches (Perrow, 1979). Thus in the case of home care coordinator there is high confidence in the job specification (i.e., we know the kind of person we are looking for), but little understanding of how this person will actually accomplish the work. The implications for selection procedures are clear; similar to jobs in cell 1, it is necessary to ensure that successful applicants meet the job specifications. Understanding that successful performance would require some relatively "unspecifiable" skills, it is essential to place additional emphasis on interviews and tests to assess assertiveness, intellectual ability, interpersonal skills, and decision-making abilities. Of particular importance in designing selection procedures for this type of job is ensuring their construct validity. In other words, if it is determined that leadership skills are critical for successful performance, as a home care coordinator, then procedures used to assess leadership must in fact measure this trait.

Cell 3 represents jobs for which there is high understanding of job-related technologies but low confidence in the specifications for effective performance of the job. An example of this type of job is the manager of an ambulatory care clinic. The technology of this job is clear and includes budgeting, supervising clerical personnel, and coordinating work schedules. Because the work of the clinic is fairly routine and predictable, these management tasks are correspondingly routine. However, in addition to the science of management, there is an "art" to management. This aspect of the managerial role thus reduces the level of confidence in objective job specifications. Therefore, for management roles, applicants may need to be evaluated through simulations, leaderless group discussions, and oral presentations. Note the distinction between jobs falling in cell 2 and those in cell 3. In cell 2, we know the kind of person we are looking for (clear job specifications) and we test to determine whether applicants can do the job. In cell 3, we know the skills the job incumbent must have (job-related technology), but we test to determine which applicants we feel would be most successful.

For the final group of jobs, located in cell 4, there is minimal understanding of the job-related technologies and low confidence in job specifications. New jobs that utilize new technologies fall into this category. Other jobs, such as researcher, chief of medical staff, and chief executive officer may fall into this category. Because of the absence of knowledge about the job-related technologies and lack of confidence in job specifications, considerable emphasis is placed on techniques that analyze biographical data, as well as reports on past experiences and use of multiple interviewers.

Technology and Performance Appraisal

The evaluation of individual performance is a central management role. Performance appraisals are used for a variety of purposes, including decisions on salary increases and promotions, individual development and motivation, employee selection and placement, discipline, and personnel and organizational planning. Four recurrent problems exist in the area of performance appraisal: (a) arriving at a single overall performance evaluation based on a series of different ratings, (b) convincing managers to grant rewards on the basis of merit, (c) obtaining greater employee involvement in the performance appraisal process, and (d) reconciling the developmental and administrative requirements of appraisal systems (Teel, 1980). There are other issues as well, such as overcoming such rater problems as biases, time pressures, and inappropriate attitudes toward the performance appraisal process.

This section focuses on the first of these problems, which involves not only the issue of developing a single performance measure, but also deciding on what precisely should be measured. As we shall see, rapidly changing technologies have exacerbated the criteria-specification problem.

The specification of performance appraisal criteria is a recurrent problem in performance appraisal. One of the central issues is whether to evaluate traits, behaviors, or outcomes of work. Trait approaches, now outdated, evaluated such items as appearance, self-confidence, alertness, ambition, and initiative. Resembling grammar school report cards more than adult performance appraisal instruments (Klatt et al., 1985), trait approaches were paternalistic and subjectively judgmental. More important, they did not measure either desirable job-related behaviors, productivity, or quality of work.

There is an ongoing debate over the merits of behavioral versus outcome-oriented performance criteria. Behavioral approaches identify critical job-related activities and behaviors, and appraise employees on their performance of these activities. Behaviorally anchored rating scales (BARS) exemplify the current application of this approach. Outcome approaches focus on the results of employee performance rather than behaviors. Management by objectives (MBO) is one example of such an approach.

Many factors affect the choice of the approach in measuring performance, including organizational norms and practices and how the performance appraisal information will be used. It is also necessary to consider the nature of the job itself and the technology that will be used to get the job done. Specifically, the type of appraisal method used depends on the extent to which knowledge about the transformation process or means-ends relationships is available and whether there are valid and reliable performance measures (Lee, 1985). The matrix of Figure 4.8 serves to clarify the impact of the presence or absence of these two factors on the type of performance appraisal procedures appropriate to different types of technology.

Where there is high knowledge of the transformation process (i.e., routine technology), and reliable and valid performance measures are available (cell 1), performance may be assessed through monitoring behavior or outputs. For example, the work of a dietary aide or an admissions clerk would fall into this category. The technology required for each of these jobs is clearly defined and requisite job-related

Figure 4.8

```
                    Understanding of job-related technology
                         High              Low
                    ┌──────────────┬──────────────────┐
Availability of     │ Routine      │                  │
reliable and        │   ╲          │                  │
valid performance   │    ╲         │ Output measures  │
measures            │  Behavioral  │                  │
                    │  or output   │ (health educator,│
                    │  measures    │ community        │
   High             │              │ health worker)   │
                    │ (dietary aide,│                 │
                    │ admissions   │                  │
                    │ clerk)       │                  │
                    │         Cell 1│ Cell 2          │
                    ├──────────Cell 3┼Cell 4──────────┤
                    │              │ Extensive use of │
                    │              │ peer review,     │
                    │              │ selection, and   │
   Low              │ Behavioral   │ training         │
                    │ measures     │                  │
                    │              │ (researchers,    │
                    │ (department  │ physicians, chief│
                    │ manager)     │ executive        │
                    │              │ officers)        │
                    │              │        ╲         │
                    │              │      Nonroutine  │
                    └──────────────┴──────────────────┘
```

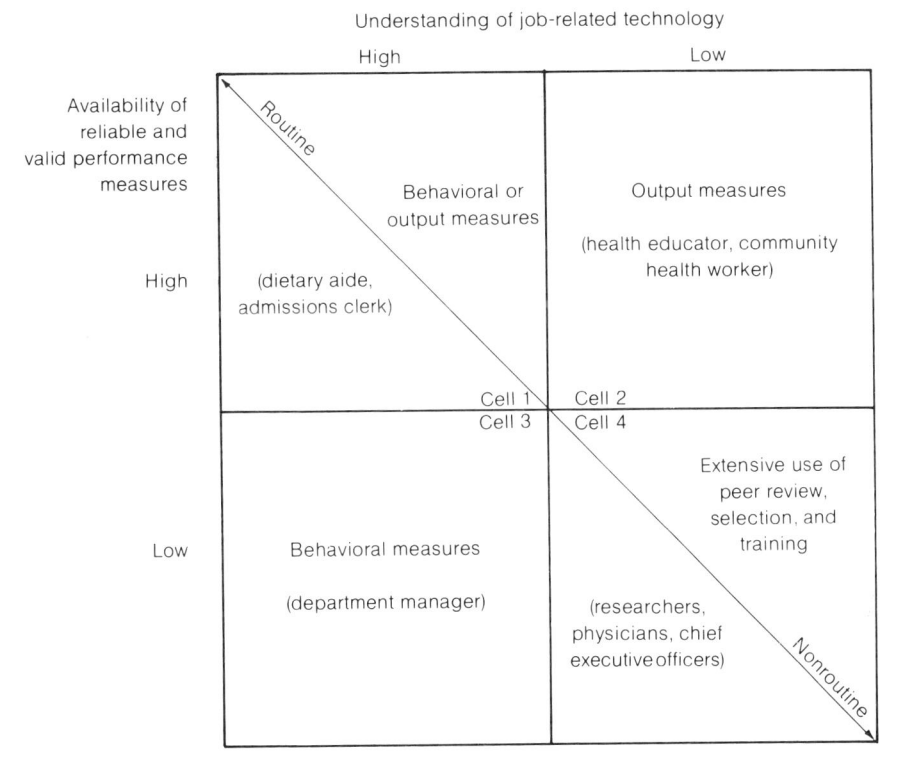

Figure 4.8. Choice of appraisal methods by understanding of job-related technology and availability of reliable and valid performance indicators.

behaviors are unambiguous. Performance appraisal criteria may, therefore, be based on behavior or output.

Cell 2 refers to jobs in which there is a relatively low level of knowledge about the technology, but reliable and valid indicators of performance exist. Here, performance appraisal should focus mainly on outputs rather than behaviors. Examples of jobs of this type are health educator and communty health worker. Although behavioral guidelines could and should be specified for these jobs, one could limit the creativity and performance of a job incumbent by linking appraisal criteria too stringently to behavioral specifications. The "technologies" used to educate patients or to relate to various cultural groups are not clearly understood, and require subtleties and sensitivities that are not easily measured. Outputs for these jobs, however, are relatively easy to measure. For example, to appraise the performance of a patient educator, one could measure the number of patients counseled, patient satisfaction with the educator, or patient behavior or attitude change. For the community health worker, it is possible to evaluate whether target groups of patients are receiving the appropriate array of services, and the level of patient and family satisfaction with the services.

Other jobs employ technologies that are well understood, although reliable and valid measures of performance are few (cell 3). Department managers generally fall in to this category. Generally, we know what managers should do, but it is difficult to develop output-based performance measures. For example, we can count the number of performance appraisals conducted by a department manager, but assessing whether the appraisal process has in fact enhanced individual and departmental performance is not straightforward. Even where department performance may be measured, it may be difficult to attribute performance to the manager. Thus, the basis for evaluating the performance of a manager is largely limited to behavioral factors.

A final job category is found in cell 4. Here, the nature of the technology is limited, and there are few valid or reliable performance measures. The work of researchers, physicians, and chief executive officers can be seen as exemplifying this type of job. Development of performance appraisal criteria is difficult, and great reliance is placed on extensive and costly selection procedures. Those involved in assessing performance for such jobs frequently rely on multiple observers and raters, as well. For jobs in cell 4, consensus among raters or observers is generally superior to behaviorally or output-oriented performance appraisal formats.

Technology and Rewards

Two basic categories of rewards have been identified by which organizations can facilitate and encourage effective performance by their members. These are named *intrinsic* rewards (i.e., the rewards are valued in and of themselves by the worker) and *extrinsic* rewards, which are provided by external agencies. Intrinsic rewards are frequently referred to as self-administered, since they are concerned with an individual's feeling about the job in terms of personal competence and accomplishment, responsibility and autonomy, and growth and development. Extrinsic rewards usually include the rewards directly controlled by the organization, such as pay, fringe benefits, recognition, and promotion. It was traditionally thought that intrinsic and extrinsic rewards were unrelated; however, recent research, although inconclusive, would suggest that in certain circumstances there may be a complex relationship between the two sets of rewards.

Four basic requirements have been identified as essential to any reward system (Lawler, 1971, 1977).

1. It must provide individuals with sufficient rewards to satisfy basic needs for food, shelter, safety, and security.
2. It must compare favorably with reward systems in other organizations.
3. It must be perceived by organizational members as fair and equitable.
4. It must be flexible, so that it can be adapted to individual needs.

The main reason for an organization's reward system is to motivate to perform efficiently and effectively. Consequently, the reward system may have a number of practical uses including attracting high-quality members to join the organization, and minimizing absenteeism and tardiness.

Lawler (1977) identified several characteristics of rewards that should ensure effectiveness.

1. The reward must be perceived to be of sufficient importance to organizational members.

2. The reward must be visible, and the relationship between performance and granting the reward must be explicit.
3. The reward should be such that it can be administered frequently without losing its importance.

In addition, the reward system must not cost so much that the price of using it outweighs any benefits in performance that might be derived. Some of the most common methods of reward are as follows.

Pay. The use of pay as a reward in organizations is almost universal. Wages or salaries can serve to reward membership in the organization as well as attendance. Salary increments and bonuses may be used to reward good performance.

Promotion. Although promotion has certain advantages in terms of rewarding good performance, there will be disadvantages if, for example, there are relatively few levels in the hierarchy to promote to. Also, a promotion tends to be a relatively permanent event, which is not easily retracted.

Fringe benefits. These rewards may vary depending on the position in the hierarchy. Common to most positions, however, are fringes such as a pension plan and life, health, disability, and dental insurance. Fringe benefits have the disadvantages of being relatively expensive yet many employees take them for granted.

Status symbols. Probably most frequently used in management or professional jobs, these rewards may include privileges such as a special office, a private secretary, a company car, or membership in clubs.

Special awards. Some organizations provide small symbolic awards for long-standing service and/or outstanding performance. These rewards have the advantage of being applicable to all categories of employees, and at relatively low cost.

Clearly, the extent to which rewards are useful in health services organizations is intricately linked to all aspect of human resources management, including the extent to which the specific jobs can be specified and evaluated. The level of professionalism of a particular occupational group will also influence the extent to which internal rewards will be effective in place.

What relationship exists between the nature of technology in health services organizations and the characteristics of rewards that may be effective in motivating human performance? To our knowledge there has been little research into this question. Building on the previous discussion in this chapter of the connections between technology, job design, selection processes, and performance measurement system, it would seem that similar relations may exist between the nature of the technology and the reward systems. For example, for a housekeeping aide, the tasks are routine, the job-related technology can be clearly specified, the confidence in the specification is high, and there are reliable and valid indicators of performance; thus extrinsic rewards will have the greatest effect. In these circumstances the reward, such as pay, can be directly linked to measurable performance of relatively routine tasks.

In contrast, a physician must make decisions with a great deal of uncertainty (i.e., perform nonroutine tasks), the job-related technology cannot be specified, and there are no reliable and valid measures of performance; thus he or she is more likely to be responsive to intrinsic rewards. These rewards will come from the individual's sense of "doing a worthwhile job" and particularly, in health services organizations, from feelings of providing a quality service. Clearly, the nature of the technology and the requirement for highly professional staff in health services organizations influences the types of rewards that will be effective in motivating performance (Chusmir, 1986).

Technology and Occupational Development

From an individual worker's perspective, the opportunities for a job or career within health services organizations are plentiful. As health services organizations become more diverse in response to changing environmental conditions, the positions within the organizations become more specialized. This has resulted in a proliferation of new occupational groups of people who work in health services organizations.

Most of the literature on careers in organizations focuses on the life cycle of an individual career, beginning with the development of career choice and identity, evolution of personal values and career choice, early career issues, mid- and later career issues, and retirement. However, the movement of an individual through a career life cycle will vary not only because of individual preferences but also because of the type of occupation(s) chosen. For example, occupational groups that are highly professionalized, such as physicians, nurses, and social workers, are likely to be strongly influenced in their career development by their professional associations, which provide extensive socialization to professional values not only during the early stages of career development but also through time as the profession changes. Standards and expectations of performance by professionals within health services organizations are more likely to be determined outside the organization by professional associations (see Mills and Posner, 1982). Similarly, ongoing technological change, especially in medicine, is likely to affect career choices and development, mediated, however, by the influence of the profession. The higher the initial investment of an individual in education for a particular occupation, the less likely the person is to voluntarily leave that occupation. Accordingly, the profession plays a very important role in ensuring that its members keep abreast of new technological developments.

In contrast, occupational groups that are relatively unskilled or nonprofessional, such as unit clerks, may approach career choices from a different perspective. The investment in the preparation for a particular position of this type is relatively low, and the major source of socialization is likely to be the organization itself. Accordingly, in situations of routine technology, a health services organization will play an important role in influencing the value system of new employees. Organizational socialization may include an in-house training program to teach routine tasks; in addition, standards of performance and expectations are likely to be well defined. Individuals who are employed to complete relatively routine tasks, requiring few specialized skills, are more likely to change jobs frequently during their work careers because any retraining costs are at the expense of the organization rather than the employee.

In summary, the characteristics of the technology in health services organizations determine the numbers, types, and levels of professional and other occupational groups necessary for the organization to meet its goals. Different occupational groups vary in their degree of professional identity and socialization and these will set the framework for career development. Where the position is responsible for routine, nonspecialized tasks then the organization itself must assume greater responsibility for socializing individuals for these positions.

GUIDELINES FOR ASSESSING THE POTENTIAL IMPACT OF TECHNOLOGY ON HUMAN RESOURCES MANAGEMENT

Throughout this chapter we have developed ideas about how new technologies or changing technologies may influence the ways in which human resources are managed in health services organizations. Health services managers of the future will need to assume a more proactive stance in identifying and assessing the potential impact of technological developments on the human resources function in the organization.

In a conclusion, we present some guidelines to assist health services managers in assessing new technological developments.

1. *Identify the technological change.* It is important that health services managers be constantly attuned to any form or variety of technological change that may be occurring within their own organizations, in other competing organizations, or in society as a whole. As noted previously, technological innovation is proceeding at a rapid rate in relation to medical practice, other health occupations practice, management processes, and information processing.

2. *Isolate the technological change from other changes.* Although it may prove difficult, this step is essential if the boundaries of the technological change are to be identified and the potential ripple effect of the change explored. The goal is to isolate the extent to which the change is relevant at different levels in the organization, including its potential effect on individual positions, work groups, the whole organization, a network, or a system.

3. *Begin to describe the nature of the technology.* Working with employees and colleagues most familiar with the technological change, managers should attempt to identify the key characteristics of the technology. One useful framework presented in this chapter is to describe the technology in terms of its routineness. Such a description should include an overview of the nature of the raw materials (patients, clients) as well as the techniques or search processes involved. As we have noted previously, in health services organizations the question of routineness is one of *degree*, not absolute routine or nonroutine.

4. *Develop a decision tree for decision making about human resources management.* This step is essential if the human resources strategy for the organization is to be appropriately matched to technological changes. The sample decision tree shown in Figure 4.9 reveals a clear relationship between the nature of the technology and the principal methods for achieving the human resources function in the organization. Where the technology is more routine, it is likely that job design will be achieved through the specification of highly specialized tasks. Techniques of job rotation and job enlargement may be useful to alleviate the boredom workers experience in completing these tasks. Selection process for jobs when the tasks are routine will rely on ability testing, as well as the assessment of skills in preparation for the job through demonstration of ability. Performance appraisal will be achieved through output measures, since these can be directly linked to job performance. Reward systems for employees whose jobs entail routine tasks will come from extrinsic sources, that is, from others, and will include such tangibles as pay, improved working conditions, and fringe benefits. The organization itself will assume the greatest responsibility for socializing individuals to fulfill these roles.

In contrast, where jobs entail the completion of nonroutine tasks and the continuous

Figure 4.9

Human resources methods	Routine	Some routineness	Nonroutine
	Description of the technology		
Routineness?	Routine	Some routineness	Nonroutine
Job design	Job rotation / Job enlargement	Degree of professionalism?	Job characteristics model
Selection	Ability testing	Role-playing presentations / Personality testing	Multiple interviews / Past experience
Performance appraisal	Output measures	Behavior and output measures	Peer review
Rewards	Extrinsic	Degree of professionalism?	Intrinsic
Career socialization	The organization		The profession

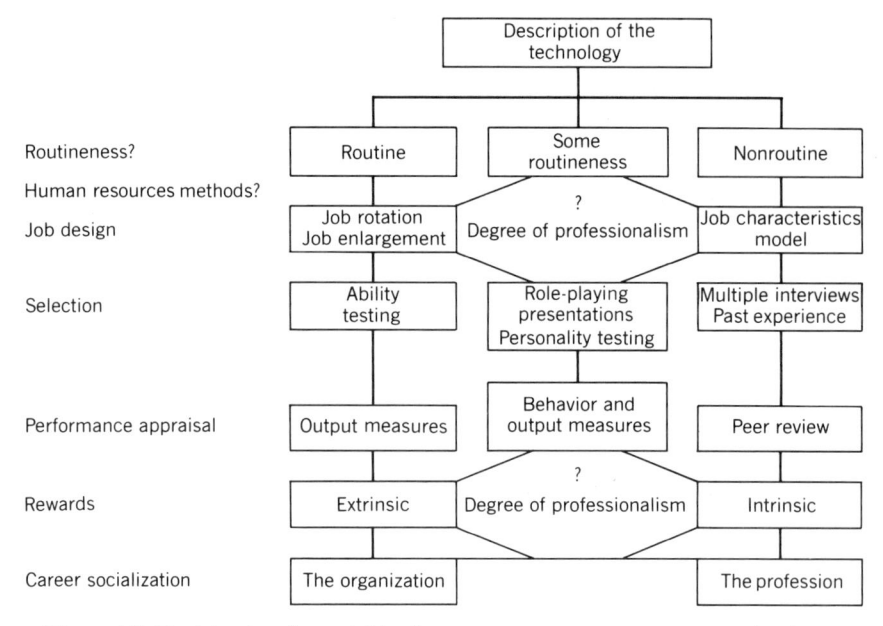

Figure 4.9. Decision tree for matching human resources management to technology.

handling of exceptional cases, the job characteristics model will be most appropriate. Selection processes will be more difficult and will incorporate several methods, including personality testing, multiple interviews, and careful examination of applicants' prior performance. Performance appraisal methods will be difficult because the expected behavior in response to nonroutine tasks cannot easily be specified or standardized. The most likely approach will include techniques for peer review. Persons occupying these positions will experience rewards that are intrinsic, that is, stemming from personal feelings about their jobs and the knowledge that they are doing well. The main influence on career socialization can be expected to come from the professional association of the particular group, especially in relation to the standards of practice. The profession itself will be the most important reference group for persons occupying positions with nonroutine tasks.

SUMMARY

In this chapter, discussion focuses on the nature of technology in health services organizations and the effect of technology on various human resource management functions. Technology is defined broadly to include managerial as well as medical technologies. Various theoretical perspectives of technology are described, and Perrow's model is then used to classify health services organizations. Additional discussion then moves to the relationships among an organization's environment, structure, strategy, and technology. The importance of specifying the unit of analysis is discussed because technology may differentially affect an organization depending upon whether one focuses on individual positions, the work group, clusters of work groups, networks of organizations, or the larger health system.

The technologies employed by an organization have important effects on the manner in which the various human resource functions are performed. Discussion centers around the impact of technology on job design, personnel selection, performance appraisal, reward systems, and occupational development. The chapter concludes with a set of guidelines for assessing the potential impact of technology on human resources management.

REFERENCES

Aldrich, H. E. *Organizations and Environments.* Englewood Cliffs, NJ: Prentice-Hall, 1979.

Alexander, J. W., and W. A. Randolph. "The fit between technology and structure as a predictor of performance in nursing subunits." *Academy of Management Journal,* 28(4):844–859 (1985).

Chandler, A. D. *Strategy and Structure: Chapters in the History of the American Industrial Enterprise.* Cambridge, MA: MIT Press, 1962.

Child, J. "Organizational structure, environment, and performance: The role of strategic choice." *Sociology,* 6:2–22 (1972).

Chusmir, L. H. "How fulfilling are health care jobs?" *Health Care Management Review,* 11(1):27–32 (1986).

Comstock, D. E., and W. R. Scott. "Technology and the structure of subunits: Distinguishing individual and workgroup effects." *Administrative Science Quarterly,* 22:177–202 (1977).

Dubin, R. *The World of Work.* Englewood Cliffs, NJ: Prentice-Hall, 1958.

Dunnette, M.D. *Personnel Selection and Placement.* Belmont, CA: Wadsworth, 1966.

Feldman, D. C., and H. J. Arnold. *Managing Individual and Group Behavior in Organizations.* New York: McGraw-Hill, 1983.

Ferris, G. R., and D. C. Gilmore. "The moderating role of work context in job design research: A test of completing models." *Academy of Management Journal,* 27(4):885–892 (1984).

Ford, J. D., and J. W. Slocum, Jr. "Size, technology, environment and the structure of organizations." *Academy of Management Review,* 2:561–575 (1977).

Fry, L. W., and J. W. Slocum, Jr. "Technology, structure, and workgroup effectiveness: A test of a contingency model." *Academy of Management Journal,* 27(2):221–246 (1984).

Glueck, W. F. *Personnel: A Diagnostic Approach.* Plano: Business Publications, 1978.

Hackman, J. R., and E. E. Lawler, III. "Employee reactions to job characteristics." *Journal of Applied Psychology* (monograph), 55:259–286 (1971).

Hackman, J. R., and G. R. Oldham. "Motivation through the design of work: Test of a theory." *Organization Behavior and Human Performance,* 16:250–279 (1976).

Hackman, J. R., and G. R. Oldham. *Work Redesign.* Reading, MA: Addison-Wesley, 1980.

Hage, J., and M. Aikers. "Routine technology, social structure and organizational goals." *Administrative Science Quarterly*, 14:366–377 (1969).

Harvey, E. "Technology and the structure of organizations." *American Sociological Review*, 33:247–259 (1968).

Hasenfeld, Y. *Human Service Organizations.* Englewood Cliffs, NJ: Prentice-Hall, 1983.

Herzberg, F. "One more time: How do you motivate employees?" *Harvard Business Review*, pp. 53–62 (January–February 1968).

Hickson, D. J., D. S. Pugh, and D. C. Pheysey. "Operations technology and organizational structure: An empirical reappraisal." *Administrative Science Quarterly*, 14:378–397 (1968).

Hickson, D. J., C. R. Hinings, J. M. Pennings, and R. E. Schneck. "Structural conditions of interorganizational power." *Administrative Science Quarterly*, 16(2):216–229 (1971).

Hoare, G. A. "Effects of retrenchment on organizational structure." In Larry Hirschhorn and Associates, Eds., *Cutting Back: Retrenchment and Redevelopment in Human and Community Services.* San Francisco: Jossey-Bass, 1983, pp. 199–226.

Kimberly, J. R. "The life cycle analogy in the study of organizations." In J. R. Kimberly and R. H. Miles, Eds., *The Organizational Life Cycle.* San Francisco: Jossey-Bass, 1980, pp. 1–14.

Kimberly, J. R., and R. E. Quinn. *Managing Organizational Transitions.* Homewood, IL: Irwin, 1984.

Klatt, L. A., R. G. Murdick, and F. E. Schuster. *Human Resource Management.* Columbus, OH: Merrill, 1985.

Kovner, A. "The nursing unit: A technological perspective." Ph.D. dissertation. Pittsburgh, PA: University of Pittsburgh, 1966.

Lawler, E. E., III. *Pay and Organizational Effectiveness.* New York: McGraw-Hill, 1971.

Lawler, E. E., III. "Reward systems." In J. R. Hackman and J. L. Suttle, Eds., *Improving Life at Work.* Glenview, IL: Scott, Foresman, 1977.

Leatt, P., and R. E. Schneck. "Technology, size, environment, and structure in nursing subunits." *Organization Studies*, 3(3):221–242 (1982).

Leatt, P., and R. E. Schneck. "Sources and management of organizational stress in nursing subunits in Canada." *Organization Studies*, 6(1):55–79 (1985).

Lee, C. "Increasing performance appraisal effectiveness: Matching task types, appraisal process, and rater training." *Academy of Management Review*, 10(2):322–331 (1985).

Levine, E. L., R. A. Ash, H. Hall, and F. Sistrunk, "Evaluation of job analysis methods by experienced job analysts." *Academy of Management Journal*, 26(2):339–348 (1983).

Meehl, P. E. *Clinical Versus Statistical Prediction: A Theoretical Analysis and a Review of the Evidence.* Minneapolis: University of Minnesota Press, 1954.

Mills, P. K., and B. Z. Posner. "The relationship among self-supervision, structure and technology in professional service organizations." *Academy of Management Journal*, 25(2):437–443 (1982).

Mintzberg, H. *Structure in Fives.* Englewood Cliffs, NJ: Prentice-Hall, 1983.

Mohr, L. B. "Organizational technology and organizational structure." *Administrative Science Quarterly*, 16:444–459 (1971).

Odiorne, G. S. *Strategic Management of Human Resources*. San Francisco: Jossey-Bass, 1984.

Perrow, C. "Hospitals: Technology, structure, and goals." In J. G. March, Ed., *Handbook of Organizations*. Skokie, IL: Rand McNally, 1965.

Perrow, C. "A framework for the comparative analysis of organizations." *American Sociological Review*, 32:194–208 (1967).

Perrow, C. *Organizational Analysis: A Sociological View*. Belmont, CA: Wadsworth, 1970.

Perrow, C. *Complex Organizations: A Critical Essay*, 2nd ed. Glenview, IL: Scott, Foresman, 1979.

Pfeffer, J., and G. R. Salancik. *The External Control of Organizations*. New York: Harper & Row, 1978.

Porter, L. M., E. E. Lawler, III, and J. R. Hackman. *Behavior in Organizations*. New York: McGraw-Hill, 1975.

Provan, K. G. "Technology and interorganizational activity as predictors of client referrals." *Academy of Management Journal*, 27(4):811–829 (1984).

Randolph, W. A., and G. G. Dess. "The congruence perspective of organization design: A conceptual model and multivariate research approach." *Academy of Management Review*, 9(1):114–127 (1984).

Teel, K. S. Performance appraisal: Current trends, persistent progress. *Personnel Journal*, p. 297 (April 1980).

Thompson, J. D. *Organizations in Action*. New York: McGraw-Hill, 1967.

Weick, K. E. *The Social Psychology of Organizing*. Reading, MA: Addison-Wesley, 1969.

Whistler, T. L. *Information Technology and Organizational Change*. Belmont, CA: Wadsworth, 1970.

Woodward, J. *Industrial Organization*. London: Oxford University Press, 1965.

CHAPTER 5

Structuring for Human Resources Management

James W. Begun
Ronald C. Lippincott

Central to the smooth and effective functioning of any organization is a set of rules, informal norms, and roles that govern the distribution of tasks in the organization. These rules, roles, and norms determine the way in which the organization differentiates and integrates work activities, and together they are referred to as an organization's structure. Typically, organizational structure is formally defined by an organizational chart and by organizational policies and procedures, including job descriptions. Power relationships among organizational members and unwritten rules also contribute to defining an organization's structure. These formal documents and informal relationships determine who performs what organizational task, and how the tasks are integrated in the production of a product or service.

Organizational structures vary along a number of dimensions—structures may be loose or tight, tall or short, rigid or flexible, centralized or decentralized. These variations have implications for human resources managers because structures influence the kinds of selection, appraisal, reward, and development mechanisms that are best suited for a particular organization. In addition, structures are tools that can be manipulated to change employee morale and productivity and overall organizational effectiveness. An understanding of an organization's existing structure and of possible alternative structures is essential to the human resources manager.

In this chapter we attempt to give the human resources manager an appreciation of why particular organizational structures emerge, how they vary, and how they affect the human resources management function. We pay particular attention to the structure of hospitals and the growing variety of complex organizational arrangements found among hospitals and other health care organizations. To demonstrate the effect of organizational structure on the human resources management function, we assess aspects of human resources management in health care organizations of three different types—solo medical practice, medical group practice, and the hospital—showing how structural differences in these organizational forms cause job design and employee selection, appraisal, and reward and development to be performed quite differently.

First, we outline a common vocabulary for describing organizational structures. In doing so, we establish an understanding of the origins of different structures and the range of possible structures.

CONCEPTUALIZING ORGANIZATIONAL STRUCTURE

Dimensions of Structure

Organizational structures can be examined at several levels, especially in large, complex organizations. First, most organizations are linked to other organizations formally through contracts, buying-selling relationships, or interlocking boards, and less formally through communications, liaison personnel, or social contacts. *Interorganizational* structures define the ways in which tasks are differentiated and coordinated among the multiple organizations. Structure at the *organizational* level refers to the division of tasks within the focal organization, while at the lowest level, *subunits* of organization have structural characteristics of their own. Structures at each level may vary substantially, and it is important to specify the level being assessed. In this chapter, we address the management of human resources at the general organizational level, with occasional reference to particular organizational subunits.

Additionally, at the outset we distinguish between formal structures, which are reflected in written documents, and informal structures, which may complement or conflict with the formal structures. It is critical for managers to recognize that both types of structure exist, and that changes in formal structures often are insufficient to produce an expected outcome unless the formal changes are accompanied by changes in informal structures. For example, many organizations have a problem with employee pilfering of office supplies. An informal norm that supports employee use of office supplies for personal needs may be more powerful and thus more important to address than written policies about employee access to and use of supplies.

In this chapter we refer to three key attributes of organizational structures in order to characterize them: formalization, centralization, and complexity. Formalization refers to the degree to which rules and procedures are written. Unwritten rules and procedures do sometimes have the same effect as written ones, as is the case for dress codes in some companies, but most attempts to enforce rules and procedures across large numbers of employees (thereby standardizing their behavior) involve writing down those rules and procedures. In addition to standardizing employee behavior, formalization generally promotes efficiency and coordination. To the degree that organizational activities can be formalized, the activities are more predictable and easier to coordinate. Formalization has other advantages as well. It ensures equity in the treatment of individuals in the same categories (e.g., workers in the same job classification category are evaluated in the same time frame) and across time (e.g., new employees are offered the same fringe benefits available to existing employees). Formalized policies and procedures also serve as an organization's "memory bank," thus eliminating the need to rely on individuals' memories and avoiding the need for recurring debates about informal norms.

Centralization is the degree to which discretion over decisions in an organization is concentrated at a single point, usually high in the organization. Centralization can be evaluated at the organizational or subunit level; for example, an organization may be decentralized, with authority dispersed to subunits, but its subunits may be centralized, with authority retained by one person.

Complexity is the degree of differentiation within an organization—the number of different activities being performed. Complexity can be introduced vertically, by adding hierarchical layers of management, or horizontally, by adding new roles or activities to the organization. Horizontal complexity is indicated by job specialization. Although necessary for efficient production of services, specialization makes it more difficult for organizations to coordinate the different tasks and requires increased attention to coordination in the organization. Vertical differentiation refers to the number of hierarchical levels in the organization. The more hierarchical levels, the more difficult it is for top management to control work activities. Often, vertical and horizontal differentiation accompany each other, because as horizontal differentiation increases, more coordinating managers (vertical differentiation) are required.

Horizontal complexity can be introduced in different ways in organizations. Workers generally are divided into groups horizontally based on function or product. Hospital personnel typically are divided mainly by function, with different departments for nursing services, dietary services, housekeeping, and so on. Functional divisions of labor are common where workers are specialized. Particular products or services also can be used to group workers; coronary care units or cancer rehabilitation units are examples of product departmentation. Most organizations use both functional and product grouping, and large organizations often add a layer of differentiation based on geography. A large multihospital system, for instance, might have East, South, Midwest, and West regional divisions, with functional and product departments within each division.

Another form of grouping workers is the matrix structure, which combines grouping by function and product. Workers in a matrix structure may have two bosses—their department head and a product or project director. Authority is shared between the two managers. Matrix structures allow flexible use of human resources across products but require extensive communication and coordination.

Determinants of Structure

We observed earlier that organizations vary substantially in their structural characteristics. A nursing home, for example, is typically structured more rigidly and tightly than a medical group practice. Matrix structures within subunits are likely to be found in large teaching hospitals, but not in smaller community hospitals. What explains the variation we observe among organizational structures?

Four major types of determinant can be identified. First, forces in the external *environment* of organizations can influence their internal structures. Complex and unstable environments create uncertainty for organizations, forcing them to create internally complex structures that are highly differentiated and require integration. Uncertainty also leads to structures that are not too formalized, centralized, or standardized, so that the organization may remain flexible. Another environmental trait influencing structure is resource dependence. Dependence on particular resources or organizations in the external environment forces the focal organization to establish favorable interorganizational linkages with the outside organizations through devices such as joint ventures and interlocking directorates.

Second, the *technology* used by an organization to transform inputs into outputs influences the organization's structure in important ways. Technologies that are fairly routine, such as those needed to produce housekeeping services, can be managed by highly formalized and centralized, and less complex, structures. The technology of much of health care delivery itself is quite nonroutine, calling for decentralized, more complex, and less formal structures. Chapter 4 covered the wide diversity in health service technologies and their human resource implications.

A third determinant of organizational structure is the *size* of the organization. Generally, larger organizations are more complex, particularly in terms of vertical differentiation. In addition, formalization often accompanies growth in size. This in turn allows organizations to decentralize, so that larger organizations frequently exhibit greater decentralization.

Organizational structures are not automatically created or totally constrained by the external environment and the organization's size and technology. Managers in the organization make decisions about how to structure tasks and relationships in the organization as they develop and implement the organization's strategy and as they assess the capabilities of the human resources within the organization. Managerial decisions about structure clearly are an element in determining how organizational tasks are differentiated and integrated.

Structural Forms of Organizations

Organizations can vary structurally on several different dimensions, and every organization is unique in some ways. *Patterns* of structural variation can be observed, however, which simplies the task of identifying major structural types and dealing with the implications of those differences. Typologies of organizational structures convey the essence of major variations. Typologies are simplified images designed to summarize and communicate the structural attributes of different organizations.

A very general typology is the classification of structures into markets, quasi-firms, and hierarchies (Eccles, 1981; Williamson, 1975). Market relationships have no formal structure at all, being governed by spot contracts created for every exchange relationship. At the other extreme, exchange relationships can be formally encompassed within a hierarchical organization, through the employer-employee relationship. In the middle are quasi-firms, relationships structured by contract, long-term buying-selling patterns, or other means less formal than hierarchical organization but more formal than market relationships. This chapter describes human resources management within hierarchical organizations.

The concept of "bureaucracy" captures several structural traits of a particular type of organization. Bureaucracies are highly formalized organizations exhibiting complexity through horizontal and vertical differentiation, with other nonstructural characteristics: separation of members' personal and organizational lives, merit-based hiring and promotion, impersonal application of policies, and career tracks for employees. Bureaucracies stress standardization and efficiency, and they are a common and effective way of structuring large organizations, particularly in stable environments (Daft, 1986).

Perhaps the best known among recent typologies of organizations is that of Mintzberg (1983), who delineates five major organizational types based on their structural characteristics. We review these briefly, so that we can use the typology later in discussing the effect of organizational structure on human resources management.

The *simple structure* is centralized, simple, and informal. Many small organizations fit this mold, with decision-making power retained at a strategic apex. This centralization enables the organization to move quickly in response to change.

A second structural type of organization is the *machine bureaucracy*, which is similar in concept to the bureaucracy described above. Machine bureaucracies have highly formalized structures, high complexity, vertical centralization, and limited horizontal decentralization. Labor is highly specialized, and a wide variety of rules and procedures serve to coordinate operations throughout the organization.

The machine bureaucracy can be contrasted to the third type of organization, the *professional bureaucracy,* which as the name implies has many professional members, requiring a high degree of both vertical and horizontal decentralization of decision making to specialists. The general hospital, with its vast array of skilled occupational groups, is an example of the professional bureaucracy. Control over the quality of work is vested in the occupational groups themselves, rather than the firm's management.

A fourth form of organization outlined by Mintzberg is the *divisionalized* form, which introduces decentralization to the division level within large organizations. Much power is retained by a central office, however, which monitors the performance of divisions. Divisions themselves (subunits of the organization) commonly are structured like machine bureaucracies.

The *adhocracy* is the last of Mintzberg's forms, and it is the least common of the five among health organizations. The adhocracy has informal structures, is complex, and is relatively decentralized. Specialists do their work in small project teams. This form of organization encourages innovation and rapid response to change, and is best suited for operation in a complex and unstable external environment.

Summary

We have introduced several ways to conceptualize the structure of organizations, ending with a typology of structures that summarizes patterns commonly found in the population of organizations. As discussed in Chapter 1, structures are linked to other major dimensions of the organization, including the organization's strategy and its human resources. In the next section we illustrate the effect of structural differences in organizations on the operation of the human resources management function.

INFLUENCE OF STRUCTURE ON HUMAN RESOURCES MANAGEMENT

Human resources managers must be familiar with the major structural characteristics of an organization and each of its subunits. Using the concepts presented earlier, the manager can classify the structural attributes of any organization. This information will have implications for human resources management. For example, performance appraisal systems will be quite different in the simple structure and the professional bureaucracy. It would be mistake to impose a highly formalized and centralized appraisal system in a professional bureaucracy. Similarly, attempts to create job descriptions in a relatively undifferentiated structure would prove fruitless.

We pursue the general point that organizational structures can affect the human resources management function by discussing selected examples of human resources management issues in three types of health care organization: the solo medical practice, the group medical practice, and the hospital. The examples are used because much is known about internal management in these organizations and because of their important roles in the health care delivery system. The three example settings illustrate the even more general point—that structure and human resources management are interdependent—and are not selected to be particularly representative of all health care organizations. They are illustrative, however, of the wide range of structural traits found in health care organizations. Table 5.1 shows the range of structural traits found in these three organizations. Table 5.1 is an oversimplification, particularly of group practice and hospital structures, since subunits or particular functions of those organizations may have different structural attributes. Patient billing in a hospital may be highly centralized, for instance, whereas employee recruitment may be delegated to individual departments. The structural traits given in Table 5.1 reflect overall organizational averages, relative to the other organizations in the table.

In terms of Mintzberg's structural typology of organizational structures, the solo medical practice represents a simple structure, while the group medical practice adds aspects of both machine and professional bureaucracies to the simple structure. The hospital, as we have pointed out, is a professional bureaucracy, and multihospital systems, which we also use in the discussion, bring in characteristics of divisionalized structures. We do not discuss examples of adhocracies, which are rare among health service organizations because the technical requirements of delivering clinical care necessitate a stable operating core of skilled workers.

Within each example organization, we draw on the human resources management functions of job design, recruitment, training, compensation, and evaluation to demonstrate the relationship between structural traits and human resources management. Because of the hospital's relative complexity and size, the discussion of the hospital setting is lengthier. More comprehensive and detailed discussions of each of these areas of human resources management are found elsewhere in this book: recruitment in Chapter 10, selection and placement in Chapter 11, training in Chapter 12, performance appraisal in Chapter 13, and compensation in Chapter 14.

TABLE 5.1. Structural Attributes of the Solo Medical Practice, Group Medical Practice, and Hospital

	Formalization	Centralization	Complexity
Solo practice	Low	High	Low
Group practice	Moderate	Moderate	Moderate
Hospital	High	Moderate/Low	High

Human Resources Management in the Solo Medical Practice

In the context of the health system, a solo practice is a relatively simple organization, characterized by three important structural attributes. The first is a high degree of centralization. The physician-owner is the "strategic apex" of the simple structure. The physician who owns the practice performs many of the core operational functions—which involve the production of medical services. In addition, basic administrative decisions are usually centralized in the practitioner. Since human resources management is an administrative function, the physician directly oversees job design, recruitment, compensation, training, and evaluation.

A second structural attribute of the solo practice is relatively low complexity. Depending on the specialty—which influences the number of and nature of work tasks—there typically are two to four employees. Horizontal complexity is low, as indicated by the low number of job titles. There is usually some combination of positions, such as receptionist-secretary, medical assistant, nurse, or technician, all roles that enable the organization to perform basic practice functions. In turn, vertical complexity is low because the small number of employees makes it possible for all members of the organization to fall under the physician's span of control as a supervisor.

A low degree of formalization is the third structural attribute of a solo practice. The small number of employees in the practice provides the basis for a high degree of informal face-to-face interaction between the members of the organization. In this situation, personal communication through direct contact vitiates the need for formal documentation such as written rules and regulations. A wide variety of informal norms affect the distribution and evaluation of work tasks.

Job Design

The low complexity of a solo practice has an important influence on a basic human resources management task—job design. With few job titles in the organization, it is unsurprising that roles exhibit low specialization, a structural condition in which organization members perform a wide variety of tasks rather than a few specific tasks. For example a single individual may handle patient phone calls and cash receipts, schedule appointments, transcribe medical dictation, and maintain medical records. Moreover, individuals often perform tasks that might typically fall under the job title of another employee. For example, a medical assistant—whose job title implies clinical specialization—might answer the telephone if the receptionist is busy making a cash collection.

Another structural attribute of the solo practice—low formalization—also influences job design. Documentation, such as written job descriptions and job specifications, is often ignored (Laetz, 1985). In this situation, confusion and inefficiency can easily arise, as when each employee claims that "bill collection is not my responsibility, with the result that an important task receives inadequate attention. On the other hand, too many different employees making appointments for patients invites scheduling conflicts, a consequence of too much attention to a task.

Efforts to resolve the job design problems usually involve the concepts of "primary" duties and "role interchangeability." Each employee usually performs certain basic tasks—typically the ones implied by the job title—which are considered to be his or her "primary" responsibilities. However, since the staff is so small, it is expected that each individual will assist other employees as the need arises. This "role interchangeability" usually involves routine administrative tasks, such as making appointments or accepting payments, which do not require great technical skills. In this situation, it is not unusual for clinical personnel who are trained in technical medical tasks to resist the performance of administrative functions. It is thus important that the expectation of role flexibility be communicated to the staff, either through an office policy manual or verbally by the physician.

Recruitment

Centralization of the human resources management functions in a solo practice potentially affects the quality of another function—recruitment. Despite the centralized responsibility of the physician for recruitment, this task would appear to be relatively simple. There are a small number of job titles, so relatively few individuals need to be recruited. Moreover, the market supply of workers for unspecialized roles tends to be large.

To the physician, however, recruitment can be burdensome. A variety of factors—greater interest in practicing medicine, unfamiliarity with recruitment techniques, a heavy appointment schedule—can interfere with the administration of an optimal recruitment process. A formal job description and job specification are basic elements in the process, but if these documents do not exist, the ad hoc job design that results may not adequately represent the position. This may create personnel problems later, which arise from misunderstandings about the nature of the role.

Additional aspects of recruitment are affected by the centralization of the supervisory administrative tasks. With few employees, there is little organizational slack, often making the need for additional personnel particularly urgent. Under this pressure, the sources of recruits—newspaper ads, colleagues, state medical society—may be too limited. Also, there is implicit pressure to minimize the time cost of the search by limiting the number of candidates. Moreover, screening can be too superficial, particularly with regard to scrutiny of references and depth of the candidate interviews.

A greater degree of formalization would clearly facilitate performance of the recruitment task. A formal job description—written by the former position holder and modified by the physician if necessary—would clarify the initial stage of the process. Establishment of a formal rule concerning adequate "notice of departure" would help to alleviate the urgency of the search. On other aspects of recruitment, particularly the number of candidates examined and their screening, a solo practitioner must be prepared to make an investment of administrative time to assure quality results.

Training

One dimension of complexity in an organization is the level of knowledge and skills required to perform jobs in the organization. Relative to other organizations in the health care system, the solo practice is low on this aspect of complexity. A number of *administrative* work tasks (answering the phone, patient reception, making appointments) do not require high levels of knowledge. Other tasks, such as bookkeeping, can be more complex, but they can be routinized into a series of relatively simple steps that can be readily learned. Also, certain *clinical* tasks, including reviewing records, ordering tests, administering medication, and giving instructions to patients, do not require a high level of medical knowledge.

The low complexity of many job tasks, and another structural attribute—the relative informality of the organization—have an impact on the way that training is undertaken in a solo practice. For a task such as making patient appointments, employee observation and self-instruction (e.g., examination of the appointment book) can be the basic modes of training. It is more likely, however, that personal "on-the-job training" by members of the practice is the prominent mode of training (Lavin, 1981). For example, how to address patients is an important element of patient reception, and while personal observation may indicate an office norm, another person, often the practitioner, will tell an employee the office policy. A physician who wants tests ordered a certain way or medications administered in a particular manner will personally instruct the medical assistant. This form of personal training is particularly prominent because the small number of employees makes the *informality* of face-to-face interaction a viable way to train employees.

Two additional aspects of training are influenced by the structure of a solo practice. First, through on-the-job training, employees are "cross-trained" in the routine, simple work tasks of different job titles. With few employees, such training enables individuals to substitute for each other in the event of absences due to vacation or illness. Paradoxically, continuity in the functioning of the organization is enhanced by this *decrease* in specialization. Second, the lack of horizontal and vertical complexity, which limits the possibilities for internal promotion or transfer, is related to the degree of investment in external employee training (e.g., continuing education courses). At a certain point, additional training will cause employees to exceed the skill requirements of their job titles. With the limitations on *internal* promotion, the more highly trained employees may seek opportunities for advancement elsewhere. Solo physicians thus have a rational incentive to limit their investment in employee training so that it meets the needs of their practice's relatively simple structure. This means that the employees seldom need a level of training more advanced than that necessary to perform largely standardized administrative duties and simple to moderately complex clinical tasks.

Compensation

Generally, the formality and sophistication of a compensation program is related to the complexity and size of an organizational structure (Stolhanske, 1977). Given the lack of complexity of the solo practice, it is unsurprising that the typical solo practice often has a relatively simple salary program. Typically, a pay range, which specifies a minimum and maximum salary, is established for each position in the practice (Owens, 1981). An elaborate process of pay grading, which evolves from the organizational complexity of many diverse roles, is unnecessary.

Despite its simplicity, the compensation program of a solo practice can have several potential problems. First, with centralized responsibility for personnel functions, solo practitioners may think that it is efficient to forego formal job evaluations of the positions in their organizations. Instead, they may make the pay scales identical to those of other organizations in their area that employ similar kinds of personnel. However, difficulties arise when the jobs in the interorganizational comparison are really *dissimilar*, a real possibility given the high degree of role interchangeability in solo practices. For example, the medical receptionist in one office may very well be performing tasks different from those handled by the person with the same formal job title in another office. The result can be a fundamental flaw in the compensation program, one that causes either overcompensation or undercompensation of personnel.

As suggested above, role interchangeability creates a second problem in the compensation program of solo practices. Role interchangeability and cross-training tend to "homogenize" jobs. As a result, it is more difficult to establish salary differentials that are based on a clear and distinct valuation of each position. If *everyone* does many *different* tasks, distinctive job valuation becomes more difficult. Ultimately, the "primary" tasks of each position are used to justify salary differences between job titles; to legitimize their compensation programs, practitioners must be careful to emphasize this point.

A third compensation problem arises from the use of a salary *range* in a structure with such low complexity. Adherence to the upper end of the range is intended to prevent paying an employee more than the market value of the job title. However, the upper end of the range also implicitly sets a limit on the salary advancement of an employee in that position. This problem is particularly severe in a solo practice, which lacks a hierarchical basis for promoting individuals to higher salaried positions in the organization. With advancement in position and salary blocked, personnel difficulties, such as loss of initiative or high turnover, are not unusual. Given this situation, increasing the top of the salary range is usually justified on two grounds: (a) performing the existing job tasks more efficiently and (b) increasing the range or extent of tasks for a particular job title.

A final problem with the compensation program is that important aspects are often administered haphazardly, a situation that reflects the lack of a formalized compensation process (Rusley, 1984). For example, a common complaint of office assistants is that physicians often forget to review salaries. On the other hand, the informal interaction in a solo practice sometimes enables employees to exert direct personal pressure concerning salaries. The result is that physicians may make unsystematic salary decisions to alleviate the pressure.

In the face of these possible circumstances, it becomes important to formalize the compensation process by stipulating the salary range of each position, the basis for salary increases (e.g., cost of living, merit), the frequency of salary review, and fringe benefits. While a personnel manual offers a way to clarify some of these matters, such formal documentation often does not exist in a solo practice. By establishing a minimum set of procedures (e.g., salary review every June), it is possible to avoid lapses in the compensation process. Moreover, the standardization of compensation decisions through regularized organizational procedures helps to reduce the personal pressures that can arise around this human resources management function.

Evaluation

Evaluation of employees is influenced by the small scale and high centralization of a solo practice. With so few members in the organization, solo physicians can either observe or participate in the work activities of the employees. This informal process provides the basis for reaching judgments about such things as attendance, interpersonal relationships, quantity and quality of work, and job knowledge. Written work, such as insurance claims and charge records, is less amenable to evaluations rooted in informal interaction, so physicians sometimes "spot check" these work tasks. Use of formal evaluation instruments is often seen as unnecessary, particularly since performance decisions are centralized in a supervisor with whom employees have constant personal contact.

The predominance of informal interction in the solo practice can create two basic difficulties in the evaluation of employees. First, a fair evaluation process requires that an employee understand both the tasks and the standards of performance that will be the basis for evaluation. It is assumed that an employee will acquire much of this information during socialization into the office routine, but this is sometimes an unreliable way to communicate the basis of performance review, particularly in an organization that emphasizes informal norms and role interchangeability. For example, even if it is assumed that everyone answers the telephone, a nurse who does not understand the informal norm and ignores the ringing telephone might object to being called "uncooperative" in an evaluation. To assure a fair evaluation process, it is important to make both the job tasks and performance standards clear to the employees.

A second difficulty in the evaluation of employees involves the process of evaluation. An information evaluation that entails some undocumented, brief oral comments provides no benchmark data for future reference. The physician might forget the discussion, and the employee, who expects a reward due to future changes in work behavior, might be disappointed. Written performance reviews help to avoid this kind of situation, and they are particularly helpful to a physician who may incur unemployment liability because of inability to document the basis for the dismissal of an employee.

Human Resources Management in the Group Medical Practice

In this section, the structure under consideration is a relatively small (three to seven physicians), incorporated single-specialty group practice. Reflecting a relatively large percentage of group practices in the United States, this type of health organization exhibits important structural differences in comparison to the solo practice (Williams, 1984). Traits of Mintzberg's machine and professional bureaucracies begin to emerge in group practice. There is an increase in horizontal complexity, as manifested by larger numbers of specialized support personnel. Also, a position—the practice manager—emerges, which is particularly important for human resources management functions.

Group practices also exhibit greater vertical complexity. In the professional corporation form of group practice, the top level of the organization is formed by those physicians who are shareholders. The board of directors, which is usually composed of all the physicians in a small group, composes the next level in the hierarchy. Beneath the board is the executive director, to whom the practice manager reports. In some practices, however, the manager reports to the founding or senior member of the practice. Support staff report to the practice manager or to particular physicians.

As suggested by the greater degree of complexity, group practices generally are more decentralized than solo practices. The small number of physicians in the group usually means that decision making about basic issues is a consensual process involving all members. In long-standing groups, however, "founding members" or "senior members" tend to exercise more authority, whether assigned or assumed. For an administrative function such as human resources management, authority is delegated to the practice manager because the physicians often lack the interest, expertise, or time to perform the task.

Finally, small groups tend to have greater formalization in comparison to solo practices. It is important to emphasize that the increase in formalization is a matter of degree. Since there are still relatively few physicians in the group, they may decide many issues informally over the telephone, at lunch, or in "corridor conferences." As described below, however, the delegation of many human resources management tasks to a specialized role is a major source of greater formalization.

Job Design

Although some solo practices and small groups may make one of their senior nonphysician staff members a practice manager, the growing business and administrative complexities of practices with about six physician members has created pressures to hire professional managers (Hamity and Gauss, 1982). Generally, the manager has responsibility for three broad functional areas—finance, systems (e.g., collections), and human resources. A manifestation of increased horizontal complexity, this specialized role marks the beginning of the differentiation of medical and administrative tasks. This development has a direct impact on the job design of the physician. Unlike the solo physician, group physicians are relieved of the central supervisory responsibility for many human resources management tasks. As a result, practitioners are able to specialize more in the core operational activity, the practice of medicine.

Whereas the position of practice manager affects the job design of the physicians, the character of hierarchical authority in a small group can create problems for a particular aspect of the manager's job design—the reporting relationships. The formal structure of corporate governance exists, but the group is still small enough that a "horizontal democracy of partners" tends to exercise significant decision-making authority (Sweeney et al., 1983). However, if each physician feels that he or she is "the boss," the command relationship of the practice manager can become ambiguous, creating potential confusion about work priorities. One physician may emphasize account collections, another supply changes, and still another personnel matters. Resolution of the problem requires clarification of the practice manager's reporting relationships by establishing the position of executive director with "real" authority or by having the administrator report to the informal leader of the group (e.g., the senior member).

With its transition from an informal to a more formal structure, the small medical group practice can pose an even more fundamental problem for the practice manager. In the vertical hierarchy, nonmedical personnel usually report to the administrator. But in a small organization with a minimal hierarchical structure, members of the organization still have relatively high personal interaction with one another. As a result, there may still be a tendency for physicians—who feel that they "own" the organization—to directly intervene in personnel matters delegated to the manager (e.g., consistent application of the compensation program). If this occurs, it poses a challenge to the legitimacy of the practice manager's position (Schatz, 1982).

To be effective, the decentralization of authority to the administrator must be formalized and supported by the physicians. A clear chain of command—best formalized in an organization chart—helps to clarify reporting relationships, but physicians must support the formal system of authority. The "practitioner-owners" must emphasize that nonmedical personnel report to the practice manager and that this individual has authority over basic human resources issues. Without such an understanding, performance of human resources management functions can subtly shift back to the physicians.

The horizontal complexity of the group practice also influences the job design of another type of personnel—support employees. In response to the need to accommodate a greater work volume (e.g., patient load), larger groups must expand the types and number of basic support personnel such as nurses, medical assistants, receptionists, billing clerks, and medical secretaries. Inherent in this dynamic is greater specialization. For example, receptionists, who may perform a wide range of tasks in a solo practice, may find that the volume of telephone calls and patient intake begins to restrict their activities to meeting these responsibilities. In effect, the greater horizontal complexity of the structure means that the tasks of particular job titles become more narrow, with a concomitant reduction in role interchangeability.

The structure of the group practice, particularly the specialization and decentralization of decision-making authority, also begins to cause greater formalization of the job design task. Unlike the solo practice, the group has a specialized practice manager who has authority over many human resources management tasks. This individual is thus able to devote more time to the formalization of the job analysis process. Through oversight of general office functions, the manager acquires knowledge of practice procedures and operations, which provides a strong basis for rationalizing tasks with particular job titles. Job descriptions are developed by having employees describe their job activities, sometimes on a preprinted form. While several different persons in the same role may have an "aggregate job description," there also are likely to remain a number of positions for which there is a single description for each particular employee. This development is an example of movement from informal, personal control over behavior toward the more impersonal regulation of activities through written documentation.

Recruitment

Because of the decentralization of human resources management tasks in the group practice, the recruitment of personnel tends to become more systematic than in a solo practice. Efforts by the practice manager to formalize job descriptions and job specifications help to decrease the possibility of inadequate definition of positions, an event that occurs too often in solo practices. Since there are a greater number of specialized support personnel, they can substitute for a departing employee. This lowers the immediate pressure to fill a vacancy and provides the slack necessary to search a greater number of potential recruitment sources.

The selection process for personnel also tends to grow in rigor. A basic expression of this development is that the practice manager has more time to interview candidates and to scrutinize their references. For medical support personnel, the manager may do initial candidate screening, since some physicians want final control over the selection of employees with whom they directly work.

Finally, the new employee's orientation to the organization, which is usually composed of informal interaction with other employees in a solo practice, often takes the form of a one-hour discussion with the practice manager. At this time, basic organization procedures and rules, such as work hours, overtime, time off, and fringe benefits, are explained. As an expression of the increased formalization of the group structure, much of this material is documented in a personnel manual.

With regard to the recruitment of physicians to the group, the group structure can create an ambiguous situation. A careful self-examination of group needs, philosophy, and style are crucial to the successful recruitment of a physician. In a small, single-specialty practice, particularly one that has survived for many years, there is usually a strong consensus concerning these factors. Because the number of physicians in the group is small, individual members tend to understand the expectations of their colleagues and the diversity of interests tends to be minimal.

At the same time, however, the time cost of the search can have a negative effect on the recruitment process. As a specialist in personnel, the practice manager can assist with certain aspects of physician recruitment (e.g., advertising the position, co-ordinating and scheduling candidate visits, outlining the history of the practice, explaining the compensation formula, and describing the community). Although the senior member of the group may take the lead, participation by physicians in the process is *not* specialized. All the members must interview and screen the candidates, a process that requires a significant commitment of time, particularly if the group wants to increase the likelihood that selection is consistent with the practice needs and philosophy. Rather than commit the necessary time, there is pressure to restrict the recruitment sources (e.g., to the "old boy" network, or a particular medical journal), to limit the number of candidates, and to conduct superficial interviews, particularly if the need for a new member is due to an increase in patient volume (Kurtz, 1984).

Training

Training in a group practice reflects the influence of the decentralization and specialization of personnel tasks. While greater specialization lessens the need for cross-training in a group, the method of employee development is essentially the same in both solo and group practices—informal, on-the-job training. In the group, however, the training is either directed by the manager, who is quite knowledgeable about office procedures and operations, or delegated by the manager to an employee who is competent in the tasks to be mastered. As in the solo practice, physicians still assume responsibility for instructing medical support staff in their particular style and procedures. Finally, given the limited opportunities for promotion, the small group, like the solo practice, typically does not make a significant investment in training.

Compensation

In developing a compensation program for a group practice, the practice manager's efforts are affected by the increase in the horizontal complexity of the structure. The increase in specialization among support service jobs results in a decline in the role interchangeability so prominent in solo practices. The content of job titles becomes more distinctive, providing a clearer basis for using a simple ranking method to evaluate the relative worth of jobs (Stolhanske, 1977). For example, skill and mental factors provide the basis for distinguishing between the value of a receptionist and medical secretary, two positions whose tasks are most likely combined in a solo practice.

A small group is a transitional structure in devising a compensation program. Some groups may still adhere to the format of the solo practice and establish a salary range for each job. However, the growth in the number of jobs begins to establish the basis for rudimentary pay grading. Pursuing this method to develop a salary structure is difficult because a simple ranking method to evaluate jobs does not reveal natural divisions. In turn, it becomes difficult to determine the appropriate number of grades. Generally, this situation leads to arbitrary decisions about the divisions, with relatively few grades being established.

In pricing the jobs, there is a basic similarity between the small group and the solo practice, and a basic dissimilarity. The method of pricing in both structures is similar; that is, salary ranges are based on a combination of personal judgment and information from colleagues and journals. In a small group, however, the greater precision in the job descriptions—which arises from the increase in horizontal specialization—helps to lessen the compensation problems that can arise from comparing similar job titles that have different job content.

This description of the compensation program suggests that as in the solo practice, there is still relatively little opportunity for internal promotion and transfer. Greater specialization may create some opportunities for advancement. However, the limited number of pay grades indicates that horizontal complexity is still relatively modest, limiting the number of alternative opportunities in the organization.

The increase in formalization and administrative specialization in the group practice is particularly important in the administration of the compensation program. Certain basic compensation matters—vacation, fringe benefits, and sick leave—are more likely to be explicitly stated in a personnel manual. The salary review process becomes more regularized than in a solo practice. The practice manager usually has a regular conference with nonmedical support employees, and written records of the results are likely to be kept in personnel folders.

The compensation of physicians in a small, single-specialty practice also is affected by the structure of the organization. Unlike the solo practitioner, the individual physician in the group generally does not have complete control over income. Although becoming less prominent, an equal allocation of net income to full-time group members has historically been the common method of physician compensation in single-specialty groups (Smith et al., 1984). Single-specialty groups minimize income differences that arise in groups composed of different specialties. Moreover, other potential sources of differences, such as malpractice premiums, types of equipment, and collection ratios, are less prominent in single-specialty groups than in multispecialty practices.

Evaluation

The specific arrangements of reporting may determine the exact process of assessment, but the evaluation function reflects the emerging horizontal complexity of the small group. In response to the differentiation of medical and administrative activities, physicians typically evaluate medical support staff, while the practice manager reviews nonmedical staff. Since professional practice managers often have a background in business administration, this specialized division of labor enhances the quality control of such business matters as the currency of accounts receivable and the accuracy of records for patient charges.

The process of evaluation also is affected by the increase in the formalization of the group structure. Evaluation often involves a standardized form, which is used to assess personal traits and the particular tasks of each role. Reviews are more likely to occur at regular intervals, and a written assessment in personnel folder can provide benchmark data for determining the relative change in an employee's performance.

Although this formalization can reduce some of the difficulties associated with evaluation in a solo practice, the structural complexity of the group is a potential source of problems for the evaluation function. In some groups, a status differential emerges between "medical" and "nonmedical" employees. The core operational activity (i.e., the delivery of medical services) is the most important work to most physicians in a group. Since physicians work closely with medical support staff who contribute to these core activities, a subtle favoritism toward these employees can arise. The result can be a systematic bias in the evaluation process, which creates inequity in the merit rewards of the compensation program. The practice manager must work closely with the physicians to avoid this problem; otherwise, morale on the "administrative side" will be undermined.

Finally, evaluation of physician performance in a small group reflects the modest hierarchical structure of the organization. With few members, there is no real need for a formal quality assurance structure. Instead, one of the physicians in the group usually is delegated this responsibility on an informal basis. Moreover, evaluation commonly arises as an informal "reactive" process (Reddick, 1977). For example, events such as patient complaints, a malpractice suit, a reduction in income, and a decrease in patient load are likely to result in a discussion between the "quality assurance" physician and the alleged offender. Since the scale of the organization is small, it is easy to communicate with other members and to convene a formal meeting of the board if necessary.

Summary

The differences between solo and group practice structures are illustrated in the following vignette, which describes actual problems that arose when one practice failed to implement the formalization that must accompany larger size and complexity in the transition from solo to group practice.

Illustration 1

Daley and Associates is a six-person medical group practice with a thriving business and serious management problems. A quick tour of the office provides ample evidence that all is not well. Three clerical workers, seated at desks piled high with charts, code books, insurance forms, and the like, are trying to code insurance claims in between answering the telephone and processing patients.

Billing problems are a major source of concern to the physicians. Interviews with billing staff reveal that no protocol for filing insurance claims exists—"We just do everything the way Dr. Daley tells us to." Dr. Daley designed the antiquated charge ticket that is used by all physicians to record procedures for billing.

When asked for the personnel manual, the office manager replies that policies are Dr. Daley's responsibility, not hers. And job descriptions do not exist—"Everyone just pitches in where they are needed." The confusion extends even to the chart filing system, in which a given chart might be filed under the patient's name, the head of the family's name, or the name of the first person in the family to visit the practice.

What is the root of the group practice's problems? The practice had been started 5 years earlier as a solo practice by Dr. Daley, with one nurse and a clerk-receptionist, who now serves as the office manager. As the practice grew, the office manager hired more workers but never trained them to be anything more than general-purpose office help. With no clearly defined job descriptions, employees never develop the skills necessary to master their duties. Dr.

Daley is a dominant personality, continuing to run the six-person group as though it were solely his own. The practice clearly has increased in complexity, but the requisite formalization and decentralization of authority have not accompanied the growth in complexity. (Adapted from Whittington, 1985.)

Human Resources Management in the Hospital

Hospitals traditionally have represented the most complex type of health care organization, a tradition superseded by the recent growth of multi-institutional health systems. Multi-institutional health systems add vertical complexity and impose some formalization and centralization across member units. Nevertheless, the hospital remains one of the most complex organizations in existence. Within its structure, a combination of routine and unscheduled tasks is performed by a diverse conglomeration of professional and other workers to produce health care services. The high degree of occupational specialization in the structure requires greater efforts to formalize and coordinate most human resources management functions, and hospitals create human resources management departments to meet these needs.

The hospital exemplifies Mintzberg's professional bureaucracy, discussed earlier. The hospital relies on the skills and knowledge of its operating professionals to function, and professionals are given considerable control over their own work. Managers and professionals work with each other, rather than working in strictly hierarchical relations (Begun, 1985). Multi-institutional health systems add another layer of complexity, becoming divisionalized structures, often encompassing a variety of structural types within the organization. Hospitals in a system may approximate professional bureaucracies, but a wellness center or laundry service belonging to the system may resemble a machine bureaucracy. Thus it is important that many human resources management functions be tailored to the subunit level.

Human resources management functions are not centralized in the hospital's strategic apex, as is the case in the simple structure of the solo medical practice. Earlier in Table 5.1 we classified the hospital's overall degree of centralization as low to moderate. Recruitment and job satisfaction, for example, often depend on the efforts and reputation of the different professional departments, such as physical therapy, nursing, and surgery. Clinicians and professional managers share power throughout much of the structure of the hospital.

We describe some of the major human resources management issues in the hospital to further illustrate the effect of structural characteristics on the problems faced by human resources managers and the range of solutions available to them. Examples are drawn both from free-standing hospitals and hospital units in multi-institutional systems.

Job Design

The high degree of horizontal complexity, a distinctive characteristic of hospitals, has a direct impact on the design of jobs. In comparison to other health organizations, there is a relatively high demand for a large number of different tasks. A rational adaptation to this situation is to narrow the range of tasks performed by employees, a process of specialization that results in a highly elaborate job classification system. There are broad categories of personnel—administrative, professional, technical, clerical, service—and numerous jobs are included within these categories, ranging from registered nurses and laboratory technicians to guards and building service workers.

The matter of medical staff privileges provides a more specific example of the impact of specialization on the design of jobs. Reflecting the high degree of formalization in a hospital, the medical staff bylaws define how physicians and other clinicians secure privileges at a hospital. A variety of factors, such as licensure, character, qualification to treat patients for which privileges are sought, availability of staff support, and bed capacity, enter a staff appointment decision. Privileges are formally granted by the hospital, with the clinical staff making a recommendation on each decision. In the larger, more complex hospitals, privileges tend to be more precise and narrow than in smaller hospitals. For example, a family practitioner who does general medical care, reads electrocardiograms, handles obstetrics, and performs various types of surgery in a small hospital might be restricted to narrowly defined family practice privileges in a large, urban hospital.

While specialization has a major effect on the complexity of job design in hospitals, formalization influences the delineation of the job classification system. In large hospitals, job analysis has become a highly formalized process. In comparison to group practices, where job descriptions are written by employees and the practice manager, hospitals use questionnaires to collect information, a more efficient technique that yields more specific data. Typically, the incumbent in a position fills out the questionnaire, with final development of the job description and job specifications a product of collaboration between the department head and the human resources management department. Depending on the position, the documentation can become quite elaborate. Moreover, given the high degree of change in hospital jobs, some hospitals attempt to periodically audit all positions.

Although the elaborate process of job analysis in hospitals would appear to clarify the design of jobs, especially in comparison to group practices, this human resources management function is not without problems. Specialization has been associated with such task characteristics as low autonomy, lack of identity, and lack of variety (Bechtold et al., 1980). In turn, these attributes influence the level of employee satisfaction, and ultimately the retention of personnel. The implication is to "redesign" jobs at the expense of specialization, that is, to *expand* task autonomy, identity, and variety in an attempt to make the work more inherently meaningful.

However, the reversal of specialization through the redesign of jobs in turn can create problems. If employees who are affected by the redesign are organized, unions may resist the task changes, particularly if existing work rules are threatened. The inherent limitations in certain jobs (e.g., housekeeping), lack of employee desire for redesign, and the impossibility of redesigning all hospital jobs pose additional obstacles to restructuring roles to promote job satisfaction.

The relationship between structural factors and job design concerns issues even more fundamental than job satisfaction. For example, there is a structural conflict between the two major types of care giver in the hospital: nurses, who typically represent more than half the modern hospital's human resources, and physicians. Historically subservient to physicians, nurses have generally increased their level of formal education and training since World War II. In turn, this upgrading has intensified the claims of nurses for greater autonomy.

In addition to its influence on the content and scope of specific jobs, structure is important in grouping the myriad roles and establishing relationships between them. As in some larger group practices, functional departments are the major way to group a large number of different hospital jobs. The departmentation in hospitals is extensive, as indicated by departments ranging from dietary and housekeeping to nursing and the various specialized clinical services.

Further differentiation is visible in the various systems of reporting relationships. Neuhauser (1972) notes that physicians use a loose organic structure, less skilled workers are under hierarchical supervision, and nurses and skilled technicians fall in between. The differentiation in hospital authority systems is much more complex than in the typical group practice, and it is essentially an adaptive response to the different levels of training and variation in task complexity among employees in the larger institution. For example, collegial authority is appropriate for physicians, given the complexity of their tasks and their high degree of professionalism. On the other hand, dietary personnel are less skilled and perform simpler tasks, making hierarchical authority a more appropriate form of supervision.

The need for coordination across departments raises basic operational issues. Whereas group practices can emphasize traditional hierarchy and rules as the basic means of coordination, the organizational complexity of hospitals usually creates pressure to develop additional bases for departmental relationships. Unit managers, scheduling departments, and management committees all exemplify structural elaboration to meet the greater need for lateral coordination in hospitals, but the best known mechanism is probably the patient care team. Consisting of the physician who is responsible for the patient and the other personnel (e.g., nurses) who assist, the teams form and dissolve as patients enter and leave the hospital. However, formalization remains important, especially since the patient's record is a crucial coordinating device for the different team members.

As Neuhauser (1972) notes, this is basically a matrix organizational structure. Vertical coordination, through the departments, exists simultaneously with horizontal coordination, through the clinician who heads the team. Nurses and other health personnel are located in departments, but they are brought together to attend particular patients under a physician's supervision. The unity of command principle notwithstanding, effective coordination makes this "two bosses" approach appear to be functional.

Interestingly, however, proposals have been made to organize the patient team in a form analogous to the approach used in group practices. Rather than having a myriad of teams forming and dissolving with the patient flow, this would make for greater continuity in the work group personnel could be assigned to the same patient units, with physicians always using the same unit (Wright and Wright, 1984). Moreover, an informal structure might develop that would supplement and support the formal mechanisms for coordination. The logical extension of this idea is a "product" structure wherein nonphysician personnel would be located in the different clinical departments of the medical staff. The structure would explicitly recognize that physicians have authority over basic operational activities, and it would group personnel by background and specialty interests rather than on a functional basis.

Recruitment

The recruitment process in hospitals exhibits a high degree of formalization, providing these institutions with a way to systematically process a large, complex work force. As in other health organizations, hospitals use job descriptions and job specifications to define the type of personnel being sought. Newspapers and "walk-ins" are major sources of candidates, particularly for nonspecialized personnel (Abbott et al., 1981). However, hospitals typically need more specialized personnel than group practices, so they make greater use of recruitment sources that reach broader markets. For example, advertising in specialized national trade publications and contact with academic medical centers occurs more often. Moreover, hospitals draw on nonstandardized recruitment methods, such as use of a nurse registry, to meet their distinctive needs. In general, hospitals are more likely to investigate a broader range of sources (e.g., unions, employment agencies, retired military personnel) than smaller health care organizations.

The selection stage of recruitment also reflects the effects of formalization in hospitals. Unlike group practices, large hospitals often decentralize human resources management functions, with professional departments handling their employees and the personnel department processing nonprofessional staff. Even without such decentralization, the personnel department's operations at "intake" often reflect the high degree of horizontal specialization in the institution (Metzger, 1981). For example, there may be employment forms tailored to fit the degree of specialization of different positions. Moreover, another level of documentation, a preliminary short form, may be used to elicit key information for an initial screening. This process enables the organization to more efficiently process a large number of applicants.

The interview aspect of screening can also become quite formalized. If many applicants are competing for the same position, screening interviews are often used to acquire information about the candidates' general qualifications. When the general credentials have been established, a placement interview, which examines technical qualifications and personality considerations, may be held. Typically, the personnel department performs the screening interview, while the placement interview occurs in the department with the opening. Also, any written or operational testing of the candidates will usually occur at this screening stage of recruitment.

The final step in an elaborate recruitment process varies, depending on the position opening. A physical examination may be required, particularly in light of the health and physical standards required for certain jobs. Furthermore, security checks may be conducted. Also, hospitals generally send out formal reference checks, although the resulting information often is not used.

In comparison to a group practice, the complexity of a hospital usually necessitates a more elaborate and formalized orientation for new employees. Indeed, the orientation occurs at three levels: hospital-wide, departmental, and job-specific (Abbott et al., 1981). At the hospital orientation, the employee receives information about such matters as the institution's philosophy and general principles, physical layout, payroll procedures, benefits, and essential fire and safety procedures. Departments will orient the employee to their particular procedures, policies, work schedules, and training sessions. The job-specific orientation usually involves a discussion with the new employee's supervisor about the work expectations and evaluation procedures for the position.

The foregoing description outlines the elaborate formalization of the hospital recruitment process, but recruitment of two clinical providers—physicians and nurses—requires special comment. Many hospitals must continually recruit nurses, a situation that can be influenced by structural factors. For example, formalization of a variety of work schedules has become a common recruitment incentive, and a formal "open-house orientation" has been suggested as a way to recruit potential candidates.

At a deeper level, high turnover among nurses is a basic reason for the continuing need to recruit these personnel. Retention of nurse employees requires constant attention. For example, too many "nonnursing" tasks is a general complaint about the design of nursing roles, and the burnout of nurses who work in specialized units (e.g., coronary care) presents a more specific structural problem (O'Donovan and Bridenstine, 1983). Retention of nurses involves another basic structural issue. With their increasing professionalism, nurses desire greater job autonomy and more involvement in hospital and patient decision making. Joint nurse-physician management of patient care units, nurses as voting members of the hospital governing board, and a voluntary nursing staff are among the structural solutions that have been proposed to meet the professional concerns of nurses (Shortell, 1982).

As for physicians, some strategies for their recruitment also involve structural factors. For example, the guarantee of a net income is increasingly used by hospitals as a way to recruit specialists (e.g., orthopedic surgeons) who are likely to generate hospital patients. Another strategy is an "internal joint venturing," which assumes a variety of forms. A typical internal joint venture by for-profit hospitals involves leasing a certain number of beds to a physician for lower than usual costs. The difference between the insurance reimbursement at full cost and the lower leasing cost is retained by the physician, becoming a financial incentive to stay with the hospital. Finally, acquisition of well-established physician practices is a growing recruitment strategy. This suggests that hospitals may be moving toward a more divisionalized structure, wherein quasi-autonomous units are linked by a central administrative office, in their efforts to gain physicians for their medical staff.

Training

In general, compared to group practices, hospitals are more involved in employee training, and this situation is associated with structural factors. Formal rules are partly responsible for the hospital's greater focus on training. For example, the Joint Commission on the Accreditation of Hospitals (JCAH) is placing greater emphasis on employee training activities in its standards. The numerous internal policies and procedures of hospitals, which function to standardize the behavior of many persons, often require careful introductory training. In this vein, new nurses often are assigned to an internship program for several weeks so that they can be socialized into the numerous unit rules and procedures (e.g., medication protocols). Finally, the complexity and dynamism of knowledge and technology is particularly great in hospitals, thereby prompting an increased need for training.

Like group practices, one way that hospitals have traditionally met their need for training is through the personnel department. Unlike other health organizations, many hospitals recently have also developed a specialized structure—the training and education department—to perform training functions. Another manifestation of the existence of complexity and decentralization in hospitals, this department has been delegated responsibility for activities such as nursing in-service training, audiovisual services, and patient education.

Specialization is one reason for the greater involvement of hospitals in training programs than is the case with group practices. The division of the hospital medical staff into departments according to specialty provides a logical organizational basis for transmitting specialized knowledge. As a result, most large hospitals have formal medical residency training programs. Nursing departments, which have historically provided training programs in technical areas, also have performed a similar function.

Hospitals are more likely to be involved in developmental programs in areas such as patient education, community health education, undergraduate and graduate medical education, and continuing medical education. In effect, the diversity of educational programs reflects the more elaborate horizontal differentiation of the hospital.

Hospitals and medical groups exhibit structural differences in the training functions, and there also are certain contrasts in the method of training. As in other health structures, individual on-the-job training is prevalent in hospitals. However, given the sizable number of employees in a particular functional area, there is a greater tendency to use a formal instructor and a systematic format. In addition, a growing number of hospitals have begun to use multi-institutional cooperative structures as a way to efficiently share educational services in a time of tight budgets.

Despite these differences between hospitals and group practices with regard to training, it is important to note that hospitals have traditionally underinvested in training. Shortell (1982) mentions two structural factors among the reasons for this situation. First, the intensive horizontal specialization of labor has undermined the incentive "for hospitals to 'cross-train' employees through job rotation or related programs." Second, although many hospital diploma schools are now closing for economic reasons, historically these institutions have had a significant role in training nurses. With subservience to administrative and medical staff needs as a basic orientation, the nurses so trained were an "all-purpose resource," performing functions that ranged from housekeeping activities to unit management administration to diagnosis and treatment responsibilities traditionally reserved for physicians.

Changes may be occurring in the hospital perspective on training. In response to the increasingly competitive health market, hospitals have begun to emphasize training that meets their operational needs. In turn, this focus has links to structural factors. For example, hospitals have recognized that allied health personnel can make significant contributions to productivity, so hospital training programs increasingly focus on such staff. Indeed, Robbins and Rakich (1981) anticipate that certain allied health personnel will be cross-trained as "supertechs," a development that would mark the reversal of greater horizontal complexity in the hospital division of labor.

Another change concerns the possibility of lifetime employment relationships with hospitals, particularly in multiunit systems. This possibility varies for managers, physicians, nurses, and others, but a decrease in the turnover of any group of personnel might prompt hospitals to invest in training that meets their particular needs. Multiunit systems, particularly investor-owned organizations, already have begun to provide a variety of management training courses. The importance of managers and supervisors in meeting operational needs and increasing productivity—especially in an organization with elaborate vertical complexity—is an additional factor underlying the trend toward management development.

Compensation

The centralized salary system in hospitals that has emerged over the past two decades is more highly formalized than the typical salary program in group practices. Job descriptions are the foundation of a rationalized salary program, so it is essentially necessary to perform a job analysis on the institution's complex division of labor. The human resources management department usually develops a standard format to complete this task.

Since it provides a systematic and logical basis to distinguish between a large number of positions, job evaluation is usually more quantitative in hospitals than in group practices. The point system is the most common approach in the performance of this task. Abbott et al. (1981) state that after detailed job descriptions have been developed,

> [t]he description is then evaluated in relation to the predetermined job factors (e.g., education, experience), which are assigned the point values. The salary grade is based upon the total point value; thus salary ranges are systematically determined for each salary grade. (P. 29)

The outcome is generally a system of salary grades more elaborate than exists in the typical group practice, a reflection of the greater horizontal complexity of hospitals.

In discussing compensation, it is important to note that there is an increasing tendency to establish a *separate* salary program for managerial employees (i.e., department heads and up). An outgrowth of the vertical complexity of hospitals and the more demanding roles of the managers, the program is based on a formalized process involving job analysis, job evaluation, and an apparent emphasis on merit review. In comparison, group practices lack such an elaborate system of administrative specialization, so a distinctive program of managerial compensation is typically unnecessary. Moreover, compensation for many practice managers has usually been determined informally, although there is a trend toward formal contracts.

The relationship between structure and managerial compensation is also exemplified by the use of incentive systems (e.g., bonus, stock options) as a method of managerial reimbursement. Group practices rarely use incentive programs, but these are becoming more widespread among proprietary hospitals, especially the multiunit chains (Cleverley and Mullen, 1982). The comparative absence of incentive programs in group practices is linked to the nature of their administrative structure. There is only a modest vertical hierarchy, and physicians—who are owner-employees—hold some of these positions (e.g., executive director). Moreover, productivity depends heavily on the behavior of the physician members of the group, constraining the contribution of the practice manager.

A highly formalized compensation program tends to centralize managerial control over salaries in hospitals, a shift from the historical power of hospital departments in this area. On the other hand, the unionization of hospital personnel, which is particularly prevalent among guards and professional/technical employees, represents a decentralizing force in hospital salary administration. In bargaining with union representatives, hospital management must share power to some degree over the hospital's compensation program, a situation that does not exist in physician-owned group practices.

Evaluation

Historically, hospitals and group practices have used similar approaches to evaluate their nonphysician members. Specifically, a supervisor rates an employee on task and behavioral traits (Smith and Elbert, 1981). Unlike group practices, however, in hospitals the complexity of work often means that the formal evaluation document becomes more elaborate. For example, hospital nurses can be assessed on several dimensions—clinical, patient/family and staff education, professional development, and other factors. Within each category there may be numerous items ranging from documentation of patient condition on admission to the unit to attendance at in-service continuing education programs. In addition to the checklist response (e.g., "exceeds standards") to each item, there is often a request for supplementary written comments. As in a group practice, this material is generally reviewed in a meeting between the supervisor and employee.

At present, hospitals—more so than smaller health care structures—are beginning to use other methods of evaluation. For example, some hospital nursing departments have begun to use behaviorally anchored rating scales (BARS) that measure "employee effectiveness through specific behaviors anchored to a scale that ranges from ineffective to effective performance" (Smith and Elbert, 1981). Basing the evaluation on specific behaviors is less subjective than the traditional trait approach, but the use of BARS has an important structural prerequisite—a large number of employees who perform reasonably similar jobs.

Another performance appraisal technique, management by objectives (MBO), has gained increasing popularity in hospitals. However, it has been most frequently used in departments such as central supply and bookkeeping, where it has been relatively easy to develop quantifiable standards of performance. In general, MBO is particularly useful in a large complex organization like a hospital because it provides a direct link between the organization's goals and individual performance expectations. This relationship frequently is unclear to employees in large organizations, which have numerous departments and hundreds of employees.

Since a supervisor's evaluative judgments are a common source of employee grievances, it is important to note the character of grievance procedures in different health structures. In group practices, the procedure often takes the simple form of an "open door" policy by the practice manager. The hospital grievance process, however, is likely to be more formalized. Written documents, a specified route of appeal, and final review, often with formal employee representation, have become common in hospitals.

The foregoing description of evaluation mainly pertains to nonphysician members of health organizations. A comparison of the way physician activities are evaluated reveals significant differences between hospitals and group practices. In groups, evaluation of physician activities is informal, possibly being the function of a quality assurance (QA) committee if one exists. The JCAH, however, requires as a formal principle that each registered hospital have a complete quality assurance progam. Hospitals have typically responded to this requirement by establishing a variety of medical review committees (Utilization Review, Medical Records, Credentials, Tissue, Quality Assurance, etc.). Moreover, each clinical department has quality assurance responsibilities. Not only is the QA function more decentralized, but these structures generally operate according to routine schedules and apply formal criteria when examining the activities of medical staff members.

In comparison to the QA program of a free-standing hospital, the evaluation of physician activities in hospital chains is even more complex (Brown and Kloss, 1984). The board of directors is obligated to oversee the QA programs of its member facilities, but to fulfill this responsibility, the board must deal with the vertical complexity of the chain structure. The board, management, and medical staff of each member hospital must design and agree on their own QA plan, subject to the corporate board's approval. Formal rules and procedures usually are established concerning the frequency of reports and the appropriate channels of communication. To enhance the efficiency and regularity of the process, the corporate and local boards may develop committees to structure the QA relationship between the corporate board and all the other member hospitals. Also, some multiunit chains have taken steps to standardize the QA documentation, permitting the corporation to undertake interhospital QA evaluations.

Finally, the degree of formalization in the evaluation of managers in group practices and hospitals tends to be different. With a relatively small number of administrators, group practices often do not formalize their evaluation process (Sweeney et al., 1983). In contrast, one aspect of the increasing formalization of the salary program for hospital managers is the use of explicit performance appraisal techniques (Abbott et al., 1981). This level of formalization is a logical consequence of a vertically complex structure incorporating many more managers who are responsible for the productivity of the organization.

The relationship between structural factors and evaluation is even more visible in multihospital chains. With two boards—a system corporate board and a local hospital board—a manager in a hospital chain faces greater vertical complexity. Indeed, the individual may be evaluated on the basis of different goals, which are established by the different levels (Hoium, 1984). At the local level, a CEO may be held accountable for such functions as planning and organization, management of resources, and promotion of the hospital, while the corporate office may evaluate the individual based on coordination of multiunit services, effective promotion of the system, and participation in system meetings.

Summary

The hospital clearly is a significantly more complex organization than the two structures discussed earlier. The wide range of service activities conducted within the hospital requires not only a high degree of formalization for coordination purposes, but a recognition of the autonomy of the diverse professional groups involved in service delivery. In the illustration that follows, a hospital CEO fails to recognize that the hospital is a professional bureaucracy, as opposed to a more centralized machine bureaucracy. She suffers the ultimate negative consequence of the evaluation process.

Illustration 2

Donna Wherry was hired as chief executive officer of a 250-bed community hospital after screening and review more than 200 candidates for the position. For 18 months, she devoted her efforts to turning around an institution mired in declining occupancy and out-of-date facilities.

While Wherry was out of town on business, a small group of doctors and nurses called for a mass meeting of hospital employees to discuss concerns about the CEO's management style. Alerted by the president of the board of trustees, Wherry hurried back to confront a hostile situation: The mass meeting had been held and a petition against her had been signed by half the medical staff and by half the hospital employees as well. A leadership committee of four doctors and nurses demanded Wherry's immediate resignation because she was "incompetent, devious, lacked leadership, had shown unprofessional conduct, and had committed negligent acts."

The charges were considered at a meeting of the board of directors, which was also attended by 8 physicians, 18 registered nurses, 5 department heads, a laboratory supervisor, a dietary aide, and the medical staff secretary. During the discussion, complaints focused on the CEO's insensitivity to the needs of the various hospital departments. A nurse complained that after Wherry's arrival, "It took her 10 months to call a meeting of the head nurses. Problems were left unsolved." A department head reported that "Wherry said Dr. Black would have to sign for an X-ray equipment request for $10,000. That is poor leadership." Another nurse stated that "Wherry was evasive and showed a lack of concern. Nurses were not present at administrative meetings." According to the president of the medical staff, "Department heads should be on board committees. No one came around and told department heads they were appreciated. Wherry has a repressive style."

After more discussion, the board voted unanimously to dismiss Donna Wherry with 2 months' severance pay. The basic problem? Wherry had managed human resources as though the hospital were a traditional, hierarchical bureaucracy. She had failed to allow participation in relevant decisions by the different professional groups in the hospital, particularly those with the most power: physicians and nurses. (Adapted from Kovner, 1981.)

SUMMARY

We have ranged widely over the diverse human resources management functions in three different health organization settings: a solo medical practice, a group medical practice, and a hospital. These organizations vary substantially in the key structural attributes of formalization, centralization, and complexity. The organizations grow in complexity and formalization from the solo to the group medical practice to the hospital. Centralization is highest in the solo medical practice. These structural differences exist in general because of the more complex technologies and environment and the larger size of the group practice and the hospital. Structures of specific practices or hospitals may differ, of course, depending on the local environment in which they operate and the strategic choices made by organizational managers.

Human resources management is affected by differences in structure. Table 5.2 summarizes the relationships discussed in this chapter. Highly formalized and more complex job design, recruitment, training, compensation, and evaluation procedures exist in most hospitals. Specialized individuals are delegated responsibility for human resources management functions in many group practices, and specialized departments serve this purpose in hospitals, reflecting the decentralization of decision-making power in larger, more complex organizations.

Structural differences also have implications for the work environment of human resources management personnel. The functions of workers in human resources management will tend to be more specialized in the more complex structures, for example, and more guidance in the form of policies and procedures will be available in more formalized structures.

All the examples discussed in this chapter reinforce the need for human resources management personnel to understand the organizational structures in which they work. With this understanding, they can be more confident of the rationale for particular management actions, and better able to match organizational settings to their own career development needs.

REFERENCES

Abbott, Ralph F., Jr., Brian E. Hayes, Jay M. Presser, Martin E. Skoler, and James N. Trono. "Personnel administration in the health care field." In Norman Metzger, Ed., *Handbook of Health Care Human Resources Management.* Rockville, MD: Aspen, 1981, pp. 7–67.

TABLE 5.2. Selected Characteristics of Human Resources Management Functions in Three Structural Settings

Function	Solo Medical Practice	Group Medical Practice	Hospital
Job design	1. Low specialization 2. Role interchangeability 3. Little documentation	1. Moderate specialization and formalization 2. Professional manager role common 3. Physicians delegate administrative functions 4. Manager-physician conflict over control of employees	1. High specialization 2. Job classification systems 3. Overspecialization can decrease morale 4. Coordination problems
Recruitment	1. Often haphazard and informal	1. Moderately formalized	1. Highly formal and elaborate process 2. Wide range of sources of employees 3. Formalized orientation 4. Problems with nurse and physician recruitment
Training	1. On-the-job training 2. Cross-training 3. Limited external training	1. On-the-job, but administrative training supervised by manager 2. Limited formal training	1. Greater investment in training activities 2. Possibility of lifetime employment
Compensation	1. Problems due to role interchangeability and cross-training 2. Salary important because of limited opportunities for promotion 3. Often haphazard and informal	1. More related to job description 2. Moderately formalized	1. Highly formalized 2. May have separate salary program for managers 3. Impact of unions
Evaluation	1. Conducted by physician 2. Often informal	1. Manager reviews nonmedical staff 2. Moderately formalized 3. Potential for favoritism toward clinical employees	1. Highly formalized (e.g., MBO)

Bechtold, Stephen E., Andrew D. Szilagzyi, and Henry P. Sims. "Antecedents of employee satisfaction in a hospital environment." *Health Care Management Review*, 5(1):77–88 (Winter 1980).

Begun, James W. "Managing with professionals in a changing health care environment." *Medical Care Review*, 42(1):3–10 (Spring 1985).

Brown, David L., and Linda L. Kloss. "Systemwide QA leans heavily on board talk." *Multis*, 2(2):42, 44, 48, 50 (June 1, 1984).

Cleverley, William O., and Robert P. Mullen. "Management incentive systems and economic performance in health care organizations." *Health Care Management Review*, 7(1):7–14 (Winter 1982).

Daft, Richard L. *Organization Theory and Design*, 2nd ed. St. Paul, MN: West Publishing Company, 1986.

Eccles, Robert G. "Bureaucratic versus craft administration: The relationship of market structure to the construction firm." *Administrative Science Quarterly*, 26:449–469 (September 1981).

Hamity, Gail I., and James W. Gauss. "A profile of the group practice administrator." *Medical Group Management*, 29(4):30–34 (July–August 1982).

Hoium, Vernon S. "Evaluating CEOs rates a concerted look." *Multis*, 2(1):51–52, 54 (March 1, 1984).

Kovner, Anthony R. "Whose hospital?" In Anthony R. Kovner and Duncan Neuhauser, Eds., *Health Services Management: A Book of Cases*. Ann Arbor, MI: AUPHA Press, 1981, pp. 222–241.

Kurtz, Michael E. "Selecting the right physician for your group." *Medical Group Management*, 31(1):24–30, 70 (January–February 1984).

Laetz, Catrien Ross. "Five steps to successful staffing." *Physician's Management*, 25(1):309–325 (January 1985).

Lavin, John H. "Now that you've found her, you have to train her." *Medical Economics*, 58(7):65–68 (March 30, 1981).

Metzger, Norman. "Recruitment, screening and selection." In Norman Metzger, Ed., *Handbook of Health Care Human Resources Management*. Rockville, MD: Aspen, 1981, pp. 85–110.

Mintzberg, Henry. *Structure in Fives: Designing Effective Organizations*. Englewood Cliffs, NJ: Prentice-Hall, 1983.

Neuhauser, Duncan. "The hospital as a matrix organization." *Hospital Administration*, 17(4):8–25 (Fall 1972).

O'Donovan, Thomas R., and T. Patrick Bridenstine. "The handmaiden revolt—Nursing staff crisis." *Health Care Management Review*, 8(1):75–79 (Winter 1983).

Owens, Arthur. "What your colleagues are paying assistants." *Medical Economics*, 58(7):9–17 (March 30, 1981).

Reddick, Frank A., Jr. "The medical staff." In Francis Foote Manning, Ed., *Medical Group Practice Management*. Cambridge, MA: Ballinger, 1977, pp. 77–95.

Robbins, Stephen A., and Jonathon S. Rakich. "Hospital personnel management in the late 1980s: A direction for the future." *Hospital and Health Services Administration*, 31(4):18–33 (July–August 1981).

Rusley, Roger L. "How to develop a sensible salary policy." *Medical Economics*, 61(12):236–243 (June 11, 1984).

Schatz, Peter S. "Should physicians be involved in personnel administration?" *Medical Group Management,* 29(1):30–40 (January–February 1982).

Shortell, Stephen. "Theory Z: Implications and relevance for health care management." *Health Care Management Review,* 7(4):7–21 (Fall 1982).

Smith, Howard L., and Norbert F. Elbert. "An integrated approach to performance evaluation in the health care field." In Norman Metzger, Ed., *Handbook of Health Care Human Resources Management.* Rockville, MD: Aspen, 1981, pp. 173–180.

Smith, Howard L., David J. Ottensmeyer, and Derick P. Pasternak. "Physician incentive compensation in group practice: A review with suggestions for improvement." *Health Care Management Review,* 9(1):41–49 (Winter 1984).

Stolhanske, James G. "Personnel management." In Francis Foote Manning, Ed., *Medical Group Practice Management.* Cambridge, MA: Ballinger, 1977, pp. 209–245.

Sweeney, Dorothy R., Geoffrey R. Anders, and J. Thomas Martin. "Why administrators sometimes fail." *Group Practice Journal,* 32(3):34–40 (May–June 1983).

Whittington, Harold. "The practice that outgrew itself." *Medical Economics* 62(13):66, 68–71, 74 (June 24, 1985).

Williams, Stephen J. "Ambulatory health care services." In Stephen J. Williams and Paul R. Torrens, Eds., *Introduction to Health Services.* New York: Wiley, 1984, pp. 135–171.

Williamson, Oliver E. *Markets and Hierarchies: Analysis and Antitrust Implications.* New York: Free Press, 1975.

Wright, Sally, and Arnold Wright. "A cooperative organizational form for hospitals." *Health Care Management Review,* 9(2):7–19 (Spring 1984).

CHAPTER 6

Information Systems for Human Resources Management

Anne L. Martin

DEFINITIONS

Information systems in health care institutions may be identified by two broad categories: clinical information systems, which support the day-to-day clinical aspects of patient care, and management information systems (MIS), which are management application systems. MIS may be defined as computer-based information systems that coordinate separate data bases to support management decision making in planning, control, and operations. MIS make available to managers uniform information for consistent long-term planning and decision making, as well as support for day-to-day operations, and this is a critical feature. For more detailed definitions, see Veazie and Dankmyer (1977).

An essential component of these management support information systems is accessibility of information about the organization's human resources. A human resources information system (HRIS) serves this need by providing a personnel data base to be used for planning and operational purposes. According to Walker (1980), the HRIS should provide the health institution with a systematic procedure for collecting, storing, maintaining, retrieving, and validating certain data needed about its human resources, personnel activities, and organizational unit characteristics.

In this chapter we discuss two key aspects of HRIS:

1. The role of HRIS in organization-wide planning and operations
2. Specific functions of information systems for human resources management (HRM)

THE HISTORY OF HOSPITAL AND HUMAN RESOURCES INFORMATION SYSTEMS

The need for collection, analysis, and management use of human resources information has outpaced the development of human resources information systems in health care institutions. Even hospital information systems, which have made significant progress over the past two decades, continue to evolve 8–12 years behind other industries. This lag is changing, as indicated by the increases in health expenditures for computer applications plotted in Figure 6.1. Increases are largely attributed to the expanding functions of information systems especially in patient care and strategic management application—two areas experiencing pressure for management accountability. To further enhance this evolution, technological developments have greatly improved the ability of managers to have direct access to multiple data bases, including human resources variables. For the most part, however, there remains much scope for improvement, and managers remain frustrated by the lack of accessibility to human resources information.

The use of information systems in hospitals began during the 1960s and has evolved in five stages (Bex, 1985).

Stage 1 (Fig. 6.2). Early hospital information systems had traditional module capabilities to process business office paperwork such as patient billing, accounts receivable, payroll, and general ledger. Information systems were not used in personnel departments. In other industries, large organizations began installation of HRIS in the 1960s.

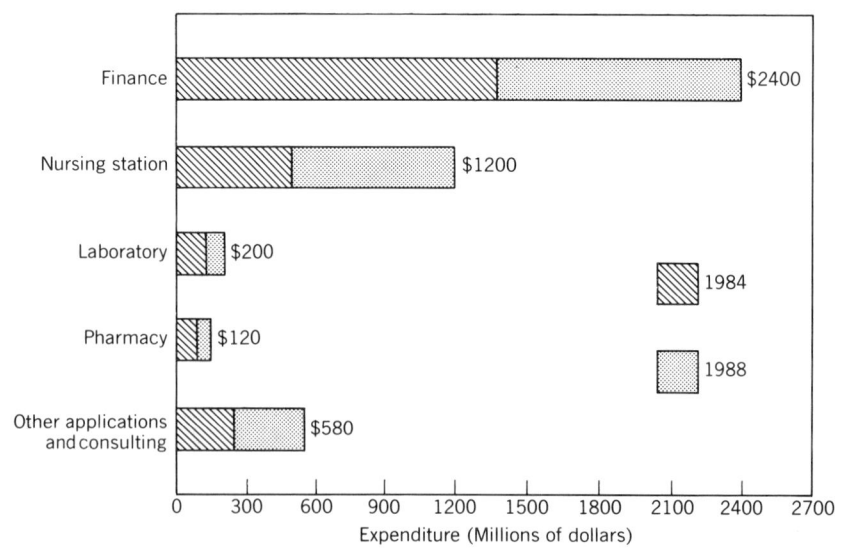

Figure 6.1. 1988 expenditures for computer related products and services segmented by application. From S. I. Dorenfest, *Hospital Information Systems: State of the Art,* 1985 Edition. Northbrook, IL: Sheldon I. Dorenfest & Associates, 1985.

Stage 2 (Fig. 6.3). By the 1970s about a third of community hospitals had begun to use computers. Limited data collection systems were developed, which incorporated the capabilities of financial information systems. In the early 1970s many larger hospitals implemented computer systems for productivity, planning, and analysis.

By the early 1970s commercial software products became available that offered multifunctional capability—a single system to serve the needs of all human resources activities. By the late 1970s most organizations in other industries had begun to install HRIS. One important impetus for the growth of human resources information systems in hospitals during this period was the increased number of employee protection laws, which grew out of the social and legislative environment of the 1960s and 1970s. These acts called for complex personnel reporting systems to state and federal governments and led to automated data bases on recruitment, placement, transfer and promotion history, job description, performance review, and vested and accrued benefits.

Stage 3 (Fig. 6.4). During the 1970s there was limited development of medical information systems. Systems were implemented to meet the operational needs of hospitals: scheduling, reporting, and management of the work load. Advances were made in laboratory and radiology reporting, pharmacy systems, scheduling of systems or operating rooms, and systems for admitting and locating patients. After the 1970s, a central feature of the progress of computer information services was the automation of medical records and the development of intelligent work stations serving integrated systems. Human resources information remained isolated in the personnel department. When analysis occurred, data had to be retrieved from multiple, independent sources. Output was independent from the rest of the information system. Interface between data bases remained a low priority.

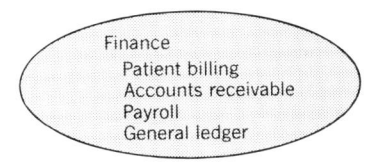

Figure 6.2. Financial information systems. From M. Bex, MIS: definition and states. *Computers in Healthcare*, February 1985, pp. 26–34.

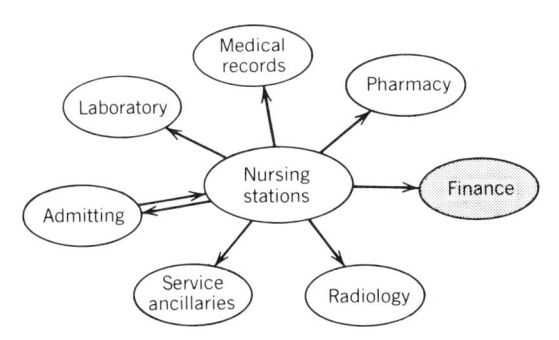

Figure 6.3. Data collection systems. From M. Bex, MIS: definition and states. *Computers in Healthcare*, February 1985, pp. 26–34.

144   Structural Systems

Figure 6.4. Medical information systems. From M. Bex, MIS: definition and states. *Computers in Healthcare*, February 1985, pp. 26–34.

Stage 4 (Fig. 6.5). Developing hospital information systems to support day-to-day operations was the focus of the 1980s. The increasing demand for improved employee benefits and financial security was a major contributory factor in the development of HRIS during this period. Benefit plans in particular became increasingly complex, necessitating automated recordkeeping.

Stage 5. Figure 6.6 illustrates the current application of computers in hospitals.

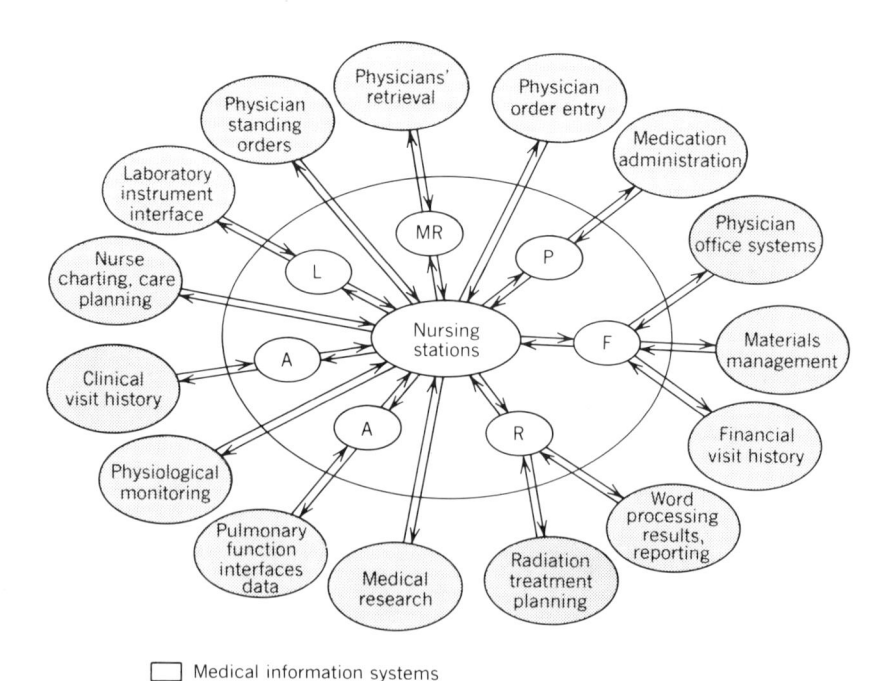

Figure 6.5. Hospital information systems. From M. Bex, MIS: definition and states. *Computers in Healthcare*, February 1985, pp. 26–34.

Almost all short-term acute care hospitals rely on computerization for at least one financial system. As many as 84 percent rely on automation for some sort of patient care or strategic management function. More than 90 percent of hospitals with capacities exceeding 500 beds have automated admitting and pharmacy systems, and 85 and 68 percent, respectively, have laboratory systems and nursing stations automated.

Table 6.1 presents the state of the art in microcomputers with software for on-line personnel application and identifies specific functions available. Software for specialized applications such as recruiting, Equal Employment Opportunity Commission reporting, job analysis, salary budgeting and planning, and succession planning are readily available. However, the application of these packages in health care organizations is not as prevalent as in other industries, where 90 percent of companies have computerized HRIS, with more than half of these being fully automated systems.

The trend in health care information systems is increasingly toward the development of integrated hospital information systems (see Fig. 6.7). Hospital computer systems are now being programmed for interrelating information from different data bases with computer and telecommunication systems physically linked into a common network. Few organizations in health care have, however, achieved the goal of an integrated approach to information gathering and communication. Few attempt to coordinate software acquisition or to force data base compatibility. It is not uncommon to find large organizations in which all the departments use different vendors. One reason may be that no single vendor is currently able to handle most needs of a complex health care system. The explosion in telecommunications development is rapidly helping to

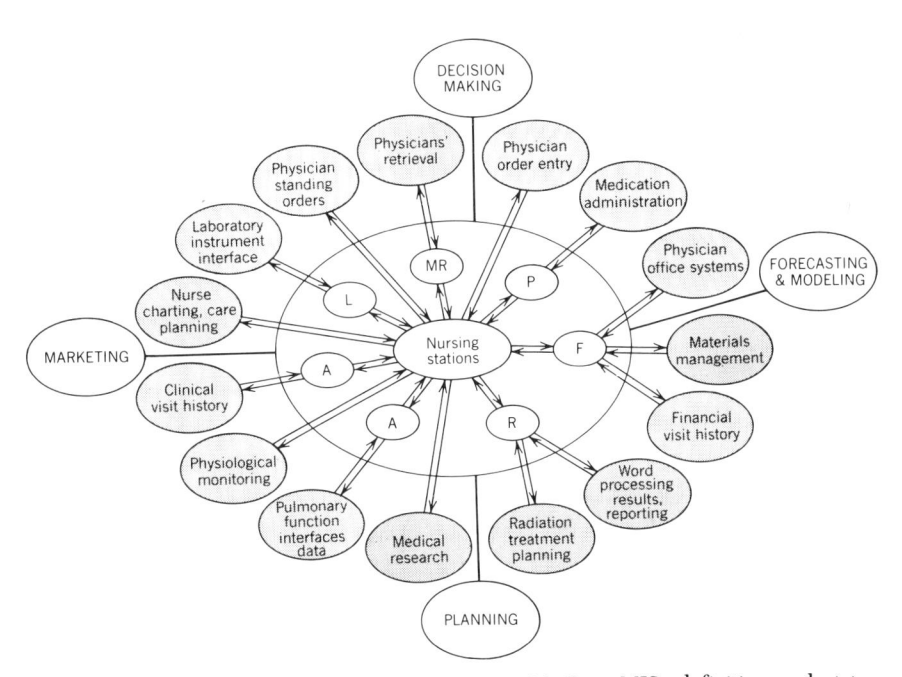

Figure 6.6. Management information system. From M. Bex, MIS: definition and states. *Computers in Healthcare,* February 1985, pp. 26–34.

TABLE 6.1. State of the Art in HRIS Software Packages

Vendor and Software Package	Strategic Level — Manpower Planning	Management Level — Management Control	Management Level — Staffing	Operations Level — Scheduling	Operations Level — Performance Evaluation	Operations Level — Wage and Salary	Operations Level — Benefits	Operations Level — Time and Attendance	Personnel Data Base — Personnel Files	Personnel Data Base — Payroll	System Features — Integrated/Networked
Administrative Strategies, Inc. ADAPT Nurse Management Systems		X	X		X						X
Compucare Distributed Management Applications Systems (DMAS)						X	X	X	X	X	X
Control Data Healthcare Services HELP Financial Management Systems					X	X	X	X	X	X	X
Cyborg Systems, Inc. Cyborg Payroll/Human Resources System	X		X				X	X			X
HBO & Company Trendpac 1 Mediflex	X	X								X	
Healthware Human Resources System		X									
Janna Medical Systems (Division SAI) JANNA PLUS		X	X								
JS/Data Payroll/Personnel System						X	X	X	X	X	
Management and Planning Services Resources Monitoring System Management Reporting Package	X		X								
Medical Information Technology, Inc. MEDITECH			X								
Medicus Systems NPAQ		X	X			X	X	X	X		X
Price Waterhouse INNOVATE-2001									X	X	X
Systems Associates, Inc. (SAI) SAINT Financial System						X	X	X	X	X	X
YRAM Applications Software STAFF SCHEDULING			X								

Fully Integrated Hospital Information System

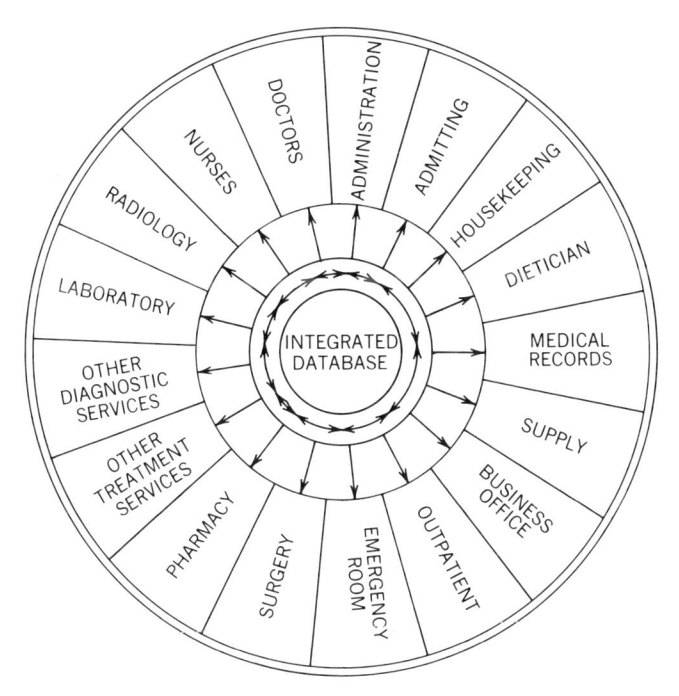

Figure 6.7. Fully integrated hospital information system.

alleviate data incompatibility problems. The major problem in the future will be "data overload," as a result of the lack of careful and selective planning about what information to collect and analyze. Most institutions today face inadequate information retrieval and modeling capability and have isolated systems that cannot communicate with corporate systems. For example, few organizations have the capacity to provide employee skill information throughout the institution for purposes of manpower planning, recruitment, or hiring.

Most human resources information systems continue to be used for record retention and report generation. Murdich and Schuster (1983) found that employee identification, payroll, equal employment opportunity compliance data, and compensation functions are the most widely computerized human resources functions. Little or no forecasting, modeling, or trend analysis is applied to these systems. These weaknesses critically hinder an institution's ability to compete.

Other trends in the evolution of automated information systems in health care facilities include the following.

1. Increased budgets for internal MIS
2. Increased capacity and reliability
3. Improvement in business-oriented software

4. Growth in communications (i.e., centralized computer systems with interactive terminals at department levels)
5. Microcomputer technology providing distributed processing

Traditionally, however, hospital personnel departments functioned solely for operational levels of management, providing support for such activities as wage and salary administration, benefit administration, recruitment, and government reporting. Recordkeeping was minimal and was applied directly to these areas of operations. Personnel departments were not perceived as having a role in organizational planning or decision making. Information available in the personnel department—often collected manually—was typically not used by management.

In the 1970s, new technologies in information systems allowed for automation of personnel data and coordination of previously separated data bases. Personnel departments began to automate operational procedures, saving staff time for other activities. Computerized reporting provided the capability to offer data on staff activities and trends to department heads and management.

This capability has paralleled a growing demand for information about the human resource in health institutions. As they evolve from social-based organizations to competitors in the world of big business, most health institutions are struggling to maintain or expand their position in the marketplace. With the current environment of competitive pricing and heavy regulation, positioning in the marketplace is not as stable as in the past but is open to challenge from resourceful and imaginative organizations.

Strategies for coping with this environment vary but generally include some attempt to reduce costs without compromising the quality of care to patients. Whatever their strategy, managers must answer two key questions:

- Are the institution's costs competitive?
- How can the quality of service provided be maintained?

Answers to these key business problems contain essential human resources elements that must be addressed as part of the organizational strategy. Furthermore, all decisions made or actions taken will affect the nature of the relationship between the organization and its employees.

Consider a business strategy to compete by becoming a low-cost producer of health care services. The internal plan may be to increase productivity and reduce staff levels while making an effort to improve performance. Managers would need to scrutinize labor force levels, productivity, skills, and employee commitment. Alternatively, a new technology introduced into a hospital, a merger between facilities, or the approval of new services would also involve important human resources decisions. These decisions and the manner in which they are implemented can reinforce or undermine the effective use of personnel.

In this environment, and given the greater availability of information technology, human resources departments will increasingly play an active role in providing information and analytical skills to assist management in achieving its strategic objectives.

This broadened role of HRM is reflected in the use of information about human resources that no longer will be viewed as a separate function unique to the personnel department but will contribute to many different departments and levels of the organization. Consider a typical personnel data base of the past, which consisted only of a payroll system that produced paychecks and labor reports. Today the HRIS would provide added management support by consolidating several personnel variables into a forecast of work loads, staffing, and scheduling. Or, consider the role of employee skill inventories, which has broadened to include use in matching applicants with jobs, planning career development with employees, performing succession planning, and making manpower projections. Or, consider employee files, which contain names of supervisors and their subordinates. This listing, probably not used in the past, is increasingly used for scheduling training for supervisors, contacting supervisors about training opportunities for subordinates, tracking the performance appraisal process, or mailing notices about policy or procedural changes. These examples point out that human resources data gathered for one purpose increasingly serve a number of purposes and are accessed throughout the organization.

A MODEL OF HUMAN RESOURCES INFORMATION SYSTEMS

So far we have emphasized the application of human resources information systems for planning the business strategy of the organization. Our perspective has been that of the planner or top manager whose decisions may affect the overall organization. In this section we broaden the approach to provide a model for an HRIS that reaches other management levels.

Understanding the roles HRIS can play throughout the organization requires taking a systems perspective of health care institutions. Systems theory provides a view of the health organization as a system made up of separate but united subsystems, which work together to serve the goal of the total system (Austin, 1983). In taking this view we see how information systems can become an important tool for supporting individual care delivery subsystems, while also integrating these subsystems for purposes of corporate policy. The rapid development of complex health systems and diversification of health structures has forced health care managers to look at their organizations as subsystems of larger systems, rather than as isolated units with discrete structures and information needs.

Figure 6.8 illustrates health care services provided by a multi-institutional system. A vertically integrated system may be composed of heterogeneous organizations such as acute care hospitals, outpatient units, long-term care units, and other chronic care units. Each contains subsystems that could be further divided into a series of smaller subsystems. For example, the acute care hospital will contain a patient care subsystem that is composed of smaller subsystems (diagnostic, therapeutic, rehabilitative, etc.). The entire network of systems and subsystems works synergistically in a structured way to comprise the total multi-institutional system.

In a multi-institutional system the coordination of each subsystem with the others is central to providing continuity of care for patients. This requires integrated information that allows managers to access comparable financial and personnel data, to be able to exercise planning and control functions for all the system components. The information system should be viewed as an independent subsystem that cuts across departmental lines to provide assistance to managers in all components of the total system. In this

150 Structural Systems

```
┌─────────────────────────────┐
│  Major service components   │
│                             │
│   Acute care hospital(s)    │
│   Chronic care hospital     │
│   Skilled nursing facility  │
│   Wellness program          │
└─────────────────────────────┘
                                    Human resources
                                 information subsystem
┌─────────────────────────────┐
│  Ambulatory surgi-center    │
│  Emergency services         │
│  Outpatient services        │
└─────────────────────────────┘

┌─────────────────────────────┐
│   Patient care services     │
│   Diagnostic subsystem      │
│   Therapeutic subsystem     │
│   Rehabilitative subsystem  │
│   Support-service subsystem │
└─────────────────────────────┘
```

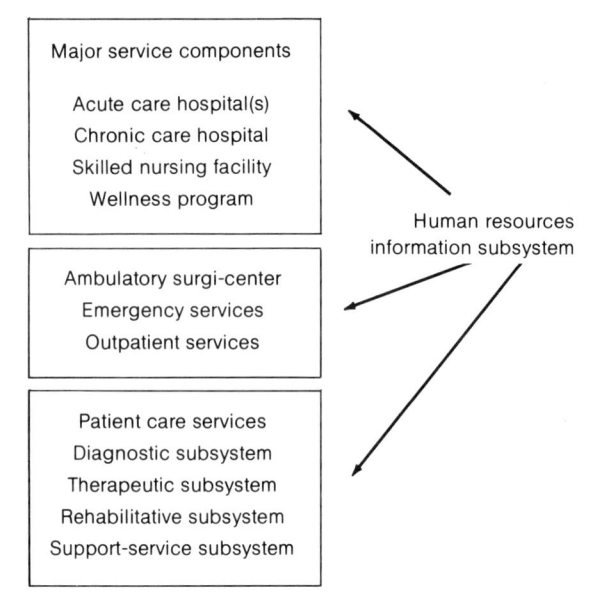

Figure 6.8. Multi-institutional system.

regard, information about human resources will be increasingly coordinated with other data bases to support a more detailed view of the organization. The effectiveness of the organization's information system will be directly linked to the extent to which its subsystems can be integrated.

Consider how the separation of acute hospital patient care, human resources, and financial systems has limited organization-wide management control over physician productivity. These three information systems have traditionally been disconnected, with little coordination between their data bases.

Clinical data systems have been used to manage the treatment of individual patients and, over the past two decades, to monitor the quality of care and the use of resources. Strategies to change utilization patterns were developed using available data from patient discharge abstracts. For example, patient length-of-stay data were commonly used to monitor the appropriateness of care prescribed by physicians. These strategies did not consider the cost implications of reduced stays.

On the other hand, financial data systems focused on administrative department functions such as patient billing and did not include information on the quantity or quality of services being provided. Human resources data on physicians typically consisted of demographic information and hospital affiliation. Often missing from these files was information indicating levels of performance, such as average length-of-stay experience for patients or treatment outcome data. Management efforts to institute effective productivity control processes are hampered by this separation. In the absence of a link between utilization and cost information, management cannot use cost accounting techniques to relate physician resource input to output. Likewise, in the absence of physician-specific utilization data, the cost implications of different clinical treatment patterns or variances in the use of resources cannot be assessed. Finally, not linking such data with information about continuing education efforts and licensing or credentialing requirements limits management's opportunity to follow up on any problems identified. A linkage of cost, human resources, and utilization data as just outlined would afford a clearer understanding of the real-cost implications of specific medical practice patterns associated with individual physicians.

Given increased pressures to monitor organizational resource use and personnel productivity, the absence of a systems approach to information processing seriously hinders the effectiveness of managers. Retrospective studies examining the impact of length-of-stay reviews have found little net savings, a conclusion that better information could have detected earlier without the expenditure of large sums of money on costly review programs.

Strategic, Management, and Operational Levels of Decision Making

Equally important to providing information to the care delivery subsystems of a multi-institutional health care organization is the need to reach different levels of management. Classic management models define three distinct levels of management decision making: strategic, management, and operational (Federico, 1985). Each level has different responsibilities, functions, and control areas and requires different information about employees. To the degree that HRI is accessible to top, middle, and operational managers, the opportunity for taking human resources variables into consideration for decision making is increased.

Human resources information plays an important role in coordinating the goals, objectives, and tasks of the three levels of management. Consider a team of nurse supervisors recruiting five additional nurse's aides for a unit that has been assessed by the vice president for nursing services as having relatively low morale and productivity. In this case the nurse supervisors may view the unit's problem as understaffing, while senior management may perceive a need for corrective management intervention. The supervisors' request for hiring new personnel may subsequently be denied, creating staff-line conflict and frustration. Timely information provided to the supervisors or the results of employee surveys and productivity indicators could be used to illustrate senior management's perspective on what action is needed and to encourage an outcome that will satisfy both management and operational supervisors.

As sharing of uniform information is increased, a more uniform view of the organization should result and isolated decision making by different levels of management should decrease.

Strategic Decision-Making Level

Strategic planning may be defined as the choice of a competitive posture and the formulation of the major policies for the corporation and its critical units. Strategic decisions are concerned with the long-term direction of the organization and its interaction with the external environment. The goal is to arrive at the best plan for the operation of the organization, given the economic, political, logistical, and operational constraints (Schmitz, 1979). This level of decision making usually involves the board of trustees, top management, and department heads. It is the strategic level that determines the particular mission and objective of the organization and draws policy guidelines for other management levels. The strategic decision makers are faced with the task of identifying future trends that are likely to affect the outcome of decisions. A key human resources task for this level of management is to assess whether personnel resources are available to carry out the strategic mission and whether it is possible to gain a competitive advantage through the best use of available human resources.

Several other specific tasks in strategic decision making that require HRIS support and are listed in Table 6.2. Most important among these is the need for aggregated information about employees for policy decisions on manpower planning and control planning.

Management Decision-Making Level

Management decisions are concerned with implementing the organizational strategy, including monitoring and controlling expenditure, productivity, and revenue. These medium-range decisions require accurate, detailed information about the internal operations environment. Specific tasks of the management level requiring HRI support may include training and development, management control systems, compensation, benefits, physician education, quality control, and labor relations (Table 6.3).

Management decisions on this level are likely to be most affected by increased accessibility to information about human resources. The more information is available to managers about the organization's employees, the better they will be able to affect the efficiency of the organization through information feedback and systematic control mechanisms. With access to human resources information, managers can effectively track employee work loads, attitudes, and productivity and can evaluate the impact of programs for improvement.

Operational Decision-Making Level

The information about employees called for at the operational level is used to support decisions about day-to-day activities (Table 6.4). The operations decision maker traditionally maintains a restricted view of the organization and may not be well informed of the many interconnections among the organization's total complement of subsystems. With increasing access to information, however, managers at operational levels are becoming better informed and are beginning to take greater part in decision making, control over productivity, and quality of care.

There are two broad categories of operational decisions: patient activities and nonpatient administrative activities, such as payroll and personnel. Operational supervisors may need to track specific employee information for day-to-day decisions or responsibilities such as scheduling, monitoring productivity, performance appraisal, or personnel changes. In addition, the automation of many tedious operational activities has given operation managers more time to become involved in planning and control activities.

THE ROLE OF HUMAN RESOURCES INFORMATION SYSTEMS

The goal of HRIS is to support improved management decision making through the provision of useful and accessible information about organizational human resources

TABLE 6.2. Strategic Level

HR Task	HRIS Data Elements Required	Output
Manpower planning	Number of employees	Manpower audit/skills inventory
Internal manpower inventory	Employee age	
	Job title	
	Experience	
	Length of service	
	Education	
	List of skills	
	Attitudes	
External manpower inventory	Scan of job/staff ratio from health competitors and other industries	Labor force assessment
	External Manpower Supply	
Forecasting staffing need	Staffing work load measurements	Project staffing needs and evaluate alternative staffing levels
	Days understaffed/overstaffed	
	Staffing cost	
	Staffing options	
	Output data (patient classification)	
Turnover analysis	Employee turnover by job category	Turnover assessment
	Exit interviews	
Employee attitude assessment	Absenteeism surveys	Attitudinal
	Quits	
	Average tenure	
	Turnover	
	Exit interviews	
Recruitment policies	Competitor medical staff and other employee flat pay/benefits/perks	Medical staff recruitment strategies; policies/strategy for obtaining and retaining other human resources
Management control planning	Productivity indicators by unit/employee	Productivity analysis
Personnel cost conformity with strategic plan	Labor cost elements	Personnel cost analysis/labor cost monitoring reports
	Relative labor cost to total costs	

capabilities and opportunities. To accomplish this requires information about future needs for personnel and data about the current and past performance, skills, and interests of employees. The emphasis we have placed on the role of HRIS in organization-wide issues has far-reaching implications for the traditional role of personnel recordkeeping. This, however, should not distract us from the importance of providing automated personnel data to operational managers for day-to-day decision

154 Structural Systems

TABLE 6.3. Management Level

HR Task	HRIS Data Elements Required	Output
Labor relations	Labor cost allocations	Analysis for bargaining
	Collective bargaining units	
Benefits analysis	Competitor benefits	Benefits package
Position control		
Manpower	Departmental activity levels	Cost-effective allocation of manpower resource
	Departmental "output" data (patient classification)	Table of organization
	Departmental resource consumption	
Management control systems	Work load data	Productivity reports
	Labor hours/dollars output data (patient classification)	
	Departmental activity levels	
	Performance data	
Resource control	Measurable standards output data (patient classification)	Resource utilization/cost reports
Training and development	Education/skills levels	Training
	Career path analysis	Objectives
	Licensure/certification	
Quality control	Quality process and outcome indicators	Quality assurance reports

TABLE 6.4. Operations Level

HR Task	HRIS Data Elements Required	Output
Manpower scheduling decisions	Department activity data	Scheduling/staffing system
Birthday/anniversary greetings	Name, birth date, marital status data	Birthday/anniversary memo, card, or gift
Training and development	Employee skill and career profile	Individual employee career and training planning
Employee benefits	Benefits profile	Development of benefits to meet needs of employees

making and recordkeeping. A broad spectrum of specific personnel-related activities on all levels of the organization can be supported by an effective HRIS.

We have stressed the importance of integrating information about an organization's human resources to all its subsystems and levels of management. This section defines specific functions of HRIS and how each is used throughout the health care system. Table 6.5 is a matrix that summarizes these functions. It illustrates the three human resources data bases from which information supporting each function is typically drawn, as well as the appropriate level(s) of management utilizing the information. For example, consider the management control function of an organization. Data from the personnel file about employers can be used to support productivity control. Analysis using information about the employee demographic mix may provide insight into differences between groups that may help explain variable productivity levels. Or, the file could be used to monitor changes in productivity among groups of employees with similar traits.

Data Bases for Human Resources Information Systems

The data bases primarily used to provide the organization with information about human resources can be categorized in three ways:

1. Personnel file
2. Applicant file
3. Payroll file

All data are not necessarily contained in the human resources department. The payroll system, for example, may be operated from the finance department. Nonetheless the data from these systems provide basic information to support many management functions throughout the organization.

TABLE 6.5. HRIS Functional Matrix

	Level of Management Decision Making		
	Strategic	Management	Operations
	HRIS Functions		
Labor relations	×	×	×
Government reporting		×	×
	HRIS Information Data Base		
Personnel file			
Performance appraisal		×	×
Training and development	×	×	×
Skills inventory	×	×	
Manpower planning	×		×
Manpower allocation		×	×
Management control		×	
Position control		×	
Applicant file			
Recruitment		×	×
Hiring		×	
Payroll file			
Compensation	×	×	
Benefits	×	×	×

Personnel File

Personnel files provide the central HRIS data base. The basic file provides employee information such as present job category or current position; age; skills, including areas of knowledge and experience; educational level, including degrees and certificates; geographical location (if the corporation is geographically decentralized); race; sex; length of service; retirement status; time in present position; potential for development and interest in promotion; wage and salary history; and performance rating and supervisory evaluations.

Access to personnel files can be made available to the strategic, management, and operational levels of the organization through the use of integrated personnel management systems or automated data base networks. At the strategic levels of the organization, aggregated information from personnel records as well as the analytical tools available to HRM is useful for manpower planning and control. At the management level, the information may be used to analyze specific personnel problems such as turnover or absenteeism. Reports for analyzing these problems are becoming a common component of personnel information system software packages. Other profiles useful for all levels of management can be generated, including the following.

1. *Personnel profile.* This profile may contain basic employee data (e.g., name, sex, race, age, marital status). From these reports operation managers can track pertinent information such as birthdays or wedding anniversaries.
2. *Career profile.* This profile is used to chronicle significant events in a employee's career: performance appraisal, job title changes, salary changes, or promotions. From this operation management decision makers can assess the development or potential for advancement of employees.
3. *Skill profile.* This profile contains data about employee education and training background to assist in considering placement and advancement within the organization.
4. *Benefits profile.* This profile contains relevant information on employee benefits to track benefit use and cost.

Applicant File

The applicant file lists all job applicants and available jobs; data typically include applicants' work histories, skills, salary requirements, and education histories, as well as job descriptions to be used in filling vacancies throughout the organization.

The applicant file supports general employment system functions (see the section on HRIS functions), including recruitment, selection, transfers, job posting for internal recruitment, skills inventory, replacement pool, employment records, and applicant testing.

Payroll

Health care organizations considering an HRIS will most likely already have some sort of computer-based payroll system or payroll service. To reconcile basic data and to minimize duplicate processing, the new HRIS must interface with the existing payroll system. The strongest advantage of this interface is that it minimizes redundant data collection, duplicative storage, and other data costs.

FUNCTIONS OF HUMAN RESOURCES INFORMATION SYSTEMS

The three data bases—used with information from other data sources—support many human resources management decisions. The section that follows discusses specific functions for which HRIS are used. This list is not exhaustive, but it provides insight into the many possible applications of information on human resources.

Manpower Planning and Allocation

Manpower planning and allocation is the critical link between the human resources strategy plan and the organizational business plan. Planning for future manpower needs will play a key role in determining the institution's successful implementation of a business strategy. Consider a hospital that is developing new service lines, including a long-term care unit and an ambulatory surgi-center, over the next 2 years. Consideration must be given to manpower levels and skills mix before a physically completed unit can be staffed.

To assess future manpower needs requires information about current employee levels and skills, along with projections of patient needs in the new facility. It requires information to assist management in anticipating employee replacement needs due to quits, promotions, or other personnel changes. Consider nursing turnover, which is a significant problem in many long-term care facilities. Data on current nursing turnover levels will need to be analyzed and projections made for recruitment targets, not only to staff the new facility but also to compensate for projected turnover rates. Close analysis of just this one job category might turn up unexpected human resources issues, such as recruitment outside the local market or even the need to increase pay scales to attract applicants.

Also required are analytical tools for asking "what if" questions when exploring alternative service strategies. Information about current staffing work load, along with forecasting or simulation techniques for analyzing data and predicting staffing needs will be useful. Computer simulation can offer the manager a way to obtain more precise comparisons of the outcomes of various personnel decisions when there are certain knowns. Consider a nurse executive using computer simulation to examine the outcomes of various staffing decisions. Output from the simulation programs might include information on the number of days each unit was understaffed or overstaffed, the yearly wage costs for each of several staffing plans, and other measures of the outcomes of staffing decisions. Or, consider a 1000-bed hospital where plans are being implemented for a 20-bed intensive care unit. Analysis for determining the nursing personnel required to staff the unit could be done by a simulation study. Output reports would permit managers to select from alternative shift work patterns while complying with standards for safe levels of care.

Once the level of manpower has been defined, management can use employee work load and productivity information for defining the exact number of people needed in each category, establishing procedures to ensure that this limit is not exceeded, and determining exactly what jobs those employees will do. This function, called "position control," is described in the case study entitled "Mason Medical Center HRIS," below.

Equally important to human resources planning is the cost-effective allocation and scheduling of existing staff. Inefficiencies in manpower allocation have often occurred because of staffing decisions based on scheduling a constant level of human resources to meet work loads that vary over both the short term (daily and weekly) and the long term (monthly, yearly, etc.). As health systems have increased in size and complexity, this task has become more difficult. Consider nursing departments, many of which have become targets for more efficient staff allocation. As indicated in Table 6.1, computerized systems have become available which provide support in determining variable staffing levels and schedules to minimize the problem of staffing inequities. Software is available to help operational decision makers assess their unit activity level and productivity. To do this, "work load measure" systems have been developed which define and monitor work load indicators. Work load levels and trends are then assessed over a period of time. Reports of alternative methodologies for work load measurement are available through the Health Care Information and Management Systems of the American Medical Association Society, Chicago, Illinois. One methodology called "acuity" patient classification systems identifies and measures nursing manpower resources based on patient-specific data (McHugh et al., 1984; St. German and Meijers, 1984).

Acuity systems are based on the assumption that staffing needs are not tied to the number of patient days but to the demands of specific patients and that labor resources input is dictated by the type and kind of patient being serviced. The system does this by identifying patients with shared clinical characteristics and relates these patients to resource use. Information is collected about each patient relative to how much nursing care is needed daily (or shift by shift). A calculation is then made to show how many nurses are needed on each unit and each shift relative to the mix of patients admitted. Nursing resources are measured by cataloging all the activities performed by nurses and the time it takes to perform each task, and categorizing them by direct patient-related activities and nonpatient or management activities.

The output of the "acuity" systems allows operation management to make reliable and cost-effective staffing decisions. Its main purpose is to aid in the day-to-day staff management process, while at the same time acting as a guide for internal budgeting. Schedules are prepared that address core staffing needs, individual nurse work load, special requests, maximum shift time, and a host of other considerations.

Originally designed to provide information about the nursing needs of patients as a rational basis for staff allocations, patient classification or acuity systems may also be used to monitor nursing resource use and productivity at management decision-making levels. A manager can, for example, receive reports on levels of activity as measured by volumes of laboratory tests, treatments, or visits using these systems. The patient-specific nature of this approach allows the measurement of patterns of practice or treatment regimes for the patient's complete stay. Since these systems can link patient-specific labor resources utilization data with the cost of these resources, it is possible to monitor the true cost of interdisciplinary treatment programs for groups of patients.

The system is also useful for monitoring decisions on staff allocation by observing trends and shifts in the ratio of time spent in each category of work. Ratios can then be used to compare time spent per category of patient between facilities or geographic areas. These systems provide some useful input for strategic levels of decision making in forecasting nursing demand. Core nursing staff determinations, typically performed annually, can use accumulated patient data from prior years, as well as forecasts, to determine how many nurses (by type) will be needed for each unit and each shift.

Hospital laboratories are another common focus for staff allocation information systems (Pang and Saint, 1985). As with nursing supervisors, laboratory managers must be able to accurately translate units of work into numbers of employees required to perform the work. Managers in laboratories feel the same pressures to track productivity levels and allocate human resources efficiently. Since the 1960s, laboratory-specific prediction models have been developed to help forecast hospital laboratory work load for the entire laboratory as well as for individual sections of the laboratory. A key challenge to these systems rests in defining accurately the factors that affect laboratory work volume. For instance, admission levels or length-of-stay differences may not be as important as employee productivity in defining work volume. Table 6.6 lists 10 predictor variables that may be used. Once the predicted work load has been determined, managers can make staffing assessments and a productivity monitoring system can be constructed. As with the nursing activity system, the purpose is to assist laboratory managers in their efforts to utilize laboratory manpower in an efficient and cost-effective way and to provide information that can be useful in planning for future needs.

Personnel Administration

Six functions of HRIS are related to specific personnel functions used by management. Each of the functions may form an independent module, together they make up the comprehensive HRIS. The functions are as follows:

Employment systems
Compensation systems
Training and development
Performance appraisal
Employee relations
Government reporting

Employment Systems

Automated employment systems, which support the process of recruitment, screening, and hiring of applicants, typically involve retrieval and matching of applicant data from the applicant file with the needs of vacancies.

TABLE 6.6. Predictor Variables in Daily Models for Laboratory Staffing

Inpatient + newborn days
Emergency room visits
Outpatient visits
Intensive care unit days
Admissions
Discharges
Public holiday
Days since last full working day

Analysis of the data contained in the applicant file is useful at all levels of decision making. Most important is its support for more efficient employee recruitment and selection by operational managers. Analysis of data contained in the file helps manager and strategic decision makers in targeting recruitment efforts. For example, if the data reveal that the community labor force is generating a large and satisfactory nursing applicant pool, advertising or recruitment concentration may be shifted to another job category. The data can likewise be used by strategic decision makers in reviewing recruitment or selection policies. An analysis of how and where employers were recruited may assist in identifying opportunities for more effective use of recruitment expenditures. For example, a recruitment policy that emphasized the use of search firms might be changed to one promoting the use of advertisements or informing schools about vacancies.

The next section presents a case study that highlights some facets of a human resources information system that supports recruitment and selection.

Mason Medical Center HRIS

Background

Mason University Medical Center has implemented an automated employment office system that allows hiring departments throughout the facility to access the employment process on-line from their office PCs. Hiring departments are able to request referrals and hire applicants on-line. The system enables most users to complete the employment process without handling papers. This system is the first step in the human resources plan to bring the services of the human resources department directly to users through the use of telecommunications technology.

Mason Medical Center is part of a university, which is the area's largest employer. The campus and medical center combined employ approximately 17,000 people. The medical center manually processed 12,000 applications each year and filled about 2500 positions. In recent years, the medical center's growth and demands for employment services had outpaced the employment office's ability to respond. In the spring of 1984, the university made a commitment to improve Mason Medical Center's employment services by automating the employment function. An automated system that was planned would minimize manual processing of employment paperwork, cutting down on waiting and leaving employment representatives free to devote their efforts to recruitment, interviewing, and management of their applicant pools. Other objectives of the system were to reduce the time required for open positions to be filled, to ensure that an adequate number of qualified applicants were available to hiring departments, and to meet all federal and university reporting requirements.

While planning the new system, a team from Mason visited other universities to view model systems. At the same time, the design team began to meet with internal committees representing hiring departments, employment office representatives, and applicants to determine their needs. These groups worked together to develop an overall system design, which was refined by the design team and reviewed by all the committees. The basic system flow was continually revised and reworked to accommodate new features or to avoid newly discovered problems.

Concurrently with the design process, the human resources information systems department began to investigate hardware and software alternatives for the system. First a minicomputer was purchased, to eliminate processing changes that would have been incurred on the university's time-sharing mainframe and to allow for establishment of a telecommunications network to support hiring departments. The hardware decision was a direct outcome of the specifications developed by the user committees during the earlier phase of the project.

At the next stage of the system's development a project team composed of HRIS staff, analysts and programmers from the data processing department, the director and supervisor of employment, employment representatives, and vendor representatives began meeting weekly to coordinate all phases of the project, review progress to date, and discuss issues and problems. These meetings continued throughout the project.

A major objective for implementation was to phase in the new system in a manner that would cause minimal disruption of the employment office operations. Operational procedures to support the new system were implemented before the system went on-line. The system was installed initially in only one department as a test. This pilot phase was implemented when the first department was brought on-line to the system. Other medical center departments were phased in over 3 months.

How the System Works

Application process. Applicants for employment at the medical center complete a standard application and skill inventory form. Employment office representatives complete minimum qualification prescreens for each applicant, and all this information is entered into the system.

Employment requisitions. A department that has an open position submits an employment requisition by entering the necessary information through a local PC or terminal. The hiring department is allowed to select up to 10 skills from a skill list for that job code. The hiring department selects only the skills that will be required for the position.

Postings. The system generates a weekly posting of all open positions. Updates not on the weekly posting are also available to applicants through a call-in "jobs line."

Referrals. The system matches each new employment requisition with the available applicant pool by matching job codes and skills. Only applicants who have cleared a history check and pass the minimum qualifications prescreen are referred. The hiring department may review applicant information on a PC or terminal screen or may print out the applicant information on a local PC printer.

Hiring. When the department has completed the interviewing phase and has made a selection, appropriate codes are put in the system. The system sends an electronic mail message to the employment representative to complete the hiring process.

Documentation. The system maintains documentation on all referrals for all employment requisitions. This information is kept in a queriable data base that can be used for research or to provide information for compliance reviews. All applicant information is maintained in an archive file and can be accessed as needed.

Position Control

Controlling the hiring of new employees is a central element of the position control function of the HRIS. The system maintains a position control file containing each approved position by department and by shift. The file has two sections: position information and employee information. When a position becomes vacant, an employee requisition is sent to HRM and the file of the employee who is creating the vacancy is put into a suspense file of the HRIS. After the recruitment process for a new employee has been completed, a new file is begun and a new-hire notification is sent to payroll. The suspense file is periodically monitored to assure that employees scheduled to leave actually do. When an employee has left, his or her file is placed in a former employee file.

There are several advantages to this system. First, referrals are available in a *timely* manner—overnight. Immediate access to the applicant pool represents an improvement over the previous manual system. Second, minimum qualifications determination by employment representatives and skills matching by the computer provide an improved and objective screening and referral process. Third, the process encourages a more active relationship between the employment representative and the department. Management confirmed early that information available through the system was sufficient to make interviewing decisions.

For the employment department, the analysis and system design phase provided an opportunity to rationalize the employment process, which had grown incrementally over a period of years, by developing work flows and job responsibilities that could be examined and revised routinely.

Finally, up-to-date applicant information is easily accessible for accurate and timely processing of applicant and employment requisitions. The information available can help assess manpower resource availability in the community. It moves representatives out of processing roles and into roles as pool managers. This allows time for recruitment and communication with hiring departments.

There are disadvantages to the system. First, the implementation of the system changed job responsibilities of some staff members, which created a lasting resistance to the system. Second, department or employment representatives no longer hand-pick applicants. Employment representatives requested this flexibility, but it was decided that the benefits of objectivity outweighed those of flexibility, which also would have added significantly to the cost of the system. Third, posting is generic instead of giving individualized position descriptions. The system does not allow for flexibility in describing positions.

Future Plans

The system is currently available to all medical center departments for on-line use. The number of on-line users is expected to increase rapidly, with major hiring departments coming on-line first. A large amount of time has been spent training hiring departments on the system.

Over time, other human resources functions will be integrated into the employment system. A position management system and integration with the employee data base are planned.

Compensation

The compensation module of an HRIS contains regular and merit pay raise intervals, contractual factors, benefits, and other variables affecting payroll costs. Compensation systems in health care organizations have become increasingly complex and costly, requiring automated processing. Traditionally, compensation packages in hospitals were straightforward, with fixed salary scales and benefits. Today, variable methods of compensation are the norm, including complicated schemes for incentives and bonuses and position-specific compensation packages. Maintaining these complex systems demands flexible data entry, processing, and analysis capabilities to assess the cost effectiveness and competitiveness of the salary structure.

A compensation file can obtain from managers information about how much each employee is paid, overtime percentages, average hourly costs, and other employee- or unit-specific indicators. It can generate mandated labor reports and monitor compliance with various labor regulations. With a strong statistical and modeling software package, the compensation information system can support strategic decision makers in system-wide decisions about salary alternatives, union negotiations, fringe benefits, and job grading. "What if" queries are useful to project the results of a change from fixed to variable pay or to model the cost of matching pay scales from competitive institutions. This can help evaluate the effectiveness of the compensation system by, for example, relating increased compensation to increased productivity or performance. Options to include in an analysis may include fringe benefits packages, flextime, commissions, bonuses of stock, ownership plans, executive bonuses, or executive stock options.

Benefit systems are an increasingly important component of health care compensation packages. As health care institutions become more business oriented, the cost and scope of benefits provided to employees is rapidly increasing. It is becoming more common for benefit packages to be presented to employees with many options in a "cafeteria" fashion—that is, a set of benefits is offered from which a preferred plan may be selected.

To account adequately for innovative benefit packages that will attract the best employees, a benefit data base must be consulted. Since the design and processing of benefit packages has become so complex, it is wise for management to turn to existing modules for help—perhaps from insurance companies or outside services. At a minimum an automated data base should be compiled that is capable of storing and retrieving file information about the facilities benefits. This includes information about employee benefit and pension plan status including vacation and sick day accruals. The system should be capable of processing statements for individual employees so that they understand their total compensation package. Consider a cafeteria benefit plan in which employee deductions may range from zero up to the maximum benefit cost. If sufficient logic were programmed into the system, deduction amounts could be calculated for individual plans. Finally, the system should help meet the federal reporting requirements.

Training and Development

Hospitals engage in a wide range of training and development activities for employees at all levels. These may include technical training, management development, executive development, career development, and new employee orientation.

A training and development information module can support these activities in numerous ways. The data base drawn from the personnel file will most likely include the educational and training background of employees, which can be monitored and analyzed. Also from the personnel file, the skills inventory profile may be used to identify a pool of people with the required skills and interests who would benefit from additional training. A program can help track continuing education needs to assure that all professional licensure and certification requirements are met.

The data base can maintain and develop routine reports of planned instructional activities for distribution to employees. This will help make training more accessible to employees and will make it possible to avoid duplication of training efforts. A training and development module may be designed to monitor training resources to match training need with resource availability.

Finally, a career path analysis or career development component may be built into the system. This, combined with data from the applicant file, makes information available to employees about career opportunities throughout the organization. From this system, data on occupations, projects training, or job opportunities within the organization are made available. Each opportunity is coded for educational requirements, location, skills needed, and travel frequency so that employees can assess their interest and qualifications. Furthermore, managers can use the system to search through personnel files to find suitable candidates for positions. This capacity allows organizations to match employee qualifications with opportunity. One organization performed career path analysis by tracking internal employee job moves to determine the most common career advancement routes from particular jobs. The analysis revealed a number of jobs that seemed to be blocked (i.e., there were very few advancement opportunities for employees in those positions). This information was used for targeting career development efforts.

Performance Appraisal

The payroll data file can be used to support the performance appraisal system in several ways. For example, it can alert supervisors that employees' yearly reviews are due. The payroll files may be reviewed weekly to obtain names of all employees who are due to be evaluated. When an evaluation is due, the employee's file may be placed in a suspense file. Following the evaluation, the date of the appraisal is put on the employee's date-of-hire file.

When integrated with payroll, the HRIS can provide data about employee performance factors that are most affected by compensation increases. For example, some performance indicators may be based not on individual performance measures but rather on group measures such as the employee's unit. Consider an operating room that increased its patient throughput over a 6-month period: this is more likely to be due to a team effort rather than reflecting the contribution of an individual physician, manager, or nurse. In this instance strategic management levels would permit examination of compensation alternatives that rewarded the work group rather than individuals. Analysis from the system could examine employee performance after compensation changes to see if improvement had occurred. If this were done, the analysis could consider performance changes in relation to patterns of promotion or salary and benefit increases, as well as other changes such as role or supervision.

Employee Relations

The key to good employee relations and to preventing unionization is to maintain employee satisfaction. A critical and often neglected role for the HRIS is monitoring employee attitudes and behavior. For example, turnover analysis is often used as an indicator of employee dissatisfaction. Analysis of turnover rates by location, position, length of service, who the employee reports to, and how the employee was recruited may indicate where intervention is needed. This information needs to be supplemented by employee attitude surveys, and both sets of data must be available on a comparable basis.

If unions are a part of the organization, the HRIS can support the negotiating process. It can provide quick access to information addressing issues that arise at the negotiating table (e.g., incremental changes in salary, benefit, or staffing agreements). Superior information is a powerful tool in any negotiating process.

Union representation may cause additional demands on management for information. The bargaining agent may ask for compliance verification regarding changes in wages, posting of jobs, review and revision of job descriptions, seniority posting, vacation schedules, hour accrual for seniority, sick leave, vacations, or changes in status. Accurate maintenance of individual records on grievance and discipline experience is also crucial in a unionized facility.

Government Reporting

Since the 1960s state and federal recordkeeping requirements imposed by several federal laws have been major tasks for human resources departments. The amount of reporting required has put manual recordkeeping out of the question in most facilities.

Meeting these reporting mandates typically requires a computerized system. Using the information contained in the basic personnel and benefits administration data bases saves considerable time in the preparation of reports. Records must be maintained accurately and often saved for several years for protection in the event of lawsuits.

Two major areas of concern for HRM are equal employment opportunity and affirmative action. It is important that the information system maintain an updated account of the institution's equal employment status and its affirmative action program. The personnel information system should include sex and ethnic codes attached to individual employee files. Statistical counts should be routinely taken regarding the number and ratio of minority employees as compared with the total work force in various specific job categories. It is equally important to use the HRIS to help develop a proactive affirmative action program through analysis of areas where deficiencies may exist in the use of minority groups and women, as well as to calculate objectives and time frames. Time-series analysis may even be necessary to confirm that salary progression and position advancement have been discriminatory.

IMPLEMENTATION

In planning the implementation of the HRIS, its three major functional components must be considered and each component carefully integrated into of the system design (Milkovich and Glueck, 1985). They are illustrated in Figure 6.9.

Input

The input functions are illustrated in Figure 6.10. They define the procedures required to collect, enter, and validate the personnel data going into the HRI system. Detailed procedures should explain who provides and who collects the data, when data should be provided and collected, and how processing is to occur. The major task in this function is to ensure that all the required data are present and that the data are valid (no garbage in).

Data Maintenance

After the data have been edited and validated by the input function, they are passed to the data maintenance function, which is responsible for updating the HRIS file (Fig. 6.11). This is done by adding new records to the files or by updating existing data elements. This function provides the capability of maintaining historical information

| Input | → | Data Maintenance | → | Output |

Figure 6.9. HRIS functional matrix.

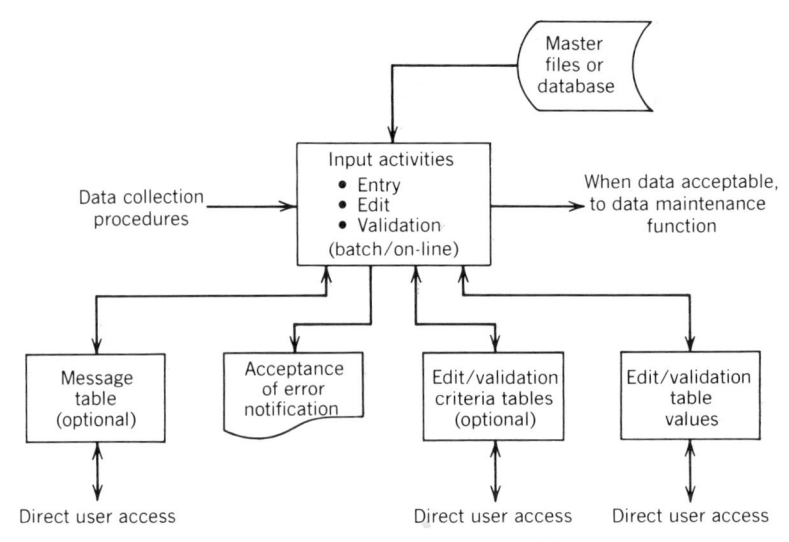

Figure 6.10. The three major functional components of the HRIS.

such as changes in salary, employee status, or performance ratings. To provide for cost-effective data maintenance, users of the information system should specify which elements need routine updating and which data should become part of a permanent history file.

Outputs

The output function is the link between the HRIS and the user (Fig. 6.12). Most users of human resources information systems are not concerned with the input and maintenance functions, but they are very concerned about how the system serves their needs for output. The output component should be evaluated carefully to be sure that it will meet user needs. The relative value of data collected and maintained will be determined by the user.

Design and Implementation of the Human Resources Information System

The first step in planning the HRIS is to clearly define its objectives. This is an essential but usually neglected step. Most often the human resources manager or an information technician simply requests potential users to list their data needs. The result is a long list of unrelated items with no direction or central thrust. Defining the goals and objectives involves specifying what the system will be required to do to fulfill the three HRIS functions. This level of specification will lead to decisions about the type of data to collect, how wide coverage should be for each element, how to collect data, and how often to make collections. For example, the objective "providing a skill inventory for managers to use in training and development programs" might be sufficient initially,

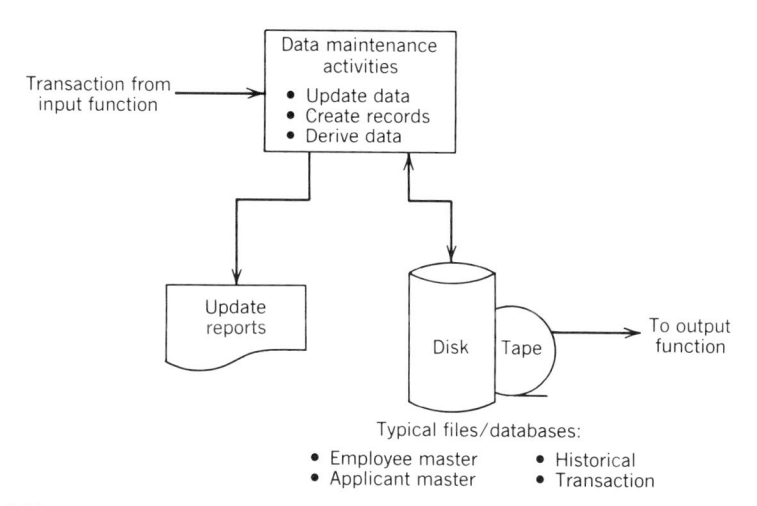

Figure 6.11. HRIS input function components. From S. H. Simon, The HRIS: what capabilities must it have? *Personnel,* September–October, 1983, pp. 38–47.

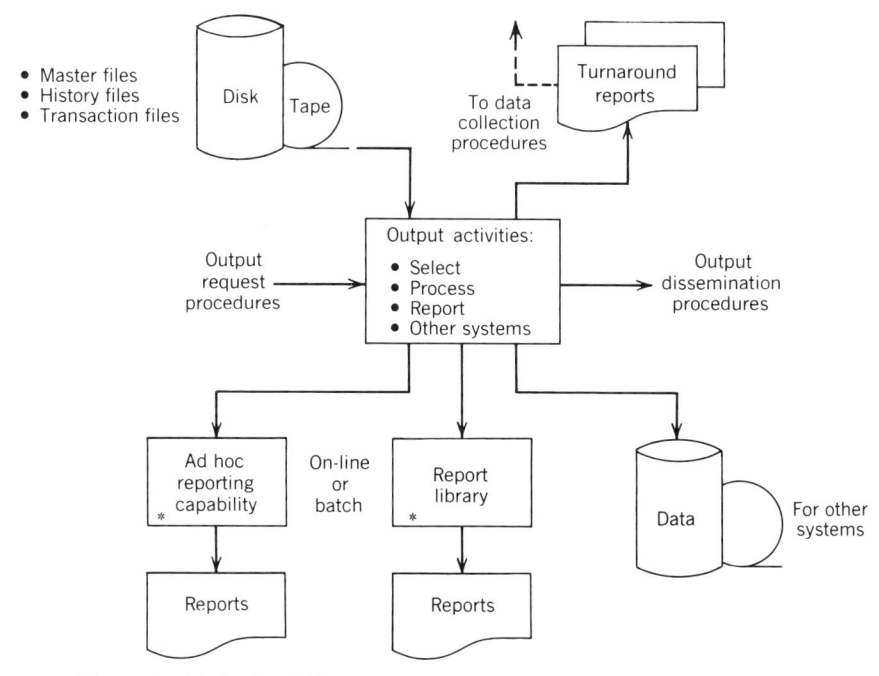

Figure 6.12. HRIS data maintenance function components. From S. H. Simon, The HRIS: what capabilities must it have? *Personnel,* September–October, 1983, pp. 38–47.

but specific data elements to guide the system technician must be defined as well. In the example of a skills inventory, the manager might define several measures of skills such as level of compensation, years of experience, and education or technical training.

The following questions may be used to guide the development of HRIS objectives.

1. What HRM variables are needed for strategic, management, and operational decision support? What information does management need? (One way to answer this may be to develop a table listing all the managers, their respective responsibilities, principal decision areas, and information needs.)
2. What are the main "business issues" the HRIS *must* address? EEO compliance? Productivity monitoring? Recruitment? Recordkeeping? (These issues should be listed and prioritized by each level of management.)
3. Who will use the system? How often? How will each user access it?

The objectives for the HRIS must be kept simple and specific. Most important, the data must not overwhelm the system or its users. If planned objectives are overly ambitious, the system will likely contain major defects and may be too much for users to absorb. A modular HRIS system with a phase-in design for implementation is strongly suggested (Milkovich and Glueck, 1985). Figure 6.13 illustrates a menu of optional functional modules, each of which is a building block of an HRIS. Selection for this menu will depend on the systems objectives and goals.

With this type of approach, each basic objective forms a module that draws on a core data set. The modules are separate but related functional components of the system, each with its own data requirements and distinctive reports tailored to a specific

Figure 6.13. HRIS output function components. From S. H. Simon, The HRIS: what capabilities must it have? *Personnel*, September–October, 1983, pp. 38–47.

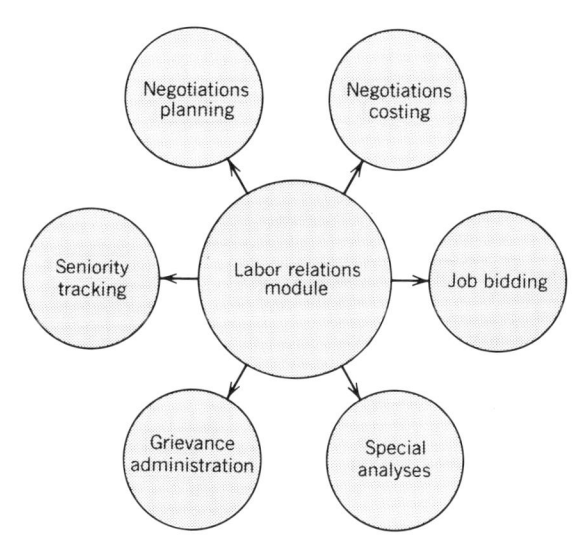

Figure 6.14. Labor relations module. From G. T. Milkovich and W. F. Glueck. *Personnel—Human Resources Management: A Diagnostic Selection.* Plano, TX: Business Publications, Inc., 1985.

Figure 6.15. Compensation module. From G. T., Milkovich and W. F. Glueck. *Personnel—Human Resources Management: A Diagnostic Selection.* Plano, TX: Business Publications, Inc., 1985.

169

function. Figures 6.14 and 6.15 illustrate modules for labor relations and compensation, respectively.

Delivering a system in pieces or modules offers several advantages. First, both the project team and the users are likely to have a higher level of confidence that the system is free of major problems, and users have the opportunity to develop proficiency in using the system gradually. Additionally, users have more opportunity for involvement in the design process, which gives them confidence before the end of the project and fosters acceptance.

With the objectives of the HRIS clearly defined, two key questions need to be addressed.

1. What data are to be collected?
2. What software is to be used?

Determining what data to collect and where to look requires careful planning. Before developing elaborate data collection methods, an attempt should be made to gather data from existing sources. There are most likely many data sets throughout the organization that can be useful to the HRIS. Existing systems for performance evaluations or payroll processing may be used through interfacing or networking the data files. Performance evaluations, for example, contain information about employees' skills and talents, present levels of performance, and potential for growth. This file may be useful for including in a skills inventory or training needs analysis. Payroll data are useful for tracking employee advancement or potential for advancement. Useful nursing human resources data are contained in records on staffing allocation histories, personnel performance, and promotional materials, which are kept in nursing departments. From the business office the operation manager can get data on supply usage by individual units, monthly budget progress, overtime expenses, and absenteeism and tardiness. Determining how to coordinate existing data bases internally is complex in health care organizations. A useful technique to simplify this process is to assess the proposed system flowchart against existing data systems within the facility.

Information should also be gathered about societal labor trends, performance patterns reported by other hospitals, and other types of data not collected internally. To do this the HRIS must have the capacity to merge data from external sources. Access to external data bases is available through many subscription services. Some provide time-series data on personnel trends from the U.S. Department of Labor and the Bureau of Labor Statistics, as well as from newspaper articles, economic forecasts, and financial reports from competing institutions. Data base retailers may charge a subscription fee plus an on-line usage fee. Several vendors offer access to multiple data bases for a single subscription fee at a considerable savings over the cost of multiple subscriptions. For example, projections about nursing staff supply may be derived from information about the number of admissions and graduations from nursing schools, the rate of nursing turnover, and the expected number and location of hospital closings. With proper hardware and software, data can easily be downloaded and merged with internal data for analysis.

Deciding what software will be needed to meet the HRIS objectives is a critical and complex step in HRIS planning. Software to support HR human resources information systems can be obtained in several ways. The choice among approaches depends on individual circumstances and objectives. For example, it is important to decide whether to buy a system, make a system, or borrow a system.

Buying Software

Most organizations purchase a package HRIS. The reasons include:

- Faster development and implementation
- More inherent flexibility
- Lack of in-house capacity

Packaged software or outside service bureaus are available that meet many of the standard human resources information needs of most health organizations. The largest advantage to the use of packaged software is that it is relatively inexpensive. Software developers are able to spread the development costs over the potential market for the package, thus reducing the cost for the individual purchaser. Another advantage is that the product is available with no waiting, but, probably more important, the user can "test drive" the package, to make sure that it performs all the functions that are required. The purchaser of packaged software can also be relatively assured that the program works and that most of the "bugs" have been eliminated. In addition, the purchaser can examine the documentation before purchasing to ensure that this material is complete and understandable.

A problem users may encounter is locating packaged software that meets individual needs. There is a large market in packaged software, however, and unless the organization is trying to do something that is unique or is not normally automated, suitable software will likely be available.

Software must also be compatible with the hardware that is currently or potentially available. In addition, since a given software package seldom exactly matches a user's specifications, each user must decide whether the differences between the software package and the organization's needs are worth the benefits of choosing packaged software. Of course packaged software can be changed, but this is usually costly.

Building Software

When an organization makes a decision to "build" its own software, either independently or in a shared system (i.e. to develop customized programs in-house or through the use of a contract programmer), it is at least possible to more closely match the organization's specifications. The software can be designed to do the job as the user has specified, within the constraints of the programmer's capability and cost. In addition, this approach ensures that there will be in-house staff who are capable of modifying and maintaining the programs.

The obvious disadvantage of the "build" option is that software development is very expensive and time-consuming. In addition, the system cannot be tested until the development process is completed. Another major problem facing organizations that try to build their own software is the difficulty of attracting and retaining qualified data

Borrowing Software

While struggling with system definition issues, it occurs to most managers that somebody somewhere has probably dealt with the same problem—Why not just borrow that solution rather than reinventing the wheel? Although it is likely that another organization has addressed the same problem with an automated solution, there are many potential stumbling blocks to outright borrowing. First, unless the other organization has developed the software in-house, the programs do not belong to the firm and are not theirs to "loan." In fact, when software is "purchased," the organization has acquired only a license to use the software. Therefore, many packages cannot legally be given away. Also, for a package to be useful to the borrower, it must be run on a hardware configuration identical to that of the borrower. In addition, an organization that "borrows" software will always have the problem of where to go for assistance—the lender is not responsible for fixing bugs or making modifications desired by another user.

Although the likelihood of successfully "borrowing" complete software packages is low, an approach that contains much merit is the "borrowing" of software concepts. By studying the software solutions others have implemented, an organization may arrive at a basic software design. Much time and expense can be saved by consulting with colleagues and patterning a system after their successes. The items enumerated below are detailed by Anderson and Robinson (1984).

Whether the organization builds, buys, or borrows the software, several steps illustrated in Figure 6.16 should be taken in determining what software is needed.

1. *Justification.* In this step the need for the software system is documented for management. It includes a general statement of requirements by module, based on the functions the new software will enhance.

2. *Software evaluation committee.* A committee is appointed to select the software; its members represent each organizational component, the potential user groups as

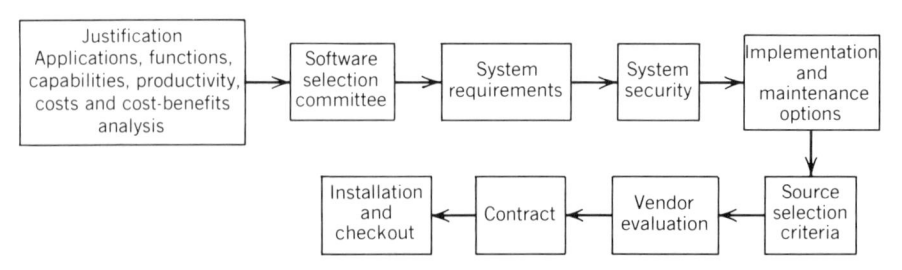

Figure 6.16. Steps in selecting software. Adapted from M. A. Anderson and S. L. Robinson. Selecting HR systems software. Reprinted from August, 1984 issue of *Personnel Administrator*, copyright © 1984, The American Society for Personnel Administration, 606 North Washington Street, Alexandria, VA 22314.

well as technical personnel. Users communicate their needs to the technicians, and their participation also fosters commitment to the software choice.

3. *Systems requirements.* The evaluation committee needs a set of requirements much more detailed than that used for justification. These guides should cover three important features:

On-line capability. Management should be able to run searches for all sorts of data and to use the data base as a planning tool.
Flexibility. Data should be obtainable in any format and combination to permit shifts in applicability.
Data content. The system requirements should be detailed in a document that also identifies the criteria for software evaluation. These should be set forth in terms of functional modules such as personnel information, compensation and benefits, training and development, and payroll.

In developing the system requirements, it is helpful to make a distinction between the current and proposed systems in the context of required inputs and expected outputs of each. This provides an opportunity to describe the shortcomings of the current service in contrast to the features proposed.

A variety of techniques can be employed at this stage. A step-by-step analysis called "scenario analysis" breaks each activity into a set of steps, each of which has its own needs. A "paper flow analysis" traces each data element or report as it enters and is processed by the system.

Decision-making tools are another important facet of the system requirements. For example, will "what if" analyses such as attrition or growth scenarios be needed? If so, retrieval capability will be essential. Also, statistical and graphics capabilities may be important.

Finally, the capability for integrating data files is important. System requirements can be set forth in terms of functions. For all the functions selected, the current system requirements that are mandatory can be described separately from the required additional capabilities of the new system. It might also be appropriate to identify a "wish list" of other data elements that would be considered if available.

4. *System security.* System security is an important consideration. A security access and control capability is essential. Access to entering and updating data must be restricted and controlled. Employees updating data should not be allowed to access all employees records, or all the data for the employees they do have access to. Security requirements must be clearly defined and capable of implementation.

5. *Implementation and maintenance options.* An early decision regarding the system output capabilities for management use is imperative. Will data be handled on the older, batch-processing basis? Is front-end on-line capability desired (i.e., the data can be viewed in real time but can be changed only by batch processing)? Perhaps a more advanced mode is required, in which interactive on-line capability provides real-time changes if desired.

6. *Selection criteria.* Next, attention can be turned to the criteria for choosing software, which can be organized into categories: applications support, hardware considerations, product support, software design and requirements, users' interface, and cost. Criteria for each category will depend on the organization.

7. *Vendor evaluation.* The qualities to be evaluated in prospective software

vendors are not significantly different from those used for selecting an organization to provide other human resources services. In addition to financial stability, professional reputation, and level and availability of services, however, certain other considerations deserve emphasis. First, since the software technology is new and still changing rapidly, vendor commitment to research and development is important. The importance of the timely availability of support staff to come on-site and assist in addressing key design and implementation problems cannot be overemphasized. This suggests that despite the relative ease of travel today, geographic location of vendor services is a relevant consideration. "Working out the bugs" is still synonymous with implementing computer programs and equipment, regardless of all the careful up-front system analysis and design.

Keep in mind that the computer field continues to be very volatile. Software companies appear and disappear overnight. Therefore, a thorough systematic evaluation is very important. This will include objective formal and informal reference checking, a financial analysis, and considerable personal interaction with the staffs of the firms being evaluated.

8. *Contract considerations.* In the excitement over completing the selection of a software vendor and the soon-to-be-acquired computerized capability, the need to have an adequate contract covering all the essential terms and conditions of the complete transaction should not be forgotten. Delivery dates should be specified. There should be a clear delineation of the responsibilities of vendor personnel and of the purchaser. There will be extensive interpersonal working relationships between the personnel of both parties in a complicated mutual endeavor, and success is in the interest of all concerned. Knowing who will be doing what helps assure this. (Source: Anderson, MA, and Robinson, SL: Selecting HR systems software, *Personnel Administration,* August 1984, 99–107. Reprinted from the August 1984 issue of *Personnel Administrator,* copyright 1984, The American Society for Personnel Administration, 606 North Washington St., Alexandria, VA 22314.)

SPECIAL ISSUES

An organization's HRIS is susceptible to failure in every step of planning and implementation. In one survey 94 percent of users of HRIS software packages said they would have implemented the system differently if they could do it over again (Walker, 1980). Why is this level of dissatisfaction so high, and is it preventable?

There are three factors most often associated with failure of an information system:

1. Improper planning before implementation, leading to poor "fit" between system and user.
2. Resistance to the system from management and/or employees.
3. Absence of top management involvement.

We have already discussed the importance of good planning and clearly defining the human resources information needs. A key determinant of the HRIS success will be the so-called fit of the system objectives and design with the organizational culture.

Most often failure is due not to technical shortcomings but to the level and quality of communication between designer and user: Deficiencies here can lead to a system that is not useful. For example, the system may be too ambitious at the onset. A system that attempts to deliver to its user the ultimate in technical capability is doomed to failure. It is essential that the design be kept simple in design and output and that it remain user-friendly. The organization should estimate a minimum of 2 to 3 years to put an HRIS into operation from conception to installation.

The second reason for failure is user or employee resistance to the system, due to lack of knowledge and exposure to the system, fear of losing management control or power, reduced opportunity for career advancement, or any combination of these.

Resistance most often develops when employees feel that the nature of their work, power, prestige, or status is threatened. Potential users of the system may feel that automation of human resources information will diminish their power and usefulness. Consider the managers who relied on their own "expert power" and were, in their respective areas, the final authority rules and regulations. Automation of data requires that all processing requirements be made explicit, which may diminish the power of these experts. Other users may fear the loss of their current positions or current pay level. In fact, some staff members may have to be transferred to other job assignments; others may stay in their present units but the nature of their jobs may change. Employees may feel uneasy about their ability to comprehend what the technical staff is doing. They may question their ability to grasp the complexity of new systems.

The process used for introducing the information system into the organization will determine the level of resistance or acceptance to the system. Users must be involved in virtually all phases of the project and must be kept informed about job security and management commitment to use automation to advance not delete jobs. Their participation will help to ensure that the system is responsive to users' needs and will secure their commitment to the system. As early as possible, technicians and staff management should identify and communicate the impact of automation on staff, organization structure, procedures, and physical arrangements. Participation should give users a voice in the system design with appropriate staff committed to the project full time as part of the project team. The representative staff should function as the communication link between technicians and the users. This will make it easier for the technicians to perform their job and to assure that the system is responsive to user needs.

To be accepted, the system must provide information that is perceived as useful, timely, and relevant. A common problem is that the users do not know how to work the system effectively and do not find it really helpful. Users often have no concept or understanding of the requirements of computer systems in general, and on-line systems in particular. Their ability to critique design documents in the planning stages will be limited. Users often give their consent to documents at the planning stage, only to point out omissions and misconceptions after the document had been translated into programs.

There are several ways other than user participation to minimize employee resistance to the HRIS. One is to initiate training programs early in the system design to orient users and other employees to its goals and objectives. Later, as the system is implemented, a thorough training program is important to show users how it works and how it can be useful to them. It is critical that users know how to access information and use the new services that are offered. Whenever possible, training programs should include hands-on exposure to the system with working demonstrations and self-instructions.

It is important to obtain a productive relationship between the human resources department and users—on all decision-making levels. Unless the human resources department is accepted as part of the organization's management team, it is unlikely that the information system will be promoted and accepted. If an assessment reveals poor relations, strategic decision makers must alter this before the system is implemented.

The active support of managers will greatly facilitate development. This support is fostered by their constant involvement, beginning with the preliminary planning and design sessions.

ACKNOWLEDGMENT

The author is grateful to L. Charest, Department of Human Resources, Duke University, Durham, North Carolina, for contributing the case study and information about assessing software options.

SUMMARY

There is a growing awareness in the health industry that information systems which contain data about hospital employees are critical for sound human resources management. Data which is routinely collected, analyzed and coordinated with the hospital's information system equips management for strategic planning and maintaining productive and satisfied manpower.

The scope of application for HRIS is broad and must be carefully tailored to meet the hospital's needs. Management can strategically project manpower levels and skills as an integral part of the institution's strategic plan. Alternative strategies may be influenced by critical human resource considerations revealed by HRIS data analysis. A more narrowly defined HRIS may focus less on the strategic and more on the management and operational levels of the institution. An effective management control program may depend on the HRIS. Labor relations, benefit programs, position control, training and development, performance appraisal system, and scheduling are some of the essential functions the state-of-the-art HRIS's perform.

The HRIS may be in one central data file or categorized in separate files such as personnel or payroll. Whatever the design of the system, the ability to network files for management and planning purposes should be assured.

The effectiveness of the HRIS depends on sound planning and implementation procedures. Objectives and data specifications need to be stated in detail. Careful consideration should be given to the software options available to the institution. Last, but most important, each employee of the organization must be brought into the planning process, if only indirectly. Commitment to and comfort with the system by employees is critical.

REFERENCES

Anderson, M. A., and S. L. Robinson. "Selecting HR systems software." *Personnel Administration*, pp. 99–107 (August 1984).

Austin, C. J. *Information Systems for Hospital Administration.* Ann Arbor, MI: Health Administration Press, 1983.

Bex, M. "Management information systems: Definition and status." *Healthcare Management*, pp. 26–30 (February 1985).

McHugh, M. L., and C. Curtis. "Automated information systems for nursing management."

Milkovich, G. T., and W. F. Glueck. *Personnel: Human Resource Management: A Diagnostic Approach*, 4th ed. Plano, TX: Business Publications, 1985.

Pang, C. Y., and J. M. Saint. "Forecasting staffing needs for productivity management in hospital laboratories." *Journal of Medicine*, 9(5–6):365–377 (1985).

St. German, D., and A. Meijers. "Nursing work load measurement: An expanded figure role." *Dimensions*, pp. 18–20 (March 1984).

Schmitz, H. H. *Hospital Information Systems.* Rockville, MD: Aspen, 1979.

Veazie, S., and T. Dankmyer. "HISs, MISs, DBMs: Sorting out the letters." *Hospitals*, pp. 80–84 (Oct. 16, 1977).

Walker, A. J. "The 10 most common mistakes in developing computer-based personnel systems." *Personnel Administration*, 25:39–42 (July 1980).

PART THREE
Behavioral Systems

CHAPTER 7

Commitment and Motivation of Professionals

Alice Atkins Mercer

The strategic management of human resources requires an understanding of the relationships that exist among various components affecting the organization, including the behavioral systems of the organization, the organizational structure, the organization's mission and values, and the environment. This chapter focuses on the strategic management of professionals, with emphasis on understanding the commitment and motivation of professionals in health services organizations, as a part of the behavioral systems that influence the strategy formulation for the organization.

Consider these two situations, which occurred in health services organizations and involve professionals.

1. Ann Jones is vice president for operations for a 600-bed hospital. In recent months, she has become concerned about the productivity of one of her management professionals, Bill Swartz, Ph.D., head of data management and information processing. Bill has been a member of the department for 15 months. His first year's performance evaluation, conducted 3 months ago, was very positive. Ms. Jones gave him the highest possible rankings in all categories. During the past few months, Bill's management reports and his scheduled projects have been late, and formation of the tumor registry has not begun. Ann feels she must determine what has affected Bill's level of motivation so negatively.

2. The inpatient hospital associated with an academic medical center has experienced a sudden decline in admissions in the past 2 months. The hospital administration has determined that the decline in admissions is not related to any of the recent changes in preadmission authorization requirements, but that the academic physicians have begun admitting their patients to another hospital located a mile away. One physician, when questioned about the change in his practice pattern, said that he saw no benefit to admitting to the university's hospital and that he was mad at the

university administration. The administration is concerned about the shift in these professionals' level of commitment to the university hospital.

Can organizations or, more specifically, can *managers* affect or influence professionals' behavior—their actions, their performance, their attitudes? How do the characteristics, the behavior, and the expectations of professionals differ from those of other employees? Should theories different from those used in industrial organizations be used to motivate employee behavior in health services organizations? Does a theory of motivation exist that can be applied in almost every situation, or with almost every type of employee? Do professionals and nonprofessionals respond differently to motivational efforts? These questions, as well as variations on them, are addressed in the sections that follow, beginning with the concept of professionals in organizations.

PROFESSIONALS IN HEALTH SERVICES ORGANIZATIONS

One of the greatest problems facing managers of health services organizations is associated with the complexity of having many different occupations and occupational groupings represented in the same institutional structure. For example, a mental health clinic may employ psychologists (at both the master's and doctoral levels of training), psychiatrists (doctors of medicine), social workers, psychiatric technicians, and counselors, to name a few. Health maintenance organizations may employ nurses, physicians, health educators, physicians' assistants, nurse practitioners, and support personnel. Large group practices may employ many of the professionals used in health maintenance organizations, plus attorneys and certified public accountants. All these health services organizations may also have affiliations with professional consultants who are not salaried by the organization. These consultants may receive their compensation as fee for service, but their influence on the organization and its other employees may be very significant: that is, considerably less tenuous than the term "consultant" implies.

A question confronting the managers of health services organizations is the issue of how to manage, control, motivate, and reward the professionals who have strong ties with the organization as well as those who are less firmly connected. This section presents the concepts of professionals in organizations and some of the factors that appear to affect the behavior of professionals within the organization.

Health services organizations require the services of professionals of many types. Defining which individuals are classified as professionals sometimes causes difficulties for managers because it seems as if many occupational groups are asking managers and society in general to treat them as professionals. In fact, these definitional difficulties have been recognized by theorists for decades and have resulted in many articles. One of the best known, entitled "The professionalization of everyone?" (Wilensky, 1964), presents the quandary most thoroughly.

In health services organizations, one would expect to find a variety of occupations represented. Many of these occupations have experienced the natural evolution toward professionalism. Members have formed professional associations, instituted codes of ethics, encouraged more rigorous credentialing and licensing of practitioners, and urged an increase in educational requirements. The health services manager must ascertain how different and how similar are these "new" professionals from the occupations that have been considered historically to be professions. Do they require or expect autonomy as a prerequisite for satisfaction with the work environment? Do they experience conditional loyalty to the organization? Will there be a difference between the expectations instilled in the "new" professional during training and the conditions and expectations of the employing organization? How does the manager identify and define the professional occupations?

Definition of Professional

Historically, the professions were defined as the ministry, law, and sometimes medicine. As educational levels increased in the United States, and as our economy moved away from an industrial base toward being service-based, more and more groups of highly trained individuals who called themselves professionals emerged in our society.

On what basis does an occupation become designated a "profession"? It is generally accepted that professionals are highly trained in an area of expertise and are certified by some institution. As a result, they possess specialized knowledge and may be considered experts in their fields. They have a commitment, considered to be enduring, to their profession. Also, many professionals feel that they are dealing with areas of social or universal concern, and they are seen as providing a service.

Professionals desire autonomy in their work environment, and many groups of professionals believe that only their peers are qualified to judge their work. Because of their specialized training, professionals expect to establish their standards and codes of ethics. Thus, autonomy and independence are conditions that help differentiate professionals from other employees in organizations (Scott, 1966).

Because of the proliferation of groups calling themselves professional, there has been an effort to find a general definition or common denominator describing all professions. Such definitions have ultimately emphasized autonomous expertise, the service orientation ideal, and a high level of technical training. The amount of autonomy the health professional is allowed to exercise becomes a very real indicator of a profession. Thus, physicians have historically been thought to belong to a profession as defined here, since they have had the most autonomy in health care settings. Members of other health-related occupations also experience autonomy within the confines of their job responsibilities and would qualify as professionals.

Individuals who consider themselves to belong to professional occupations in health organizations can be divided into four groups. Each group brings its own orientation, its own set of priorities, and its own goals, which may conflict with those of other professionals. One group consists of nurses and allied health workers. Another group consists of physicians. A third group consists of technicians. The final group consists of managerial personnel. There is some debate concerning whether all four groups possess all the attributes associated with professions. Yet it is worthy of note that each of these groups, in one way or another, identifies itself as a profession. The remarks in the section that follows apply to professionals and professional behavior in varying degrees, depending on how closely the individuals and occupations are aligned with the definitions of a profession.

Conflict for Professionals

Professionals may experience conflict working within a bureaucracy because the profession and the bureaucracy are based on fundamentally different principles of organization (Scott, 1966). In a bureaucracy, for example, certain conditions exist such as a hierarchical structure and delineated formal procedures, with rules and controls. Specialization of labor, with few workers trained or hired to produce the entire product, is another feature of many bureaucracies. Professional behavior, however, is centered on commitment to autonomy and peer review (Wilensky, 1964). The conflict arises when the professional is employed by the bureaucratic organization, and the professional's requirement for autonomy is incongruent with the bureaucracy's need for structure and control.

There is some indication that this conflict is not inevitable; professional attitudes and commitment to a bureaucratic organization are not necessarily incompatible. Health care workers such as nurses or research scientists may find that the bureaucracy supports their professional goals by supplying them with a structure that facilitates their work. In fact, there is slight evidence that physicians and their feelings of autonomy can be affected positively by bureaucracies, but researchers are going to have to investigate this condition more thoroughly in health services organizations (Engle, 1969; Goss, 1963).

Conditional Loyalty

The professional's behavior within the organization may be best understood via the concept of conditional loyalty (Scott, 1966), which occurs because the professional identifies first with his or her profession and then with the work organization. This strong form of identification is established and fostered when the professionals receive their training (e.g., in nursing school) and is perpetuated through professional groups and associations in which they maintain membership. Through the extensive, and often lengthy, training professionals receive, they develop a "professional" self-image: They value their skills highly and are more concerned with their reputations among their peers than with pleasing organizational superiors; they view the organization and its administrators as facilitators, and their loyalty to the organization may be conditional upon their evaluation of the adequacy of the organization's facilities and programs relative to others of which they are aware.

Other types of employee may also experience a conditional loyalty to the organization, especially part-time employees. This tendency is especially important for health services organizations that employ or affiliate with individuals on a quasi-contractual basis—for example, agency nurses who are available for temporary duty during periods of peak demand, physicians who come as consultants, or employees of a contracted housekeeping service. Such persons may enter the organization with fewer and different expectations concerning the role the organization will assume in their lives as contrasted with the full-time salaried employee. Employees' expectations of the organization are related to their levels of commitment and motivation, as discussed later in this chapter.

The example that follows illustrates the type of conflict that occurs among professionals health services in organizations.

Ms. Carter, director of nursing at General Hospital, a 140-bed community hospital, is disturbed that the administration has vetoed the nursing department's equipment request for a microprocessor for use in patient care planning. In addition, the cardiology medical staff has just received a denial of a $5000 request for new equipment. Both the nurses and the cardiologists feel that the denials were arbitrary and that their needs were not evaluated fairly. They have complained formally to Ms. Carter and to the chief of the medical staff, Dr. Johnson, who have asked that the senior administrator, Mr. Bells, consider their appeal. The central argument is that the organization's failure to assign priorities in the allocation of the hospital's financial resources is improper: "Whenever the nursing or medical staff requests equipment, we have to justify and rejustify our needs, but when the administration wants to renovate their offices, the money is spent without any input from us. What's the difference? Who approves their justification? Do they even have to provide a justification? Patient care needs should be met first." Mr. Bells responds by saying that he too is a professional who knows what is best for the organization and that he takes everybody's needs into account when he makes allocation decisions.

Increasing Conflict

As more and more health services organizations are pressured to be efficient producers of quality health care, and especially in this era of changing reimbursement systems, conflict between health care professionals and administrators is increasing in some environments. One of the primary reasons for the increase in conflict is the perception that what is in the best interest of the patient will become secondary to the goal of efficient delivery of that care. This conflict is demonstrated most dramatically by changes occurring to professionals such as physicians. Some physicians believe that they have lost control of their profession. More and more physicians are working on salary rather than fee for service, and they are beginning to feel like "hired hands." They are attacked in editorials in local newspapers, and in national publications such as the *Wall Street Journal* (1986) as being unresponsive to changes occurring in the delivery of health care. They feel as if they have fallen victim to a reimbursement system and to third-party payers who are controlling the future of medicine rather than being allowed to concentrate on what is best for the patient. Hospitals expanding their outpatient clinics are perceived to be in direct competition with physicians; alternative delivery systems are affecting their patient mix as well as their incomes; relative value scales bring economic uncertainty.

Health professionals are feeling forms of attack in addition to increasing competition and changing reimbursement systems. Changing liability and malpractice laws, changing Joint Commission on the Accreditation of Hospitals (JCAH) requirements, changing regulations encouraging more review of medical decisions, and greater numbers of professionals and occupations, as well as a better educated society questioning their decisions and judgment, are all contributing to a significantly changed environment for health professionals.

Professionals may see their role in the health care organization in terms completely different from those used by organizational management. The professional may view himself or herself as a scientist working for a patron, whereas the organization sees the relationship in terms of client-professional, or even employer-employee (Raelin, 1984). This role ambiguity undoubtedly will contribute to dissatisfaction or decreased levels of organizational commitment.

Health professionals may complain about administrators' interference in clinical decision making and administrators' contributing to inefficiencies in the delivery of health care. They may perceive that they are losing their professional independence. On the other hand, health professionals may find that the accomplishment of their professional goals is facilitated by the marketing and management support provided by these very administrators. If administrators are assisting the clinician in the acquisition of needed or desired resources, a form of team building between the two groups may occur. Conditional loyalty to the health services organization may become contingent on the provision by the organization (and the administration) of needed support to the physician or clinician.

Organizations that have interactions with health professionals in this rapidly changing environment are using various methods to alleviate this conflict. Some of the new forms of organizational design are presented next.

Organizational Forms Used by Health Professionals

The relationship between health services providers and health services organizations has been formalized in several ways, which were described by Scott in 1982.

Autonomous Structure

The institution may delegate to the professional group the responsibility for setting goals and standards, and for seeing that the standards are maintained. This type of arrangement is justified because of the complexity and uncertainty involved in the work performed, conditions especially applicable in health services organizations. Many professionals, because of their training, ascribe to the system of peer review (which involves assessment of performance by colleagues, not by administrators or others outside the specialty area) as a form of control and review of their performance. Only professional colleagues are believed to possess the specialized training and experience that would permit valid evaluations of performance. The administration does not have control over the quality or type of professional practice occurring in the health care organization.

Many general hospitals have an autonomous form of organization; that is, the governing board has delegated the legal responsibility for the care of patients to the medical staff. The medical staff in organizations of this type has organized formally to determine the physicians who will be granted admitting (or other medical) privileges. The degree of organization of the medical staff varies with the individual health organization. Some hospitals, for example, have a medical staff that operates as a very organized collective unit. Others have a staff comprised of a loosely connected group of individuals who do not view themselves as part of a group or a formal structure.

Various control arrangements in addition to peer review exist in organizations with this type of professional structure. The JCAH has described minimal credentialing requirements for medical staff and various committees involving physicians as a requirement of the health services organization. Federal regulations associated with reimbursement have also established controls, such as the Peer Review Organization (PRO), which reviews elective Medicare hospital admissions on a prospective basis. This type of preadmission review is being required by other third-party vendors, such as commercial insurance carriers, and by health maintenance organizations, in an attempt to encourage more appropriate inpatient utilization. The PRO constitutes a significant control on physicians, and physicians have reorganized some of their administrative and clinical procedures in response to the introduction of this institution.

Administrative Dominance of Health Professionals

In a second kind of organizational form, professionals in health organizations do not have the kind of autonomy described previously and are subjected to regular supervision and instruction. In most health service organizations, nurses are a part of this administrative dominance type of structure. Also included are groups such as physical therapists and medical technologists. The case study entitled "Professionals at West End Hospital" demonstrates this type of organizational form.

Professionals at West End Hospital

The Department of Health Services Research at West End Hospital is composed of highly educated and highly trained professionals whose goal is to provide many forms of biostatistical information to various areas of the hospital. Within this department is the Cardiac Disease Registry, a group of technical employees who gather data on all the cardiac patients in the hospital.

Much of the work of the registry was done by registered records administrators (RRAs), who possessed bachelor's degrees and were trained to abstract clinical information from medical records. They were paid relatively little for doing highly technical but routine and tedious work. The RRAs were supposed to be on an equal administrative level with holders of master's and doctoral degrees, but these professionals had difficulty imagining that anyone less highly trained and educated than themselves might perform research functions.

As time progressed other professionals, accredited records technicians (ARTs), were added to the Cardiac Disease Registry to perform follow-up functions. This information was added to the basic data base to give a full and complete picture of the cardiac patients and what happened to them over time.

All these factors began to take a toll on the registry members, and one by one as professional openings with a new outlook and challenge as well as better compensation came along, the RRAs left the registry. Not unexpectedly word of the registry's situation spread, and it became difficult to recruit new RRAs for the positions that were available.

As a part of a hospital-wide program, the job descriptions for these positions were reevaluated and rewritten for the registry personnel. After consideration of the skills necessary to perform the registry functions, a new job description was written for a technician rather than a trained, degreed professional.

The job analysts determined that trained professional skills were not needed to perform chart abstracting. The job actually required knowledge of the medical record order and catheterization report format, and familiarity with the variables to be abstracted from both these sources. The job itself was fairly routine and repetitious, and the skills could be learned on the job by performing the abstracting duties. A general knowledge of medical terminology was an advantage, but this could be assimilated over time by working with the medical records and catheterization reports. Clinical skills required were also minimal because routine information was given in the catheterization report format of history, physical and laboratory values, anatomy, procedures, findings, and diagnosis on each patient. The single most difficult part of the abstracting, after familiarity with the variables was acquired, was interpreting the physician's catheterization report. This, too, was expected to become less difficult in time and after the analyst came to know each physician's style of presenting information.

The follow-up positions in the registry involved slightly different skills, but again it was decided that professional skills were not necessary. Basically the job was to update the data base using income information from follow-up questionnaires. These skills could again be mastered over time and on the job.

From both these job analyses, new job descriptions were compiled that became almost generic. The job descriptions followed the basic format of listing job title, job summary, and duties and responsibilities in the job description portion and enumerating employee qualifications to perform the job satisfactorily in the job specification portion.

Once selected and hired, new registry employees were trained by a senior member of the registry unit. The length of time for training varied with the individual, but the goal was to enable the new employee to achieve productivity standards outlined in the performance appraisal for the position.

A productivity profile was kept on each abstract assistant. Although it is somewhat difficult to quantify the work of the abstract assistants and impossible to do so for the follow-up assistants, some figures of productivity are necessary for such a large group of employees to justify their existence to administration and to continue to be budgeted.

The performance of each employee is appraised annually, to provide information needed for determining whether individual work results are consistent with expectations, and for identifying high, marginal, and unsatisfactory performance. Information on compensation determination is also received, as well as feedback for both employee and supervisor.

At one time a type of management by objectives was accomplished by these performance appraisals in that both employee and supervisor would independently list goals to be accomplished within a time frame chosen by the employee. These objectives were discussed privately, and a common list for each employee was compiled through the joint efforts of employee and supervisor. Now, however, the supervisor lists objectives and deadlines for each employee, and the list is conveyed to the employee at the time of evaluation.

The retention of registry analysts, once trained, is affected by several factors, both positive and negative. The negative factors can be summed up under the overall topic of motivation. These employees have lower than expected motivation levels because of perceived inequity in status with their departmental coworkers, lack of opportunity for advancement because of the emphasis on degrees in the other units of the department, and poor communication associated with relations with the other members of the department, as well as unclear lines of responsibility and accountability.

The positive factors almost outweigh the negative and make retention of employees possible. Benefits within the organization are fair, but possibly the department's best benefit is its "flex time" policy. This is appealing to the personnel presently in.the registry, many of whom are mothers of young children. Because time schedules are somewhat flexible and time can be made up within a pay period, loss of pay and attendance holidays because of children's illnesses and so on is not a threat.

The registry positions also offer an entry into the organization and can help individuals to be aware of other hospital positions and possible advancement.

And how have the registry personnel achieved at least the basic esteem level of Maslow's hierarchy? Over time they have gained acceptance for their skills within the department through recognition of what their job involves and what a valuable source of information their data base is. For example, a recent newspaper article was able to quote mortality figures for cardiac patients using the data base. Also, university physicians are now using the data base for research and publication. Articles and presentations of national distribution have come from this large data base.

The merging of two different types of personnel, both of whom consider themselves to be professional, has been a challenge for the health services research department of West End Hospital. Understanding professional behavior and factors influencing the motivation of these employees has helped in providing an environment conducive to the retention of these employees.

In some environments physicians are coming more and more under the administrative dominance type of organization because of the changes that are occurring rapidly in health care and because of the new organizational forms that are emerging. There are many reasons for these changes. The need for additional coordination of the many specialized activities (and occupations) associated with patient care has become more apparent. Changing reimbursement mechanisms are putting institutions more at risk financially because of the activities of the professionals associated with those institutions. Physicians are increasingly dependent on hospitals or other institutions to provide their practice environment, and many are becoming salaried by these institutions.

The professionals working in this type of organization are not organized separately into an autonomous unit. There is much more overlapping of administrative and professional actions. However, an individual professional still has considerable discretion over clinical decision making and tasks. Administrators, many of them clinical professionals themselves, coordinate the activities of these individuals. This form of organization is seen in some academic medical centers and in the Veterans Administration health units. Teams are used frequently for the provision of patient care, forming a type of product line management focus, as is common in industrial organizations.

What form is the most appropriate for the resolution of conflict described previously? There is no general agreement. However, the section that follows describes one approach that allows for the special requirements of professionals.

Resolution of Conflict

Some health services organizations are attempting to resolve the conflict occurring when professionals (especially physicians) work in bureaucracies. They are establishing coequal professional groupings in which the medical staff hierarchy and the administrative hierarchy are roughly equal in power and importance, and are interdependent (Scott, 1982). Communication vehicles and participation in decision making contribute to this perception of equality. Under this form, health professionals are concerned with the delivery of patient care to individual patients, and administrators are concerned with the larger issues of patient care, the delivery of care to groups of patients. There is no longer the artificial and increasingly inefficient distinction of having patient care issues decided by the medical staff, and support and maintenance issues by the administrators.

Both groups, the health professionals and the administrators, need to be kept informed of the other group's concerns. For example, when the hospital must choose between allowing a nonpaying nonemergency patient to be admitted and reducing the amount of care provided to paying patients in a systematic manner, the health professionals must be kept informed of the nature and complexity of the dilemma faced by the administrators.

In addition to understanding the organizational forms being used by institutions employing and associating with many professional groups, health services administrators may be interested in attempting to increase the amount of loyalty health professionals exhibit for the organization, perhaps in the belief that a more loyal health professional is a more productive one, working for the organization rather than against it. Hospitals frequently try to increase the employees' feelings of "linkage" (Mowday et al., 1982) or ties with the organization. Some organizations are attempting to increase the degree of commitment health professionals (as well as other employees) feel toward the organization. The sections that follow present the topic of commitment.

COMMITMENT

Often one person may be spoken of as being more committed than another to the institution they both serve. What ramifications does this condition of "commitment" have for the health services organization and for the management of those participating in such organizations?

Organizational commitment can be defined as the relative strength of an individual's identification with and involvement in a particular organization. "Organizational identification" and "involvement" are terms that may be synonymous with "organizational commitment," which is characterized by at least three factors: (a) a strong belief in and acceptance of the organization's goals and values, (b) a willingness to exert considerable effort on behalf of the organization, and (c) a definite desire to maintain organizational membership (Porter et al., 1974). For example, the organizational commitment of an allied health professional could be measured partially by considering the strength of the person's desire to remain affiliated with the employing organization. Or, considering the hospital-physician relationship, is the physician willing to put in a great deal of effort beyond that normally expected to help the hospital be successful? Or, does the charge nurse demonstrate agreement with the organization's stated goal of treating each patient and visitor as a "guest" in the hospital? What degree of concurrence exists between the health services professionals' goals and values and those of the organization?

Certainly an individual can be committed to entities other than or in addition to his or her place of work. People can be committed also to family, hobbies, education, and other life goals, and these multiple commitments do not have to be in conflict with each other. The administrator in a health services organization, however, does need to be aware of these competing commitments, which affect the behavior of professionals and lower level employees alike.

Organizational commitment is thought to be based on the concept of exchange. Employees have a set of needs, desires, and career and personal goals, which they hope and expect the organization to help them to meet. In exchange for the organization fulfilling these expectations, employees are expected to contribute meaningfully to the organization, and it is hoped that they will become more committed to it. Of course, if these expectations are not realistic or are not fulfilled, employees' levels of commitment may be negatively affected. Commitment may be viewed in part as the result of both the organization and the employees engaging in a tacit relationship of quid pro quo.

Commitment to an organization is not the same as satisfaction with a job. Both research and theory indicate that commitment is a more complex process than job satisfaction, and is more enduring. Commitment is less affected by day-to-day events and involves the employee's feelings about the entire organization.

Determinants of Organizational Commitment

There are various schools of thought concerning the conditions that lead to increased individual commitment. The personal characteristics of the employee affect the degree to which he or she may be committed. Most studies have demonstrated that the older person is, the more likely he or she will be committed to the organization. Also, the longer someone has been associated with for the organization, the more committed he or she will be. Of course age and tenure do not explain absolutely the relationship with the organization. For example, the older employees become, the fewer opportunities they may have to work elsewhere.

Another determinant of commitment is level of education. The more educated person is, the less commitment he or she feels toward the organization. Because of the high levels of education of many health professionals, awareness of the effects of this condition is important. Attaining organizational commitment from its employees and from its affiliated professionals may be more difficult for Massachusetts General Hospital than for Federal Express. This condition is partially explained by the discussion presented earlier in this chapter on professionals in organizations and their conditional loyalty to the organization.

Women have generally been found to be more committed to organizations than men (Mowday et al., 1982), but very few studies of women professionals in health services organizations have attempted to determine whether gender is a predictor of commitment under these conditions.

In addition to the personal characteristics mentioned above, job-related characteristics are thought to influence commitment levels. These job-related characteristics include conditions such as the amount of challenge the job presents, and how clearly the job and the employee's role have been defined to the employee. Employees whose job roles are ambiguous have a degree of uncertainty associated with the tasks they undertake. This uncertainty affects their feelings about the job, as well as their commitment to the employing organization.

Work experiences within the organization also affect the employee's level of commitment. The feeling that the organization supports the employee in his or her endeavors and that the organization can be "counted on" when necessary contribute to feelings of commitment. Other work conditions that may affect commitment include feelings of equity and fairness, especially when associated with pay and reward systems.

Finally, the structure of the organization may influence levels of commitment, although the research is inconclusive and still in formative stages. Structural components such as degree of decentralization, participation in decision making, and formalization of rules and procedures have been found to be related to commitment.

Thus, the determinants of commitment originate from many sources, from within the individuals themselves to the structure of the organization. (Figure 7.1 depicts the major determinants of commitment during the early employment period.) Managers cannot change the individual and personal characteristics of professionals associated with their organization, but understanding these characteristics can affect their choices when future members of the professional staff are being considered. On the other hand, managers can affect the structure of the organization, and they can increase conditions such as involvement of the professional staff in the decision-

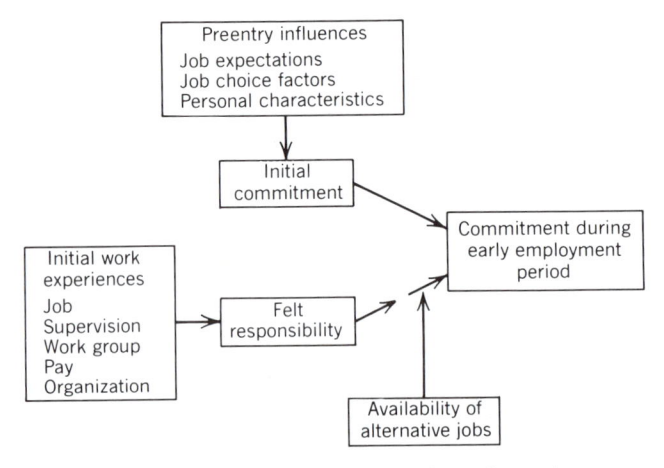

Figure 7.1. Major determinants of commitment during the early employment period. (From Mowday et al., 1982, p. 56.)

making and policy-setting events that may lead to higher levels of organizational commitment.

Outcomes of Commitment

For the manager, it is important to know what the possible effects of commitment are. Having a high degree of commitment within a med-surg unit, for example, may be "nice," but does it make any difference in the patient care delivered in that unit? Although that particular question cannot be answered with absolute certainty, more general questions about the effect of commitment can be addressed. For example, does an attitude such as commitment lead to a given behavior or to a predictable set of behaviors?

Organizational commitment can take attitudinal and/or behavioral forms. For example, employees may "say" that they are committed in many different situations; this serves as an indicator of attitudinal commitment. The notion of behavioral commitment suggests that being committed makes some difference in employees' behavior; ideally, this "difference" will have a positive impact on the organization or on its clients or patients.

The outcomes of organizational commitment that have been studied include turnover, absenteeism, tardiness, tenure with the organization, and job performance. Less committed employees and professionals tend to exhibit withdrawal from the organization. These outcomes can take the forms described in Figure 7.2.

Employees who are less committed generally have higher turnover rates than do other, more committed employees. This statement does not suggest that less commitment causes higher turnover; it simply means that these variables are related. People who are less committed to an organization tend to quit their jobs more frequently.

One way in which commitment is measured is to ask whether an employee intends

Commitment and Motivation of Professionals 193

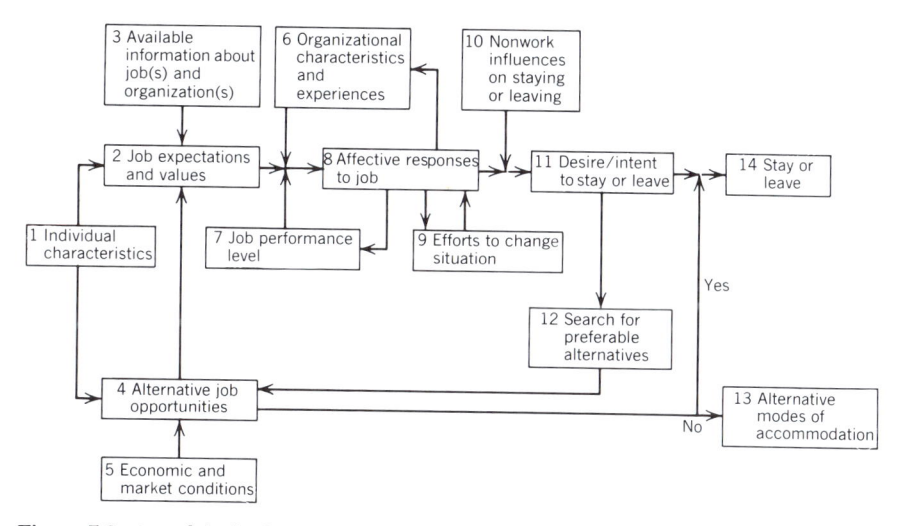

Figure 7.2. A model of voluntary employee turnover. (From R. Steers and R. Mowday, in L. Cummings and B. Staw, Eds., *Research in Organization Behavior*, Vol. 3. Greenwich, CT: JAI Press, 1981, p. 242.)

to stay in the organization. In the model of employee turnover illustrated in Figure 7.2, intent to stay is measured instead of commitment. Professionals, among others, may underreport their intent to stay if they fear reprisals emerging from such a revelation (Arnold et al., 1985). However, when commitment has been measured over a period of time, a decline in the level of commitment of an individual employee may serve to warn management that he or she will leave the organization soon.

Commitment is one of the many conditions that affect absenteeism rates. The more committed employees are absent less frequently. However, in examining absenteeism rates, a manager must be aware of different types of absenteeism and their meanings for the organization. For example, it is thought that employees who are absent one day at a time are responding to an entirely different set of circumstances and exhibiting different behaviors from employees who are absent in blocks of several days for each absentee episode. In fact, the organization may have encouraged an "absence culture" in which certain short absences, perhaps of one day at a time, are acceptable informally (Nicholson and Johns, 1985).

Tardiness is a form of absenteeism and has some of the same features. Tardy employees may be thought of, in some cases, as being absent for a portion of each day. Employees who are tardy more frequently are usually less committed to the organization. Again, however, declining commitment is not a result or a cause of tardiness or absenteeism. There is simply a relationship between the tardiness and declining commitment.

The length of time a person has spent with the organization is also related to the commitment. Although commitment to the organization is viewed by many scholars as an enduring affective response, it probably has some variability, being expressed in one form when the individual first joins the organization and in other forms in later stages in his or her association with the organization. It would be expected to mature and evolve over time. This condition can be explained in several ways, largely involving the concept of "exchange" mentioned earlier.

The level of commitment of an individual upon first joining an organization is affected by his or her personal characteristics, expectations about the job, and circumstances of job choice (Mowday et al., 1982). These conditions can be used to predict an individual's likelihood of building longer term commitment to the organization. A prospective employee whose personal value system and goals mesh with those of the interviewing organization will probably be more likely to become committed to the organization. The same is true if the prospective employee has realistic and accurate expectations about the job and what is expected of him or her as an employee. The events surrounding the job choice also affect commitment levels. If the job choice is a public decision and is somewhat irrevocable, the employee is more likely to be committed to the job. When an individual first joins an organization, he or she may initially question the decision to join. The questioning may result from cognitive dissonance, which can lead to attempts to reconfirm the wisdom of a major decision. During this period, a manager may expect to see a decline in organizational commitment. But as the person spends more time with the organization, he or she readjusts expectations about the organization and comes to know it better. After several months, the degree of commitment should increase, returning to preemployment levels.

The longer an employee or professional stays with an organization, the more he or she has invested in that organization. An individual who is paid a salary with benefits has more financial investment in the organization, such as retirement plans, annuities, or perhaps even a profit-sharing arrangement. The individual also has a greater personal investment in the organization from serving on committees and getting to know the organizational ropes. Also, the job itself perhaps has gotten better over time, resulting in stronger feelings of commitment. Moreover, alternative job opportunities decrease as individuals become older. And finally the individual has become more firmly tied to the organization and to the community, having inevitably developed working relationships as well as social relationships that would be difficult to leave. All in all, it is not surprising that there is a relationship between commitment and tenure with the organization.

There also should be a relationship between job performance and organizational commitment. One would imagine that a more committed individual, having become meshed or allied with the organization's goals and values, would be more productive. This item is especially hard to measure for professional employees, so there is no conclusive evidence (other than intuition) that more committed professional employees are more productive. A "highly productive" person who sees many patients in a day may not be delivering the highest quality of care. But, on the other hand, having many patient encounters in a day does not imply low-quality care. This situation exemplifies a recurring question in health care, namely, What is quality? The JCAH has indicated that some health organizations, using their quality assurance framework, are beginning to examine outcomes associated with health delivery such as hospital readmissions due to complications, drug or transfusion reactions, unexpected transfer from a general bed to a special care unit, and perforation during an invasive procedure (personal communications, July 1986). These outcomes could be matched with number of admissions for physicians to develop a component of a productivity index. Previous types of negative outcome examined were more general, such as deaths or unexplained surgery. It is not clear, however, whether negative events are in part the result of declining commitment, or the cause of declining commitment. Or, to phrase it differently, it is not clear whether committed individuals perform more positively and to the advantage of the organization, or whether positive actions result in more committed individuals.

Many health care executives, realizing that they are dependent on physicians for patients, express concern about the relationship of their organization with its medical staff. Increasing physician commitment to the organization is believed by many administrators to result in more productive medical staff, who are more willing to serve on hospital committees and are more positive members of the hospital organization. There are few studies of physician commitment to organizations, and even fewer studies of nonsalaried physician commitment to organizations. Like other forms of commitment, physician commitment is very complex. For example, certain organizational conditions, such as the existence of a bureaucratic structure, may not increase commitment by themselves, but in combination with other positive conditions they may result in an increase in commitment. Also, one would expect different medical specialties to respond to different conditions in relation to commitment. For example, a hospital-based radiologist would have requirements significantly different from those of a cardiologist. In general, the following components of physician commitment have been studied.

There is some indication that physicians who are more satisfied with overall hospital services are more committed to the organization. Hospital services can be measured by departments (laboratories, admitting, medical records, etc.), and the services themselves can be evaluated on the basis of quality, timeliness, cooperation of unit providing service, and comprehensiveness.

The types of reward a health services organization provides its physicians may also affect the level of commitment the physicians feel toward the organization. Rewards can be in the form of salary or enticements such as providing professional office space. Some institutions offer support for the physicians' practices in the form of marketing and communications advice. Some institutions provide prerequisites such as free parking.

The higher the quality of the nursing staff and the administrative staff, the more committed the medical staff. An administrative staff that is involved in long-range planning and remains responsive to physician needs helps to increase commitment levels. The physicians' perceptions of how satisfied their patients are with the hospital also influence how committed physicians are to the hospital. Physicians who are more active with the hospital, serving on committees and taking leadership roles, are generally more committed to the institution (Mercer et al., 1985).

Managerial Applications

The growth of professionalization in health care has been accompanied by increased formalization and expanded controls over providers, which may affect professionals' commitment levels. These control systems have been insisted on by the courts, the reimbursement systems, and accreditation and regulatory entities, all of which are requiring more accountability. The discretion allowed the individual professional is declining.

The control systems are varied. Some operate through the use of committees that examine unusual events. For example, in hospital mortality committees, a group of health professionals examines the events leading to every death that occurs in the hospital. If the committee finds that irregularities have occurred, an investigation will be conducted. Tissue samples and pathological reports may be reviewed. Protocols may be examined. All these events, however, imply that someone or some committee is looking over the shoulders of the physicians and the nurses, evaluating their decisions. This evaluation is a significant change and represents a threat to any professional.

Other types of control are being placed on professionals. The Medicare prospective pricing system initiated in 1983 has resulted in individual hospitals conducting full-scale reviews of the lengths of stay of Medicare patients treated by their medical staff. To influence physician behavior, some hospitals provide each physician on their staff with data on his or her patients' length of stay for specific disease conditions as well as information on institutional or regional lengths of stay norms. These data allow the hospital to identify lengths of stay that are significantly greater than expected. In addition to aggregate data analysis, patients who are still in the hospital are concurrently reviewed, and retrospective chart review is conducted on records of discharged patients. Management systems of these types suggest to physicians that they are losing control of their profession.

Physicians and nurses appear to fall into two groupings with respect to their reactions to increased control systems. One group of professionals assumes a broader perspective about patient care and desires the rationalization of health services. Members of the other set are concerned about individual patients, not society, and see themselves as the individual patient's advocate. Of course these differences in perspective result in different responses to control systems, in addition to other administrative concerns. Disillusionment among some professionals is leading to withdrawal from professional or clinical life. Others seek to secure or regain control by becoming more a part of the organization, attaining administrative positions, for example (Scott, 1985). Still others remain in the health care field and resist the control systems. Understanding the differing perspectives will assist the health services administrator in dealing with the increasing conflict that can be expected.

Assuming that professionals do indeed have the needs described at the beginning of this chapter—the needs for autonomy, independent decision making with peers, freedom from organizational control—what will happen to professionals working in health organizations in the future, as more and more control is placed on them while their accountability increases and their organizational commitment levels decline? One would expect the professionals sooner or later to employ adaptive or withdrawal mechanisms to make their current situation more acceptable. Nevertheless, there will be disillusionment, grounded in part on their unrealistic or unfulfilled expectations about the role of the organization in a health care professional's life. If it is important to have a match between goals and values of the organization with those of the associated professional, managers should ensure that new professionals have accurate information about conditions and expectations within the organization. This requirement is especially necessary for professionals whose training does not prepare them for organizational life. For example, if physicians are expected to perform certain committee and managerial duties with or for the hospital, this requirement (or expectation) should be delineated in the medical staff bylaws. New applicants for medical staff privileges should receive a copy of the bylaws and should be given the opportunity to meet with the chief of staff and the senior hospital administrator. Such a meeting should include a frank discussion of the goals and values of the health care organization, as well as a candid discussion of the aspirations of the professional. In this manner, professionals can develop realistic expectations about a portion of their role within the hospital.

Organizations and managers that demonstrate that they value the expertise and services of the professional may be more effective in helping the professional adapt and adjust to the organization. Often administrators who show a competence and awareness of certain professional matters are more effective in working with professionals. This is one rationale for having clinical professionals occupy senior administrative positions in health care facilities. For example, the vice president for professional services in a large hospital might be a nurse. In the case study that follows, the administration of a psychiatric hospital chooses another response to a threat triggered by an action of disgruntled professionals.

Windom Hospital Responds to Changing Commitment Levels

The field of psychiatric health care has expanded both in terms of services offered in existing hospitals and with respect to competitive facilities providing that care in a hospital setting. Windom Hospital has been the sole free-standing psychiatric hospital in a large metropolitan area for 10 years. As the competition for psychiatric care has grown, other free-standing facilities have entered the marketplace. The additional providers have changed the marketplace picture, but they have directly threatened Windom. The shifts in utilization patterns have been gradual, and the personnel changes have been random and orderly.

However, a new facility, Memorial Hospital, has brought some drastic changes to the orderliness of change experienced by Windom. The significant change began when a major group of physicians announced suddenly the relocation of their practice to Memorial. Because of existing contractual agreements, the administration of Windom was hard put to fill the services that these professionals had been providing. Yet this was only a prelude to the manpower problem that was about to beset Windom.

The exiting physicians identified key personnel and positions needed to staff the new hospital, and the administration at Memorial began to offer Windom staff members similar positions with significant salary increases. The result was that key members of the Windom staff began to leave for new positions at Memorial, and within 3 months Windom had lost 23 staff members to their competitor, several in key management positions. Nontransfer losses for the same period amounted to seven persons, making the total number of turnovers in the given 3-month period four times higher than normal.

The loss of key personnel strained the Windom Hospital staff in the midst of all-time high patient census figures, and as a result some regular patient care activities were limited. The stress on remaining Windom staff of providing additional coverage caused serious morale problems. A number of concerns were voiced by remaining staff about the meaning of the staff defections and the implications for those assuming the larger responsibilities. It was clear to the human resources director that the problem had to be evaluated and responses to the massive changes implemented.

The human resources director, after consultation with staff and administration, decided to foment something positive from the transitions. First, supervisors polled the staff about whether they were considering a change in the near future. This served to identify those who were leaving or thinking of leaving, and had been offered a position with the new hospital. It also gave supervisors a chance to hear the employees' concerns about the transitions and about overwork in the period of high census. Finally, it identified persons wanting to stay with

Windom Hospital and desiring to advance within the organizational structure, which suggested some internal promotion possibilities. Searches could begin to fill positions for which vacancies were anticipated.

Step two, the decision to address the changes and their meanings and implications, was coordinated with the regular quarterly meetings of hospital employees. It was decided that the administration, which usually spoke to employees during these meetings, would directly acknowledge and address the loss of the physicians and other key personnel. It would also give the administration a chance to thank the personnel for their additional work in response to the need to cope with the holes in the staffing and the high census demands.

A third step, the awarding of small bonuses, was decided on as a tangible sign of the administration's appreciation for the extra work of the staff.

A fourth step was the decision to publicize the internal promotions, emphasizing that there was room for advancement for persons who chose to remain with Windom Hospital and that good work was recognized and rewarded. Considerable space in the hospital's newspaper was devoted to lists of internal promotions, the quarterly meeting was used to introduce present personnel taking new positions, and internal departmental memos were circulated to acknowledge and reinforce the changes.

A fifth step, to encourage persons anticipating leaving to depart immediately was decided on as a means of imparting stability to the hospital by bringing the attrition rate back to normal levels and allowing the human resources department to focus its attention on other matters of concern, such as the replacement of departing personnel.

When questioned about the effects of these changes, the human resources director reported improved work climate, more positive feelings, improved morale, and the discovery of new talents among personnel. Although the turmoil of the mass transitions was trying and demanding for the administration and the human resource departments alilke, the initial assessment of the short-range outcome was that the negative results and impact had been minimized by the proactive stance of seeking to make gains from losses.

Professionals traditionally have required and enjoyed autonomy in their relationship with many organizations, and this requirement can be met even when organizations place controls and demands on the professionals. One way is to have professionals serving on policy-making committees for the organization. Thus they are still exercising a degree of control and have not been subsumed by the external control of the organization and nonprofessionals.

Professionals often complain, with some justification, that committee work is inefficient and wastes their time. It is the responsibility of the administration to allocate committee assignments equitably and to run efficient meetings. Also, the answer to every problem or issue is not to establish a new committee or study group. Other solutions should be explored.

Professionals also have a need for interaction with their peers. The collegial environment the health organization establishes may affect the level of commitment the professional demonstrates. This environment can be affected by the type of medical staff or the hospital's relationship with a near-by medical school. Grand rounds, medical staff updates, and continuing education courses can also be used to increase the desirability of the work environment for professionals.

Professionals in organizations pose complex situations and conditions for management. Understanding, influencing, and affecting their behavior are major challenges in health services organizations. The section that follows presents the concept of motivation, which represents one component of professional behavior and attitude that may affect professionals' interaction with the organization.

MOTIVATION

Understanding why people behave the way they do is important for the effective management of human resources—including professionals—in a health services organization. Being able to affect or influence employee behavior is a necessary challenge for the manager, who is responsible for the most efficient allocation of the organization's resources. Given the labor intensiveness of an increasingly service-oriented society, ensuring the efficient application of human resources is a major part of the manager's duties. Clearly, motivating employees is a significant component of human resources management, and in order to understand the concept of motivation, managers must develop an understanding of human behavior in organizations.

Why Is Motivation Important?

One of the manager's most pressing and significant challenges is developing human resources—encouraging employees to be more productive, to be more useful members of the production team, to be more innovative in the solving of difficult problems, to be absent and tardy less frequently, and to be more creative in the discovery of new products or processes. The function of the manager as the leader of employees is discussed in Chapter 9. But before a manager can be an exceptional leader, he or she needs to possess an understanding of human behavior, which leads to an understanding of motivation.

In the sections that follow we discuss the basics of motivation and present some of the more prevalent theories of motivation, concluding with a few remarks on goal setting as a motivational device.

The Basics of Motivation

Many of the first principles of motivation were based on the thoughts of Greek hedonist philosophers. The primary tenet of hedonism when used to explain human behavior is that people will pursue the activities that bring them pleasure and avoid the actions that bring them pain. Thus hedonists may be considered people who are pleasure seekers and pain avoiders. Although the concept appears to make sense intuitively, it does assume that all individuals are motivated similarly and it does not account for individual differences in actions and motives.

Subsequent psychological approaches to motivation constitute variations on the theme of hedonism. The roles of instinct and unconscious motivation and how they affect human behavior and reactions were considered; the concepts of drive and reinforcement theories were examined; and finally cognitive theories (later known as expectancy theories) were developed (Steers and Porter, 1979). Some of these theories became too cumbersome to be workable and understandable, but they provide elements for the concepts described below, which are often applied in work, job, and managerial settings.

Motivation: Content and Process Theories

People studying motivation theories have traditionally divided them into two groups. The first, called content theories, examine the characteristics of individuals when considering what motivates them. Examples of content theories described below are the classical theory, the hierarchy of needs theory, the learned-need theory, and the two-factor theory. All four focus on what dimensions of individuals seem to affect motivational level.

Motivation theories in the second category, often called process theories, are concerned with how individuals are motivated. The example of process theory described in this chapter, namely expectancy theory, may be the theory used most frequently to explain motivation. It is important to note, however, that none of the theories that follow has been found to be the "best," or the "most right." Each contains components that assist and contribute to the understanding of human behavior, however.

Content Theories

Classical Theory. One of the earliest motivation theories is most often represented by the work of Frederick Taylor (1919). Management scholars whose ideas were later labeled "classical" believed that employees are motivated by economic needs and desires and will become more productive as a result of financial enticements. In the simplest application of this theory, employees were placed on a piecework system. The quality of the work environment was not considered in this model of motivation, nor were variations among individual needs. Only economic incentives were considered. The purity of this system was affected later with the establishment of conditions such as unattainable quotas. This approach appears to have its weaknesses, primarily in its simplistic assumptions of what motivates individuals to perform.

There are few examples of the application of classical motivation theory to health care settings. Because of the extreme variation in the amount of care a given patient requires, it would be meaningless to pay a nurse based on the number of patients he or she cared for. This theory is applicable only when the task is clearly defined and the amount of time required to complete it somewhat fixed (e.g., stocking inventory in a department of materials management. Classical theory fits few clinical settings.

Hierarchy of Needs. The *hierarchy of needs* theory of motivation was based on the belief that an individual's needs have a natural ordering that determines the importance to the individual of discrete needs. As developed by Abraham Maslow (1943), this theory aligns human needs on a multistage hierarchy, where lower level needs must be satisfied before higher level needs can be met or influenced. Maslow identified the basic and fundamental human needs as physiological needs such as food. Once physiological needs have been met, an individual can be concerned with the second level of needs: safety and protection. The third level involves social needs, which must be fulfilled before the fourth need, namely self-esteem, can be dealt with. The fifth and final human need is the achievement of self-fulfillment or self-actualization. Thus Maslow and others maintain that humans have a need to grow and to develop, and that need serves to motivate their actions when the first four have been met.

These five levels are not independent of each other or mutually exclusive, and they may recur. Also, once a need has been mostly met or fulfilled, it can no longer serve as a motivator. To use an extreme example to illustrate Maslow's theory of human behavior, a health services organization could not expect that offering free desserts to encourage or stimulate productivity would motivate employees whose basic level need for food had already been satisfied.

Learned-Need Theory. Other researchers have identified alternative or additional needs that serve as motivators. McClelland's learned-need theory suggests that people are motivated by needs for achievement, power, and affiliation. These needs are learned from the culture we live in, and they can be influenced by managers. If a need is strong in a person, the person will be motivated to try to satisfy it.

Individuals who have high achievement needs have a need to improve their performance. Power needs can be defined in terms of the importance of reputation and influence to the individual. Affiliation needs are associated with establishing and maintaining relationships with people and coworkers. Managers who are interested in learned-need theory want to know how they can influence an individual's need for achievement. Some research indicates that high achievers like setting their own goals, that they avoid setting unrealistic goals and prefer moderate, achievable goals, and that they like receiving feedback (McClelland and Burnham, 1976).

Two-Factor Theory. In the two-factor theory of motivation Herzberg et al. (1959), suggested that satisfaction has both a maintenance dimension (dissatisfiers) and a separate motivational dimension (satisfiers). Instead of evaluating job satisfaction on one continuum, Herzberg's research suggested that we think of one continuum representing items that satisfy employees and one representing items that dissatisfy employees. Examples of dissatisfiers as suggested by Herzberg include salary, job security, and interpersonal relations. Examples of satisfiers were achievement, recognition, growth potential, and responsibility. Of these two dimensions, only satisfiers would serve to motivate employees. There are researchers who believe that the causes of motivation are more complex than the two-factor theory indicates.

Process Theories

Expectancy Theory. Expectancy theories have been advocated by researchers who support the belief that individuals will perform better if they expect their performance to result in some positive, valued outcome or reward. Outcomes that are valued may have more effect than negative outcomes such as punishment (Lawler, 1973; Vroom, 1964). Figure 7.3 illustrates an expectancy motivation model, showing the relationship between effort, performance, outcome, and the intervening effect of expectancy in determining both performance and outcome.

According to these ideas, health services organizations can affect the behavior of individuals through organizational reward systems, depending on each employee's capabilities (Staw, 1984). It is uncertain how an individual's specific needs fit with expectancy theory, but it is believed that they play an indirect role—for example, in helping define what is a "valued" outcome for a particular employee. A nurse in the oncology unit of a hospital may respond to a reward system that provides free stress management courses for high performers, whereas a clerk-typist working in administration may value increased insurance benefits.

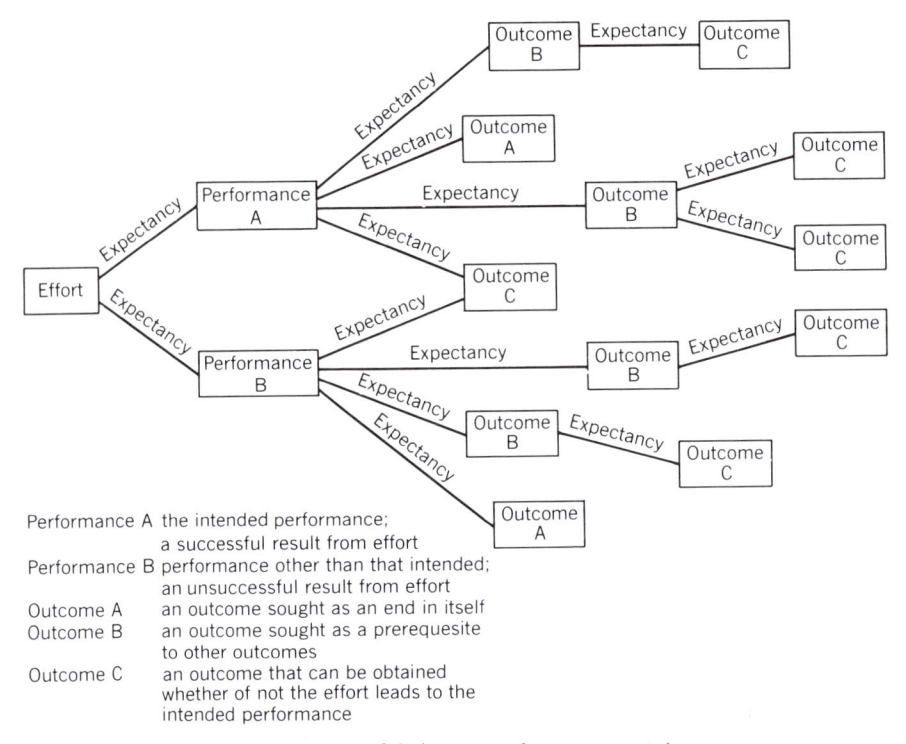

Figure 7.3. Expectancy motivation model. (From Lawler, 1973, p. 50.)

Performance A — the intended performance; a successful result from effort
Performance B — performance other than that intended; an unsuccessful result from effort
Outcome A — an outcome sought as an end in itself
Outcome B — an outcome sought as a prerequesite to other outcomes
Outcome C — an outcome that can be obtained whether of not the effort leads to the intended performance

Summary of Theories

From the brief descriptions above, it is apparent that there has been extensive research into the topic of motivation but that there is no single best motivational technique. Understanding the basic concepts of content and process theories should provide the manager with the foundation to adapt them to individual situations. People's behavior and motivational level at work are affected by a combination of their individual characteristics and perceptions, the specific dimensions of the job itself, and organizational characteristics.

It is important to acknowledge the role that *perception* assumes in understanding motivation. Attempts to motivate both professionals and employees of other types are affected by the employees' perception and interpretation of the motivational effort. For example, if an employee perceives that the organization is attempting to accomplish a certain organizational goal via its motivational devices, that perception — accurate or not — will affect his or her reaction and subsequent actions.

Motivation in Health Services Organizations

As stated above, the ability to impact employees' behavior is a function of the characteristics of the individual, the job, and the organizational environment. Health services organizations have dimensions that distinguish them from industrial organizations and consequently make the motivation of employees a somewhat different task for managers. One difference is that health services organizations usually do not have clearly defined goals. Is "health" a goal for a health services organization? Can we measure when someone is "well"?

Not only are the goals not very clear or measurable, the "product" is not well-defined. Are we attempting to encourage the proper utilization of a health facility, or are we trying to process people? The fact that our product involves "health" also compounds our measurement problem. "Health" is perceived by some people as a right, and health services employees are seen as providing a service citizens have a "right" to receive.

Furthermore purchasers of health services usually are using an agent (the physician) to make many of their decisions. This condition affects how the purchasers react to employees within the purchased environment.

The presence of extensive amounts of technology in most health care settings also differentiates this environment from organizations of other types. The existence of this technology, which requires many varied, highly skilled occupations to operate and maintain the physical equipment, often results in a great deal of interaction between the client (patient) and the employees. There is much more direct contact between client and operating personnel than is found in many traditional industrial organizations.

Finally, and perhaps most important, health care is dominated by professionals, many of whom possess the traditional characteristics of professionals addressed earlier in this chapter. The presence of clinically oriented professionals, especially physicians, is often in direct contrast to the management of health services organizations, with its bureaucratic orientation.

These conditions (lack of clearly defined product or goals, use of physician-agents, technology, and dominance by professionals) all contribute to the complexity of the health care organizational environment and affect the motivational levels of individuals working within that environment.

How does the administrator apply these theories in the everyday management of human resources?

According to Staw (1984), the issue in the future may not be how to motivate employees to be more productive, but how to motivate them to work smarter and to increase their focus on work- or task-related concerns, rather than personal, family, or nonwork concerns. Among organizations employing many professionals, such as health services organizations, inducing professionals to be more innovative and creative is the motivational focus.

One motivational technique that has received a great deal of attention in recent years is goal setting. The section that follows briefly discusses goal setting as a motivational device. Managers interested in this topic should read additional material.

Goal Setting

Most individuals and organizations engage in some type of goal setting: It may be explicit, such as the decision to attempt to increase by 20 percent the number of patients who indicate satisfaction with nursing care, room design, and food served, or it may be more broadly expressed, such as "improve the quality of care delivered in the health organization."

Encouraging employees to set performance goals appears to have positive results for the organization and for the employee. Much of the information presented above seems to suggest that individuals determine what is important to them and act on that determination. A health services organization that recognizes that many individuals are predisposed to determining goals will have the opportunity to influence employees' behavior.

Someone who has set a goal has made a value judgment about what is important to him or her and will take action based on the goal(s). Goals thus serve to motivate behavior. The health services manager must work with the employee and the organization in the setting of goals. The employee must be convinced that the environment encourages the setting of goals and will reward success accordingly. The manager and employee separately and together should determine the goals that are appropriate for the unit and for the individual employee. The goals should be ambitious but not unattainable. Simply stating that the employee should "do better" is not explicit enough. The goals should be clear to the employee and to the manager, so that there is no confusion about what is being stated or expected. And, there should be a follow-up assessment after the setting of goals, to see how much progress is being made and to catch any problems that may have arisen. Feedback of this nature appears to assure more positive results from the goal-setting activity (Kim, 1984).

Goal setting is not a magic solution to organizational problems or to poor performance. But it frequently does produce positive results for both the organization and the individual.

SUMMARY

In this chapter some of the factors that affect professionals working in health services organizations were examined. Various definitions of "professional" were presented, and the characteristics of professionals that differentiate them from other employees were discussed. We have considered the complexity of the health environment and the building of organizational commitment within that environment. The difficulties of fostering commitment in an organization that employs professionals were identified, as well as some of the characteristics of commited employees and professionals. Finally, some of the theories of motivation were discussed, and the example of goal setting as a motivational device was presented.

ACKNOWLEDGMENTS

The case studies entitled "Professionals at West End Hospital" and "Windom Hospital Responds to Changing Commitment Levels" are based on materials from unpublished class papers by Virginia Maddock and Alan B. Bell, respectively, submitted to the author in April 1987.

REFERENCES

Arnold, H., D. Feldman, and M. Purbhoo. "The role of social-desirability response bias in turnover research." *Academy of Management Journal*, 28:955–966 (1985).

Engel, G. V. "The effect of bureaucracy on the professional autonomy of the physician." *Journal of Health and Social Behavior,* 10:30–41 (1969).

Goss, M. E. W. "Patterns of bureaucracy among hospital staff physicians." In E. Freidson, Ed., *The Hospital in Modern Society.* New York: Free Press, 1963.

Herzberg, F., B. Mausner, and B. Synderman. *The Motivation to Work.* New York: Wiley, 1959.

Kim, J. "Effect of behavior plus outcome goal setting and feedback on employee satisfaction and performance." *Academy of Management Journal,* 27:139–149 (1984).

Lawler, E. E., III. *Motivation in Work Organizations.* Pacific Grove, CA: Brooks/Cole, 1973.

Maslow, A. H. "A theory of human motivation." *Psychological Review,* pp. 370–396 (July 1943).

McClelland, D., and D. Burnham. "Power is the great motivator." *Harvard Business Review,* pp. 100–111 (March 1976).

Mercer, A., S. R. Hernandez, and K. Bilson. "Factors influencing organizational commitment by physicians." In R. Robinson and J. Pearce, Eds., *Academy of Management Proceedings 1985.* San Diego, CA: Academy of Management, 1985.

Mowday, R., L. Porter, and R. Steers. *Employee-Organization Linkages: The Psychology of Commitment, Absenteeism, and Turnover.* New York: Academic Press, 1982.

Nicholson, N., and G. Johns. "The absence culture and the psychological contract—Who's in control of absence?" *Academy of Management Review,* 10:397–407 (1985).

Porter, L. W., R. M. Steers, R. T. Mowday, and P. Boulian. "Organizational commitment, job satisfaction, and turnover among psychiatric technicians." *Journals of Applied Psychology,* 59:603–609 (1974).

Raelin, J. "An examination of deviant/adaptive behaviors in the organizational careers of professionals." *Academy of Management Review,* 9:413–427 (1984).

Scott, W. "Professionals in bureaucracies—Areas of conflict." In H. M. Vollmer and D. L. Mills, Eds., *Professionalization.* Englewood Cliffs, NJ: Prentice-Hall, 1966.

Scott, W. "Managing professional work: Three models of control for health organizations." *Health Services Research,* 17:213–240 (1982).

Scott, W. "Conflicting levels of rationality: Regulators, managers, and professionals in the medical care sector." *Journal of Health Administration Education,* 3:113–131 (1985).

Staw, B. M. "Organizational behavior." In M. R. Rosenzweig and L. W. Porter, Eds., *Annual Review of Psychology,* Vol. 35. Greenwich, CT: JAI Press, 1984, pp. 627–666.

Steers, R. M., and L. W. Porter. *Motivation and Work Behavior.* New York: McGraw-Hill, 1979.

Taylor, F. W. *Scientific Management.* New York: Harper & Row, 1919.

Vroom, V. H. *Work and Motivation.* New York: Wiley, 1964.

Wall Street Journal. "Medical benefits." July 9, 1986, p. 20.

Wilensky, H. L. "The professionalization of everyone?" *American Journal of Sociology,* 70(2):138–158 (1964).

CHAPTER 8

Management of Corporate Culture

Ralph H. Kilmann

The importance of strategy in determining future direction for a health services organization and the relationship of strategy to human resources management has been illustrated in preceding chapters. In addition, the relationship of structural design features of the institution to major personnel functions was discussed. These factors, and their interrelations, make valuable contributions to the performance of organizations.

The likelihood that a health services organization will achieve success in today's dynamic and complex environment is not determined by the skills of its leaders alone, nor by the strategy, structure, and reward systems that make up its visible features. Rather, the organization itself has an invisible quality—a certain style, a character, a way of doing things—that may be more powerful than the dictates of any one person or any formally documented system. To understand the essence or soul of the organization requires that we travel below the charts, patient treatment protocols, rulebooks, diagnostic and therapeutic technology with sophisticated machines, and buildings into the underground world of corporate cultures.

What exactly is culture? Nobody knows for sure, nor will there ever be a clear definition that meets with everyone's approval. The topic generates multiple meanings. William B. Renner (1981) vice chairman of the Aluminum Company of America, highlights the dilemma of defining culture:

> Culture is different things to different people. For some, it's family, or religion. It's opera or Shakespeare, a few clay pots at a Roman dig. Every textbook offers a definition, but I like a simple one: culture is the shared values and behavior that knit a community together. It's

---

Adapted from Ralph H. Kilmann, *Beyond the Quick Fix: Managing Five Tracks to Organizational Success.* San Francisco: Jossey-Bass, 1984.

the rules of the game; the unseen meaning between the lines in the rulebook that assures unity. All organizations have a culture of their own. (P. 1)

Most definitions of culture disagree only on *what* is shared among the members of an organization. Is it rules, norms, beliefs, expectations, values, philosophies, or all these things? For most purposes, these intangibles are so interconnected that it makes little sense to argue about how each is similar to or different from the others. However, it is worthwhile to learn how these intangibles become shared among the members of any group, and why this creates such a powerful force that guides behavior (Kilmann et al., 1985). Thus, the most exciting thing about culture is discovering how it first captures and then directs the collective will of the membership.

Culture provides meaning, direction, and mobilization—a social energy that moves the health services organization into action. One has to experience the energy that flows from shared commitments among group members to know it: the energy that emanates from mutual influence, "one for all and all for one," and "esprit de corps." Can management tap this source of energy for organizational success and create a sense of unity among the diverse groups working in institutions delivering health services, or will the energy remain immobilized? Or, worse yet, will this social energy be working against the mission of the organization?

In numerous organizations that provide health care this social energy has been diffused in all directions or even deactivated. It is not mobilized toward anything; indeed, it has barely been tapped. Most members seem apathetic or depressed about their jobs. They no longer pressure one another to do well. There is inadequate concern about quality of care being delivered. Pronouncements that top management itself will improve the situation fall on deaf ears. The members have heard these promises before. Nothing seems to matter. The soul of the organization is slowly dying.

In other cases, while the energy is alive and flourishing, it moves members in the wrong direction. The organization lives in an immense culture lag or *culture gap*—the social energy pressures health services providers to persist in behaviors that may have worked well in the past but clearly are dysfunctional today. Professional behavior of physicians may be influenced, in part, by the old norms that demanded the provision of services at any cost rather than acknowledging the reality of the health services industry today, namely that resources are limited.

The gap between the outdated culture and what is needed for success gradually develops into a *culture rut*—a habitual way of behaving without asking any questions. There is no adaptation or change; only routine motions are enacted again and again even though success is not forthcoming. Here the social energy not only works against the organization but is contrary to the private wishes of the health providers. Nobody wants to be ineffective and dissatisfied, but people pressure one another to comply with the unstated, below-the-surface, behind-the-scenes, invisible culture. This rut state can go on for years, even though morale and performance suffer. Bad habits die hard.

Culture shock occurs when the sleeping organization awakes and finds that it has lost touch with its mission, its setting, and its assumptions. Many health services organizations are having this experience. The former "soft" life that existed with cost-based reimbursement, limited competition, and numerous entry barriers established in a regulatory environment has been replaced by situations that demand rapid responses and tough decision making.

Today's world has left the insulated provider institutions behind—a Rip Van Winkle story on a grand scale. Rather than experience this shock, the hosital or medical group may decide not to wake up; the health services executives simply continue to believe the myth of erroneous extrapolation: What made the organization successful in the past will make it successful in the future.

The first part of this chapter explores three interrelated questions:

What are adaptive cultures—in contrast to dysfunctional cultures?
How do cultures form—what brings them into being?
How are cultures maintained—what forces keep cultures intact?

Understanding the answers to these questions is necessary for assessing and changing cultures in health service organizations. In these discussions, we will see how cultural norms provide the leverage points for creating and maintaining adaptive cultures more directly than any of the other manifestations of an organization's way of doing things.

The second part of this chapter presents five steps for managing culture that can be used by hospital systems, medical groups, or other provider organizations: surfacing actual norms, articulating new directions, establishing new norms, identifying culture gaps, and closing culture gaps. These five steps show how the organization can gain control over its culture rather than vice versa. The health professionals can decide what new norms are needed for today's complex problems and then can proceed to energize their work groups toward the new directions they envision—thereby closing the corporate culture gap.

The value of focusing attention on the management of corporate culture in today's health services industry cannot be overstressed. Organizations are making strategic shifts to take market share from competing hospitals. Acute care facilities and medical groups vie for the right to obtain a scarce certificate-of-need for a piece of equipment that offers hope for improved treatment for some patients—and higher profitability. The values associated with cooperative behavior that existed in the days of limited interest in cost control and in attention to patient/market issues are not productive in today's environment. Health services organizations may lay the foundation for modifying their approach to the marketplace through strategic planning and structural design changes; if the organization's employees are to support the change, however, the culture of the organization must be handled properly. Thus, proper attention to the corporate culture can facilitate the accomplishment of desired objectives in strategic and human resource management.

WHAT ARE ADAPTIVE CULTURES?

Even if we accept the idea that the term "culture" will always be a bit vague and ill defined, unlike the more tangible aspects of organizations, it is still important to consider what makes a culture good or bad, adaptive or dysfunctional. Wallach (1983) provides a summary of what cultures do for or against the organization:

> There are no good or bad cultures, per se. A culture is good—effective—if it reinforces the mission, purposes, and strategies of the organization. It can be an asset or a liability. Strong cultural norms make an organization efficient. Everyone knows what's important and how things are done. To be effective, the culture must not only be efficient, but appropriate to the needs of the business, the company, and the employees. (P. 32)

Why does one health services organization have a very adaptive culture while another lives in the past? Is one a case of good fortune and the other a result of bad luck? On the contrary, it seems that any organization can find itself outdated if its culture is not managed explicitly.

A culture does, if left to itself, become dysfunctional. Human fear, insecurity, oversensitivity, dependency, and paranoia seem to take over unless a concerted effort to establish an adaptive culture is undertaken. Everyone has been hurt at one time or another, particularly in childhood. It is, therefore, rather easy to scare people into worrying about what pain or hurt will be inflicted in the future, even in a relatively nonthreatening situation. As a result, people cope by protecting themselves, by being cautious, by minimizing their risks, by going along with a culture that builds protective barriers around work units and around the whole organization.

A dysfunctional culture also helps explain some of the self-defeating behaviors that have been observed in many organizations. Such behaviors include doing the minimum to get by, resisting or even sabotaging innovation, and being very negative in general about the organization's capacity to change. Worse yet are such behaviors as lying, cheating, and stealing, and intimidating, harassing, and hurting others. These behaviors may seem unthinkable, but they do receive cultural support even though they cause difficult problems for the organization. They also significantly undermine, to say the least, both morale and performance.

The most detrimental behavior in the long run, however, is continuing to see and act out what made the organization successful in the past rather than adapting to the dynamic complexity of today and tomorrow. For example, the challenge facing Miami Valley Hospital in Dayton, Ohio, was to get out of the culture rut created by the trap of erroneous extrapolation. The hospital had been successful under strong medical leadership from the 1940s to the mid-1970s. However, it had developed a fortress mentality and believed that it could continue to focus on internal operations and on continued leadership in the community by merely strengthening the medical staff. As noted by one board member:

> When I came here, I remained silent for a while, but I was appalled at the posture of the hospital. Those who came before me built a fine institution, but they gradually developed tunnel vision. Their vision stopped at the edge of the campus. (Peters and Tseng, 1983, p. 206)

An *adaptive* culture, alternatively, entails a risk taking, trusting, and proactive approach to organizational as well as individual life. Members actively support one another's efforts to identify all problems and to implement workable solutions. There is a shared feeling of confidence: The members believe, without a doubt, that they can effectively manage whatever new problems and opportunities will come their way. There is widespread enthusiasm, a spirit of doing whatever it takes to achieve organizational success: The members are receptive to change and innovation.

HOW DO CULTURES FORM?

A hospital culture seems to form rather quickly based on the organization's mission, its setting, and what is required for success: efficiency, product reliability, quality of services provided, advanced medical technology, innovation, hard work, and/or loyalty. Generally speaking, when the organization is born, a tremendous energy is released as health professionals struggle to make it work. The culture captures everyone's drive and imagination.

As the reward systems, policies, procedures, and rules governing work are formally documented, they have a more specific impact on shaping the initial culture; however, they cannot compete with either the bold or even the more subtle actions of key individuals. For example, the objectives, principles, values, and especially the behavior of the founder of the organization provide important clues to what is *really* wanted from all members both now and in the future. Carrying on in the traditions of the founder, other top executives affect the culture of the organization by their every example.

Thomas Frist, founder of Hospital Corporation of America (HCA), is an example to his employees. Having the initiative and taking the risks necessary to create the world's largest investor-owned hospital system, he sold the strategy of building a large base of profitable hospitals without compromising on quality of care or concern for individuals. He maintains contact with individual hospitals through several methods. Photographs of the HCA hospital administrators are in his office. When Frist contacts an administrator over the phone, he is able to look directly at the photo and thereby improve his feel for having personal contact with the individual.

Employees also take note of all critical incidents that stem from *any* management action—such as the time that so-and-so was reprimanded for doing a good job just because he was not asked to do it beforehand, or the case of someone else, who was fired because he publicly disagreed with the hospital's position. Incidents such as these become the folklore that people remember, indicating what the health care corporation really wants, what really counts in getting ahead, or, alternatively, how to stay out of trouble—the unwritten rules of the game. Work games adopt these lessons as norms on how to survive and make it, how to protect oneself from the system, and how to retaliate against the organization for its "sins of the past."

As a culture forms around a recognized need, the setting, and specific task requirements, it may be very functional at first. But, over time, the culture becomes a separate entity, independent of the initial reasons and incidents that formed it. The culture becomes distinct from the formal strategy, structure, and reward systems of the organization. As long as it is supportive of and in harmony with these formally documented systems, the culture remains in the background.

In a similar vein, culture becomes distinct from the membership. All members throughout the health services organization are taught to follow the cultural norms without questioning them. By the time employees have been around for a few years, they have learned the ropes. Even a new top executive who vows that things will be different finds out—often the hard way—how the culture is bigger and more powerful than he is. Single-handed efforts by the executive to counter the "invisible (social) hand" are met with constant frustration. For example, a senior health care executive can call his subordinates into his office, individually, and get verbal commitments for some new policy or plan; however, when each person leaves the office and again becomes part of the corporate culture, the executive may find that the new plan is bitterly opposed.

Running up against the culture becomes even more apparent when management attempts to make a major strategic shift or tries to adopt entirely new work methods: The culture may not support the intended changes. Now the energy and the separateness of the culture quickly become evident. The intangible culture is revealed when management cannot pinpoint the source of apathy, resistance, or rebellion. Management is puzzled as to why the new work methods are not embraced automatically and enthusiastically by the members. To management, it is obvious that these proposed changes are necessary and desirable. Why cannot everyone else see this?

At the same time, top management is also caught in the grip of the organization's separate and distinct culture. Health professionals from below wonder why managers play it so safe, why they refuse to approach things differently, why they keep applying the same old management practices even those that simply do not work. Employees wonder why management is so blind to the world around them. They wonder if management is "mean" or just stupid.

HOW ARE CULTURES MAINTAINED?

The force controlling behavior at every level in the hospital—a force that can brainwash members into believing that what they are doing is automatically good for their organization, their community, and their families—must be very powerful indeed. That such dysfunctional and self-defeating behaviors can persist for years again suggests some powerful force at work. Is it magic, or is it the psychology of what most people will do to be a member of a group? We must understand the invisible force if we wish to control it rather than allowing it to control us. A deeper knowledge of norms and how they are enforced is essential.

Social scientists speak of norms as the unwritten rules of behavior. In a company, for example, a norm might be: Don't disagree with your boss in public. These norms are very crystallized when a strong consensus exists among a group of people concerning what constitutes appropriate behavior. If a norm is violated—if someone behaves contrary to the norm—there are immediate and strong pressures to elicit a behavior change. Consider, for example an employee who persists in presenting personal reservations about the hospital's new line of services at a group meeting—just after the boss has argued strongly for investing heavily in its marketing campaign. The person is stared at, frowned at, looked at with rolling eyes, and given other nonverbal messages to sit down and shut up. If these efforts are insufficient, the dissenter will hear about it later, from coworkers or from the boss.

The need of everyone to be accepted by a group—whether family, friends, coworkers, or the neighborhood—gives each such group leverage to demand compliance to its norms. If people did not care about acceptance at all, a group would have little hold, other than formal sanctions, over individuals. The nonconformists and the mavericks who defy pressures to adhere to group norms always do so at a considerable price.

A simple experiment conducted by Asch (1955) demonstrates just how powerfully the group can influence its deviants. The experiment was presented to subjects as a study in perception. Three lines—A, B, and C—all of different lengths, were shown on a single card. Subjects were asked to indicate which of these three lines was identical in length to a fourth line, D, shown on a second card. Seven persons sat in a row. One by one they indicated their choices. Although line D was in fact identical to line C, each of the first six persons, confederates of the experimenter, said that line D was identical to line A. As each person gave the agreed-upon fraudulent response, the unknowing subject became increasingly uneasy, anxious, and doubtful of his or her own perceptions. In fact, the subject (the seventh person) agreed with the rest of the group almost 40 percent of the time. The error rate in choosing the wrong line with no one else present was less than 1 percent. This showed quite a difference in behavior!

In this experiment, which has been replicated many times, there was no opportunity for the seven persons to discuss the problem. If there had been such an opportunity, the effect would have been stronger because the six would have attempted to influence the seventh member. It is not easy being a deviant in a group when everyone is against you. People need acceptance from others so much that they often will deny their own perceptions when confronted with the group's norms of "objective" reality. Objective reality becomes a *social* reality.

Imagine just how easily such socially defined and distorted perceptions of reality can be maintained when backed up by formal sanctions—pay, promotions, favorable work shifts, and other rewards. The group or the entire organization can reward its members so that they ignore not only the changes taking place in the health industry but also the disruptive behaviors of various troublemakers inside the organization. The members collectively believe that everything is fine, and they continue to reinforce this myth and reward one another for maintaining it. In essence, everyone agrees that the dysfunctional ways can continue without question. Anyone who thinks otherwise is treated as a deviant: severely punished and eventually banished from the tribe.

ASSESSING AND CHANGING CULTURAL NORMS

While culture manifests itself through shared values, beliefs, expectations, and assumptions, it is most controllable through norms: the unwritten rules of game. Even norms that dictate how one should behave, the opinions one should state, and one's facial expressions can be surfaced, discussed, and altered:

> Norms are a universal phenomenon. They are necessary, tenacious, but also extremely malleable. Because they can change so quickly and easily, they present a tremendous opportunity to people interested in change. Any group, no matter its size, once it understands itself as a cultural entity, can plan its own norms, creating positive ones that will help it reach its goals and modifying or discarding the negative ones. (Allen and Kraft, 1982, pp. 7–8)

A good way to assess a health services organization's culture is to ask members to write out what previously was unwritten. I have done this many times with a variety of organizations. Members are willing and able to write out their norms under certain conditions: (a) that no member will be identified as having stated or suggested a particular norm (individual confidentiality), and (b) that no norm will be documented when one's superiors are present (candid openness). Furthermore, the members must have faith that the norm list will not be used against them but will instead be used to benefit them as well as their organization. The consultants and managers who guide members to state norms, therefore, must generate trust and commitment throughout all five steps of the process.

Step 1: Surfacing Actual Norms

The first step is for all group members (generally in a workshop setting) to list the actual norms that currently guide their behaviors and attitudes. This can be done for just one group or for many groups, departments, and divisions, depending on how many people can be included and managed in one setting. Sometimes it takes a little prodding and a few illustrations to get the process started, but once it begins members are quick to suggest many norms. In fact, they seem to delight in being able to articulate what never had been stated in document and rarely had been mentioned in conversation.

For a health services organization whose culture is dysfunctional, some of the norms people list are: Don't disagree with your boss; don't rock the boat; treat women as second-class citizens; put down your hospital; don't enjoy your work; don't share information with other groups; treat subordinates as incompetent and lazy; provide the minimal attention possible to patients; look busy even when you are not; don't reward health professionals on the basis of merit; laugh at those who suggest new ways of doing things; don't smile much; openly criticize hospital policies to outsiders; complain a lot; don't trust anyone who seems sincere. Ironically, this list could not be developed without violating one norm. Don't make norms explicit!

As these norms are listed for everyone to see, there is considerable laughter and amazement. The members become aware that they have been seducing one another into abiding by these counterproductive directives. But no one made a conscious individual choice to behave this way; rather, as the employees entered the organization, each was taught what behavior was expected—often in quite subtle ways. The more cohesive the group, the more rapidly such learning takes place and the more strongly the sanctions are applied. In the extreme case, a highly cohesive group that has been around a long time has members that look, act, think, and talk alike.

Step 2: Articulating New Directions

The next step is for all group members to discuss where the organization is headed and what type of behavior is necessary to move forward. Even when a health service has a very dysfunctional culture from the past, members, as individuals, are aware of what changes are needed for organizational success. Similarly, members are aware of what work environment they prefer for their own sanity and satisfaction.

A certain amount of planning and problem solving, however, may help to articulate the new directions. In work groups that have been in a culture rut, members are so absorbed with the negatives that they have not spent much time thinking about or discussing what they desire. Sometimes it is useful to ask them to reflect on their ideal organization: If they could design their own organization from scratch, what would it be like? This generally brings out what could be different in the present organization and what should not be accepted just because it has been that way for a long time.

Step 3: Establishing New Norms

As the third step, all group members develop a list of new norms for organizational success. At this point, the members usually catch on to the impact the unwritten rules have had on their behavior. They experience a sense of relief as a new way of life is considered. They realize that they no longer have to pressure one another to behave in dysfunctional ways. The members can create a new social order within their own work groups and within their own organization. Part of this sense of relief comes from recognizing that their dissatisfactions and ineffectiveness are not due to their being incompetent or bad. It is much easier, psychologically, for members to blame the invisible force called *culture*—as long as they take responsibility for changing it.

In organizations needing to be more adaptive to modern times, some of the norms that are often listed are: Treat everyone with respect and as a potential source of valuable insights and expertise; be willing to take on responsibility; initiate changes to improve performance; congratulate those who suggest new ideas and new ways of doing things; be cost conscious, so that the organization remains efficient relative to its competitors; speak with pride about your organization and work group; budget your time according to the importance of tasks for accomplishing objectives; don't criticize the organization in front of patients or customers; enjoy your work and show your enthusiasm for a job well done; be helpful and supportive of the other professional groups in the organization.

Step 4: Identifying Culture Gaps

The contrast between these desired norms (step 3) and the actual norms (step 1) can be immense. My colleague, Mary J. Saxton, and I refer to this contrast as a culture gap. We have developed a measurement tool for detecting the gap between what the current culture is and what it should be: the Kilmann-Saxton Culture-Gap Survey (Kilmann and Saxton, 1983).

The survey was developed by collecting more than 400 norms from employees in more than 25 different types of organization. Many of these norms were also developed through culture change projects. The items included in the final set of 28 norm pairs were derived from statistical analysis of the most consistent norms that were operating in most of the organizations we studied. An example of a norm pair is: (A) Share information only when it benefits your own work group, versus (B) Share information to help the organization make better decisions. Each employee chooses A or B for each norm pair in two ways: according to the pressures the work group puts on its members (actual norms), and according to which norms should be operating to promote high performance and morale (desired norms).

The differences between the actual norms and the desired norms represent the culture gaps. There are four types of culture gap, each made up of seven norm pairs:

1. *Task support.* Norms having to do with information sharing, helping other groups, and concern with efficiency, such as "Support the work of other groups" versus "Put down the work of other groups."
2. *Task innovation.* Norms for being creative, being rewarded for creativity, and doing new things, such as "Always try to improve" versus "Don't rock the boat."
3. *Social relationships.* Norms for socializing with your work group and mixing friendships with business, such as "Get to know the people in your work group" versus "Don't bother."

4. *Personal freedom.* Norms for self-expression, exercising discretion, and pleasing yourself, such as "Live for yourself and your family" versus "Live for your job and career."

When used in numerous profit and nonprofit organizations, our survey has revealed distinct patterns of culture gaps. For example, in some of the high-technology firms, lack of cooperation and information sharing across groups has resulted in large culture gaps in the area of task support. In the automotive and steel industries, failure to reward creativity and innovation has resulted in large culture gaps in task innovation. In some social service agencies in which work loads can vary greatly, large negative gaps in social relationships are found, indicating that too much time is spent socializing, as opposed to attempting to get the next job done. Finally, in extremely bureaucratic organizations, such as some banks and government agencies, large gaps in personal freedom are evident. Here members feel overly confined and constrained, which affects their performance and morale.

The most general finding to date is the presence of large culture gaps with respect to task innovation. It seems that American industry is plagued by significant differences between actual and desired norms in this area. This is consistent with all the attention that has been given to the productivity problem in the United States. An industrial culture that pushes short-term financial results is bound to foster norms against efforts at long-term work improvements, regardless of what the formal documents and publicity statements seem to advocate.

Do all members in the same organization see the same culture gaps? Apparently not. The smallest culture gaps are found at the top of an organization's hierarchy. Health services executives believe their own publicity; they say that they reward creativity and innovation but seem to forget that actions speak louder than words. The culture gaps at the executive level are small. It is at the bottom of the hierarchy, where the gaps reveal alienation and distrust, that such discrepancies are largest. Here the work groups can explain what is meant by the norm: Don't trust management. In essence, work groups see management as being up to no good, getting caught up in fads to fool and manipulate employees, or thinking the workers are too stupid to see what is behind management's latest whim.

This sort of grass-roots culture not only describes a culture rut but suggests why developing an adaptive culture must precede any other effort at change and improvement. Without an adaptive culture, every action by top management will be discounted—ignored—by the groups below. Ironically, even top-down efforts to change the future will be unsuccessful: I have seen executives try to dictate a new culture by dramatic changes in their own behavior coupled with symbolic deeds and fiery speeches, but to no avail. Only when work group members encourage one another to be receptive to overtures by the other groups—as a result of this five-step process for closing culture gaps—will the necessary change in culture take place. For example, various work groups would include such new norms as give management another chance, and assume good intentions. Managers and consultants, therefore, have to work especially hard to encourage work groups, including the executive groups, to meet one another half way.

Just as the *size* of culture gaps can vary according to the shape of the organization pyramid, the *type* of culture gap can differ from one health professional group to another in the same organization. Professional groups have different histories, educational backgrounds, professional concerns, modes of practice, and status levels. As a result, the groups in a large, complex health services organization may require very different cultures for achieving their respective brands of effectiveness.

Large, multiunit health systems also may have different cultures within separate divisions. This possibility is illustrated by the Sisters of Mercy Health Corporation, which noted this diversity:

> Each division operates in a unique community environment shaped by different needs, expectations, talents, and resources.
>
> The development of governance and management capabilities in the division is uneven, as is the availability of human resources.
>
> Each division has developed a different combination of role and goals responsive to its local situation. (IGMP: Integrated Governance and Management Process, 1980, p 7)

Step 5: Closing Culture Gaps

How can the culture gaps be closed? How can a health services organization move its culture from the actual to the desired? Can a hospital be taken out of a culture rut and be put back on track for solving present and future problems? Will the organization survive this culture shock?

When the current culture is at least hopeful, it is almost miraculous what an impact the survey results or the lists of desired norms have on the members of a work unit. As mentioned before, there is often a great sense of relief as people become aware that they can live according to different norms and that they have the power to change. Surprisingly, some change from the actual to the desired norms can take place just by listing the new set of norms. Members start "playing out" the new norms immediately after they are discussed.

When the current culture is cynical, depressed, and very much in a rut, the response to the survey results is quite different. Even when large gaps are shown or when a listing demonstrates the tremendous differences between actual and desired norms, the members seem apathetic and lifeless. Members respond by saying that their work units cannot change for the better until the level of management above them and the rest of the company change. Members believe that it is the external system that is keeping them down.

Curiously, when I do a culture-gap survey at the next highest level, the very same arguments are heard again: "We have no power to change; we have to wait for the next level to let us change; *they* have the power!" It is shocking, after conducting the culture-gap survey for an entire organization, to present the results to the top management group only to find the same feelings of helplessness. Here top management is waiting for the economy to change. In actuality, it is the corporate culture that is saying: Don't take on responsibility; protect yourself at all costs; don't try to change until everyone else has changed; don't lead the way, follow; if you ignore the problem, maybe it will go away.

This is the perfect example of an organization in a culture rut. When the shock of realizing the discrepancy between actual and desired norms is just too great to confront, the organization buries its head in the sand and hopes everything will be sorted out by itself. Even in the face of strong evidence of a serious problem, time and time again I have witnessed this form of organizational denial—a much more powerful and perhaps more destructive force than any case of individual denial. The group's power to define reality clouds each person's better judgment. The dysfunctional culture "wins" again.

One large industrial organization asked me to present a 3-day seminar to the top executive group (the chairman of the board and chief executive officer, and the 10 corporate officers) on the topic of corporate culture. I suggested that a representative survey of culture gaps be conducted across all divisions in the company. In this way, I could report on the culture of the organization specifically, and this information could be expected to generate an interesting and lively discussion. In a couple of weeks, the vice president of human resources gave me his response: "No, we'd better not do this. I don't think the executive group really wants to know what is going on in the company. Besides, we can't take the chance of surprising them with your survey results." Who was protecting whom?

At another meeting with a major company on the same topic, I shared the foregoing anecdote as an example of the industrial culture problem in America without, of course, mentioning any names. The response to my story was: "That must be *our* company you're talking about!" It was not the same company, but the message was still the same.

A major lesson to learn from changing corporate cultures, especially very dysfunctional and depressed cultures—those in a culture rut—is that people do not have to feel powerless and inept. If managers and members decide that taking on responsibility for change and feeling the power to change should be part of the new culture, it can be so. Power and control are social phenomena, not objective, physical realities. Many times individuals and organizations have moved forward and achieved great success when everyone else "knew" it was impossible.

Merely listing and stating the new norms, however, is not enough to instill these throughout the organization. Also, norms cannot be altered just by requesting a norm change. Members have to develop agreements that the new norms will indeed replace the old norms and that this change will be monitored and sanctioned by the work groups. They must reinforce one another for enacting the new behaviors and attitudes and confront one another when the dysfunctional norms creep back into the work group. Consider the new norm: Congratulate those who suggest new ideas and new ways of doing things. If any member notices that a coworker frowns when some new product idea is suggested, the coworker should be given suitable stares and reminders of the new norms. An even stronger expression might be: "We thought you were part of the team and had agreed to make the switch. What's your problem?"

In another major organization undergoing a change program, I suggested that each new norm be written on an index card and given a number. Each member in a work group was then responsible for monitoring several norms and bringing attention to behavior that did not conform. Eventually, group members no longer cited the norms—only the numbers. Coworkers would state "You just committed a number 12," or "You pulled a 7 on me." These members found it very effective to enforce their new norms in this lighthearted manner, yet the point of adopting the norms was made unequivocally. Of course, outsiders who heard such interchanges certainly were confused; but this seemed to add to the group's cohesiveness, since the members now had their own secret code.

SUMMARY

Any health services organization that determines the extent of its culture gaps using the aforementioned steps is in a position to chart the directions for a culture change. Conducting sessions for each division, department, and professional group, including the ways in which the new norms will be monitored and enforced, will begin the change process. However, any new culture will gradually return to the old dysfunctional one if the culture change is not supported by all the formally documented systems (strategy, structure, and reward systems) and by top management behavior (Kilmann, 1984). The right leadership styles—such as risk taking, openness, and flexibility—must be enacted in each work unit to support an adaptive culture. Job descriptions must specify information sharing and cooperation, both within and between all work units. Also, the reward system must encourage members to enact the new cultural norms and to suppress the old unwritten rules. In these ways, the social energy and the formal system will be working together in an orchestrated manner—precisely what is required for organizational success in today's competitive health services industry.

REFERENCES

Allen, R. F., and C. Kraft. *The Organizational Unconscious: How to Create the Corporate Culture You Want and Need.* Englewood Cliffs, NJ: Prentice-Hall, 1982.

Asch, S. E. "Opinions and social pressure." *Scientific American,* pp. 31–34 (November 1955).

IGMP: Integrated Governance and Management Process. Farmington Hills, MI: Sisters of Mercy Health Corporation, 1980.

Kilmann, R. H. *Beyond the Quick Fix: Managing Five Tracks to Organizational Success.* San Francisco: Jossey-Bass, 1984.

Kilmann, R. H., and M. J. Saxton. *The Kilmann-Saxton Culture-Gap Survey.* Pittsburgh: Organizational Design Consultants, 1983.

Kilmann, R. H., M. J. Saxton, R. Serpa, and associates. *Gaining Control of the Corporate Culture.* San Francisco: Jossey-Bass, 1985.

Peters, J. P., and S. Tseng. *Managing Strategic Change in Hospitals: Ten Success Stories.* Chicago: American Hospital Publishing, 1983.

Renner, W. B. "The new corporate culture." Department of Public Relations and Advertising, Aluminum Company of America, Pittsburgh, 1981, pp. 1–4.

Wallach, E. J. "Individuals and organizations: The cultural match." *Training and Development Journal,* pp. 29–36 (February 1983).

CHAPTER 9

Organizational Change, Transformational Leadership, and Leadership Development

Richard Kurz
S. Robert Hernandez
Cynthia Carter Haddock

Given the rapid shifts occurring in the health services industry, a clear need exists for innovation and change within organizations providing health services. This need for innovation includes not just the development of new services or technologically sophisticated machinery for diagnosis and treatment of patient conditions, but also new social and organizational approaches that improve the manner in which the activities of the organization are conducted.

Thus, innovation and organizational change refer not only to medical advances in patient care, but also to the processes for bringing new problem-solving ideas into use by the institution. Concepts for corporate restructuring, methods for cutting service delivery costs, installation of budgeting and cost identification systems that complement hospital reimbursement under diagnosis-related groups, and improved hospital information systems are all examples of the types of innovation required for the continued viability of health services organizations.

The identification and implementation of new management systems, technological advancements, and strategic adaptations constitute a major concern for those responsible for the strategic management of the institution. If an organization is to be successful, the change process itself must be handled well by those in leadership positions (Kanter, 1983; Quinn, 1980).

The contribution that senior leadership makes to organizational change also deserves the critical attention of human resources managers. Other chapters have noted relationships that exist between strategic, structural, and behavioral system changes and selected human resources functions. Thus, changes initiated in other elements of the personnel organization obviously affect components of the personnel system. Understanding the role of leadership behavior in fostering, or inhibiting, types of change will help human resources managers to facilitate modifications in the personnel functions that are required by the institution.

Additionally, the critical role of leadership in nurturing organizational change suggests that human resources managers should devise leadership development activities that cultivate the positive behaviors needed in future health services organization leaders. Designing activities for leadership growth will help improve the possibility that the desirable corporate culture described in Chapter 8 will be realized.

A detailed discussion of organizational change and innovation is provided elsewhere (Kaluzny and Hernandez, 1987), but a brief overview of the change process and types of change is presented in this chapter. Next, the role that types of leadership play in organizational change is discussed. This material is followed by suggested approaches for leadership development in health services organizations.

ORGANIZATIONAL CHANGE

The contribution to change of the attitudes and behaviors of organizational leaders can be understood by identifying their role in the process of organizational innovation and by classifying the types of change with which they must deal. This process of innovation is seen as composed of two major stages, initiation and implementation, with a number of substages (Zaltman et al., 1973).

The Innovation Process

In the initiation stage of innovation, organizational leaders obtain knowledge about a new idea or process, form attitudes about the possible use of this idea in the organization, and reach a decision. The initiation stage begins with the perceptions of leaders that a performance gap (Downs, 1966) exists between the organization's current level of performance and its capabilities. Thus, health services executives play a pivotal role in determining that institutional performance could be improved based on criteria they believe to be valid.

These executives then gather information on options that offer potential for improving organizational performance. Of course, information about an innovation that might be used by a health services organization may create the "gap" leading to change. For example, knowledge of the benefits in terms of both financial performance and improved patient services that are possible from the use of magnetic resonance imaging (MRI) may create the perception that current performance could be improved through the use of MRI technology.

After information on the innovation has been gathered, senior managers must evaluate these options and make a decision. Their attitudes concerning the features of the new activity and the potential benefits form the nucleus for the decision that is rendered.

Types of Change

Changes in health care have affected all types of delivery system and all aspects of management. Their impact on the organization has ranged from modest departmental restructuring to a substantial reversal of incentives for managers and physicians. If health services executives are to respond to changes, they must be able to categorize them, as a means to a better understanding of their significance for the organization.

Organizational change may occur in the goals an organization wishes to accomplish, the means through which it attempts to achieve them, or both (Kaluzny and Hernandez, 1987). Changes may modify an organization as technical changes (alternations of means but not goals), as transition (alterations of goals but not means), and as transformation (alterations of both means and goals). Moreover, the impact of the alteration increases from technical change to transformation. The result of these possibilities is a three-part categorization of change, as illustrated in Table 9.1.

Technical change includes both technological innovation and structural modifications in the design of the organization. For example, the addition of new diagnostic machines in place of outdated equipment constitutes technical change, as does the use of an internal review committee to screen proposed new services. The means used within the organization change, but not the goals.

A transition, on the other hand, requires a shift in the goals of the organization, although the means remain the same. For example a hospital might decide to expand its market area from a regional to a national focus. Both the technology and basic structure would be quite similar to those that already existed in the institution.

Finally, transformation requires an alteration of both means and goals. For example, a hospital might purchase and operate a health maintenance organization to increase its control of its patient days and to stabilize referral patterns. Such a change requires the organization not only to expand its mission but also to adapt its orientation to the delivery and financing of care.

LEADERSHIP AND CHANGE

Types of Leadership

If fundamental changes are occurring in the environment for health care delivery, how can health care organizations adapt or respond to such changes to provide high-quality, cost-effective care? Answers to this question will certainly vary, including redesigning delivery systems, adding new financing mechanisms, expanding marketing efforts, and rethinking human resources systems and roles. Within the latter area of human

TABLE 9.1. Types of Organizational Change

Types	Means	Ends
Technical	Change	No change
Transition	No change	Change
Transformation	Change	Change

SOURCE: Adapted from Kaluzny and Veney (1977).

resources, one factor many scholars and practitioners agree will be crucial is the leadership of those systems. Peters and Waterman (1982) make the point in *In Search of Excellence*.

> We argue that the excellent companies are the way they are because they are organized to obtain extraordinary effort from ordinary human beings. It is hard to imagine that billion-dollar companies are populated with people much different from the norm for the population as a whole. But there is one area in which the excellent companies have been truly blessed, with unusual leadership, especially in the early days of the company. (Pp. 81–82).

Unfortunately, the empirical study of leadership and its impact on performance does not support widespread confidence in leadership. Lieberson and O'Connor (1972) investigated the influence of leadership changes along with yearly industry and company influences on three measures of performance: sales, earnings, and profit margin. They found that industry and company accounted for far more variance in two performance measures than leadership and that the effects of each factor were similar for the third indicator. A related study (Salancik and Pfeffer, 1977) using similar methods found that the impact of mayors on variations in city budget categories is relatively small. Two recent studies by Weiner and Mahoney (1981) and Smith et al., (1984) question both the methods and results of those earlier reports. Thus, the empirical evidence regarding the effect of leadership on performance remains clouded.

This apparent conflict between empirical research and common belief may be resolved, at least in part, by adopting a more contemporary definition of leadership. One approach is that presented by James MacGregor Burns. In his widely recognized work *Leadership* (1978), Burns identified two types of leadership: *transactional* and *transformational*. Transactional leadership "occurs when one person takes the initiative, making contact with others for the purpose of an exchange of valued things . . ." and transformational leadership "occurs when one or more persons engage with others in such a way that leader and follower raise one another to higher levels of motivation and morality" (Burns, 1978, pp. 19–20). Let us consider each type in turn.

The practical relevance of these abstract definitions is illustrated by the identification of patterns of behaviors associated with each type of leadership based on subordinate perceptions. Through a factor analysis of items evaluated by a group of military officers, two factors have been associated with transactional leadership and the three others with transformational leadership (Bass, 1985). These factors are important because they indicate the behaviors that subordinates perceive as significant in motivating them to accept leadership from designated persons.

Transactional Leadership

The transactional form of leadership was found to consist of two leadership behavior patterns termed *contingent reward* and *management-by-exception*. In the contingent reward approach the leader identifies tasks to be done to accomplish a goal, explains them to subordinates, and indicates the outcomes that will accrue to them if tasks are performed and the goal accomplished. Typically, the rewards subordinates receive are praise for work well done and recommendations for pay increases. Path-goal theory

(Evans, 1974; Georgopolous et al., 1957; House, 1971; House and Mitchell, 1974) attempts to explain why contingent reward works. This approach also identifies contingencies, such as worker experience and task complexity, that influence the relationship between leader and follower. Perhaps the best current description of the transactional leadership process as contingent reward is found in *The One Minute Manager* (Blanchard and Johnson, 1982).

A second transactional leadership process is management by exception or contingent punishment. The leader in this case intervenes with subordinates only when something goes wrong. This type of leadership is often advocated for use with skilled professionals such as physicians or nurses, especially by the professionals themselves. The use of management by exception without coincident use of reward does not appear to be effective (Bass, 1985). Without opportunities for positive reinforcement, relationships become negative and soon deteriorate.

The function of transactional leadership is to maintain the organization's operation rather than to change it. The development of efficient methods to carry out tasks is the role of the leader (Burns and Becker, 1987). This concept of leadership is well-suited to theories that concentrate on the formal structure of organizations and their function of goal attainment (Fayol, 1949; March and Simon, 1958; Taylor, 1911; Weber, 1947). In sum, transactional leadership is the process through which those in formal positions achieve existing organizational goals by clarifying subordinates' expectations about the outcomes of their effort and by reducing role ambiguities and conflict. In contemporary health care organizations, such leadership is important to management at all levels.

Transformational Leadership

As noted earlier, three patterns of transformational leadership behavior were identified through factor analysis of subordinate perceptions (Bass, 1985). Another study using qualitative techniques to identify the characteristics of transformational leaders found seven qualities of leaders through in-depth interviews with a dozen business leaders (Tichy and Devanna; 1986a). By presenting these positions in tandem, the relevance of each is increased.

The first major factor resulting from the quantitative analysis was *charisma,* "an endowment of an extremely high degree of esteem value, popularity, and/or celebrity-status attributed by others" (Bass, 1985, p. 39). Eighteen items loaded on this factor, three of which are listed below as illustrations:

- Makes everyone around him or her enthusiastic about assignments.
- I have complete faith in him or her.
- Is a model for me to follow.

Four characteristics identified by Tichy and Devanna appear to be related to charisma. Such leaders view themselves as *change agents.* They are not "turnaround artists" or entrepreneurs; rather, they are committed to a path of long-run change and can motivate others to move along it with them. Consistent with House's (1977) explanation of charisma, the leader's ability to motivate appears to result from three other characteristics. Transformational leaders are *visionaries.* They are able to formulate an image of the future and to make it meaningful to others. In addition, such leaders are *value-driven.* They are able to state a core set of values and then perform behaviors consistent with them. And finally, they are *courageous* in an emotional sense. They are "prudent risk-takers, individuals who take a stand" (Tichy and Devanna, 1986a, p. 271). They are willing to propose what they believe to be correct both to those who do and those who do not want to hear it. Thus, charismatic leaders create change by developing a vision, espousing it to others, and steadfastly supporting it based on shared values.

Intellectual stimulation is the second transformational factor (Bass, 1985). This dimension indicates the transformational leader's ability to create in others increased awareness of problems and their solutions. In other words, such leaders focus on and promote strategic thinking among their associates. Subordinates are not aroused to action, but rather to conceptualization of their environment. Three items loaded on this factor:

- His or her ideas have forced me to rethink some of my own ideas which I had never questioned before.
- Enables me to think about old problems in new ways.
- Has provided me with new ways of looking at things that used to be a puzzle for me.

Three qualitative characteristics (Tichy and Devanna, 1986a) provide an indication of how this capacity for intellectual stimulation of others is created, sustained, and implemented. First, the creation of this quality is related to the ability to *deal with complexity, ambiguity, and uncertainty*. This ability may result from the leader's mastery of the technical as well as the social and political aspects of an enterprise. In health care, the ability of managers to understand the significance of new treatments for organizational change requires at least a reasonable understanding of biological systems, disease processes, and medical terminology. Second, the capacity for intellectual stimulation is sustained through the quality of *life-long learning*. Such individuals are open to self-assessment and willing to adapt their thinking to new ideas. Moreover, they learn from their mistakes rather than merely burying them. Finally, the intellectual side of *courage* is evident in transformational leaders and their concern for their concern for implementation. They are able to step back from the immediate situation and see the strengths and weaknesses of the organization. This is a key element of strategic thinking which others can observe and emulate.

Individualized consideration is the third and final transformational factor identified by Bass. This factor consistently appears in factor analyses of leadership behavior (Bowers and Seashore, 1966; House and Baetz, 1979). This dimension has two aspects. One is the leader's *concern for subordinates* as individuals with unique problems, which is often expressed through one-to-one interaction focused, perhaps, on job-related or personal issues. The second aspect is the leader's *attempt to develop subordinates*. This element may take many forms such as delegation, team building, or mentoring. Examples of the items in this factor are:

- You can count on him or her to express appreciation when you do a good job.
- Makes me feel we can reach our goals without him or her if we need to.
- Treats each subordinate individually.

One transformational leadership characteristic appears related to this factor. Transformational leaders *believe in people*. This quality emphasizes the developmental aspect of individualized consideration. Although such leaders understand in many cases the concerns of their subordinates, their primary goal is empowerment of them. That is, through a clear understanding of motivational principles, they allow subordinates to use their full capacity to contribute to the organization. These authors quote John Harvey-Jones's concise statement of the process of worker empowerment.

> Industrial success is a matter of getting commitment of people and the art is to get them involved and then you can sort of ratchet people by continuously setting them or getting them to set themselves aims which are a little more achievable to them. Then every time they achieve, you can ratchet them a bit further and after a little while you can get a team which really believes it's a leading team. . . . (Tichy and Devanna, 1986a, p. 274)

The function of transformational leadership—namely, to be creative with regard to the development and establishment of organizational goals—appears to follow from a natural system perspective on organizations (Burns and Becker, 1988; Scott, 1981). The goals of the organization are not viewed as determined but rather to be determined (i.e., committed to) by all organization members at its initiation and as it continues to evolve. From a transformational perspective, leadership plays a key role in creating organizations with clear direction by developing leader-follower relationships based on inspiration, intellectual stimulation, and individualized consideration. For health care organizations, this approach to leadership is not new but is unrecognized. For example, many religious-based health care systems were founded by an individual or small group of individuals who had a vision or charisma that is reflected in a contemporary corporate mission statement or set of values. The recent emphasis on corporate culture in health care may assist these organizations in identifying individuals who have been or are transformational leaders or in developing related characteristics in others.

Relationship of Transactional and Transformational Leadership

Transactional leadership is distinguished from transformational leadership both in the behavior patterns involved and in the intellectual functioning of leadership. The factor analysis discussed previously measured leadership behavior and identified three factors labeled as indicative of transformational leadership and two as transactional leadership. However, the theoretical relationship among the factors was provided by Bass's understanding of their relation to the concepts of Burns, not by the factor analysis procedure. To ascertain the relationship among the factors empirically, Bass conducted a second-order factor analysis, using factor scores for each case.

The result was two factors, one termed proactive leadership and a second called reactive leadership. The proactive factor consisted of the charisma, intellectual stimulation, individualized consideration, and contingent reward factors from the original analysis. The reactive factor contained the management-by-exception factor. This outcome suggests that the transformational dimensions and contingent reward have effects on the respondents' perceptions of "proactiveness." A careful analysis of Bass's original factor matrix indicates that the contingent reward items loaded at approximately equivalents level on charisma. This pattern suggests that a link between contingent reward and transformational leadership occurs because contingent reward behaviors are also viewed as indicative of charisma rather than intellectual stimulation or individualized consideration.

Bass (1985) also views the intellectual functioning of leadership as differing on proactive versus reactive leadership.

> What may separate transformational leaders from transactional leaders is that transformational leaders are more likely to be proactive than reactive in their thinking; more creative, novel, and innovative in their ideas; more radical or reactionary than reforming or conservative in ideology; less inhibited in their ideational search for solutions. (P. 105)

Rusmore's (1984) study of public utility managers, which uses seven measures of intellectual ability, is suggestive of the difference. Rusmore found two factors: general intelligence (based on tests of verbal, quantitative, and abstract reasoning abilities) and cognitive creativity (based on tests of obvious consequences, remote consequences, and unusual uses). These factors correlated negatively and positively, respectively, with management level in the organization. Thus, supervisors were more dependent on general intelligence and executives on cognitive creativity.

Relationship Between Leadership and Change

Three types of change that have different effects on the organizational mission, structure, and internal operations of health service organizations have been identified. In addition, two types of leadership have been described; each depends on different patterns of behavior between leader and follower and has a specific function regarding organizational action. The relationship between the two types of leadership and the three types of change can be identified through two general propositions.

First, *technical change and transition can be successfully implemented through transactional leadership,* although transformational leadership will facilitate these changes. Technical and adjustive change requires either a change in means or a change in ends, but not both. These alterations may constitute significant change in the system but not of the system (Cameron, and Ulrich, 1986). After such change, the system is altered but remains recognizable to insiders and outsiders alike. Community hospitals will continue to add new technology and/or business procedures and they may use their existing procedures to specialize services or expand markets. Despite such significant modifications, however, they will remain community hospitals. The function of transactional leadership is to maintain the system through control of current procedures or through improvement in them when necessary. Through processes based on contingent reward, leadership can explain and implement changes in means or ends without altering workers' fundamental understanding of the organization's mission and their relationship to it.

Second, *transformation can be successfully implemented through transformational leadership if the processes of transactional leadership are in place.* Transformational leadership is fundamentally a process of motivating others to do more than they originally expected to do. Bass (1985) states that this process can alter motivational level in three ways:

1. By raising our level of awareness, our level of consciousness of the importance and value of designated outcomes, and ways of reaching them.
2. By getting us to transcend our own self-interest for the sake of the team, organization, or larger polity.

3. By altering our need level on Maslow's . . . hierarchy or expanding our portfolio of needs and wants. (P. 20)

Although the process of transformational leadership will be discussed, it is important to indicate the common elements of these methods. Each suggests that the relationship of individual members to the organization is fundamentally altered as part of the adaptive change process. Unlike transactional leadership, followers of a transformational leader do not merely recognize more clearly the relationship of their behaviors to their outcomes; they change the meaning of the outcomes and perhaps their behaviors as well. However, as the second-order factor analysis indicated, contingent reward is also necessary for proactive leadership.

Thus, both transactional and transformational leadership are significant for organizational change. The processes of transactional leadership have been the core of management studies until recently and are best summarized in the role of manager consisting of planning, implementation, and evaluation. Scholars and practitioners have elaborated these activities from both rational system and natural system perspectives, primarily focusing on internal operations.

With recent emphasis on environment-organizational relations, management writers have begun to explore an expanded role for leadership. Child (1972) and others (Lorange, 1980; Pfeffer and Salancik, 1978) recognized that the environmental-organizational relationship as well as internal operation can be socially constructed and, in doing so, have indicated why and how political and cultural considerations can influence an organization through the process that will be described as transformational leadership.

The Process of Transformational Leadership

Since much has been written about the processes of transactional leadership, this section concentrates on the process of transformational leadership and how it can result in organizational transformation. This process has been described by Tichy and Devanna (1986a, 1986b) and Cameron (1986). According to Tichy and Devanna (1986b), transformational leadership is "systematic, consisting of purposeful and organized searches for changes, systematic analysis, and the capacity to move resources from areas of lesser to greater productivity" (p. 27). This process is presented in Figure 9.1.

The mechanism is triggered by changes in an organization's environment that cannot be ignored. The changes discussed earlier in this book certainly constitute environmental triggers for most health care institutions. The description of the transformational leadership process will begin with a discussion of the organizational dynamics and conclude with a consideration of the individual dynamics of change.

The first task of the transformational leader is to create a felt need for change and to overcome the resistance to it. The fundamental problem such individuals have is referred to as the "boiled frog phenomenon" (Tichy and Devanna, 1986a). If you put a frog in boiling water it will jump out, but if you heat the water slowly around the frog, it will fail to respond, and thus will die. Changes in an organization's environment may occur so slowly that they are not recognized until it is too late for remedial action.

The transformational leader can create awareness in several ways. One method is to challenge the routine of the current leadership with devil's advocate positions. This approach ensures frequent examination of the assumptions on which current

228   Behavioral Systems

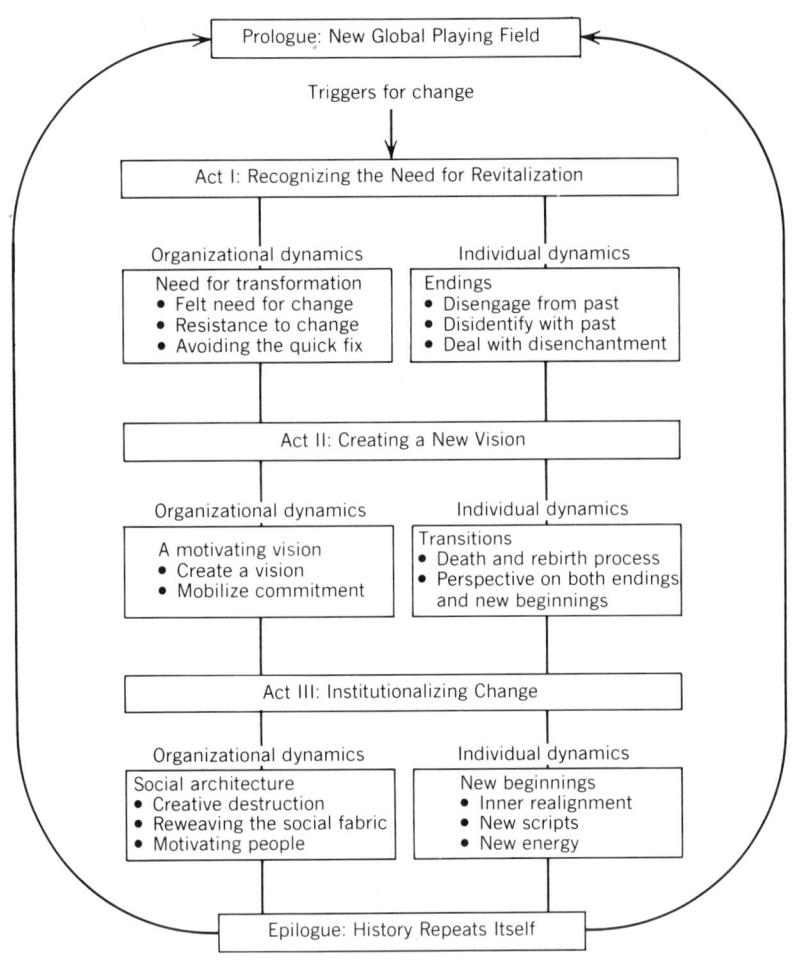

```
                    ┌─────────────────────────────────────────┐
                    │  Prologue: New Global Playing Field      │
                    └─────────────────────────────────────────┘
                                Triggers for change
                                        ↓
                    ┌─────────────────────────────────────────┐
                    │  Act I: Recognizing the Need for Revitalization │
                    └─────────────────────────────────────────┘

        Organizational dynamics              Individual dynamics
        ┌────────────────────────┐    ┌──────────────────────────┐
        │ Need for transformation │    │ Endings                  │
        │ • Felt need for change  │    │ • Disengage from past    │
        │ • Resistance to change  │    │ • Disidentify with past  │
        │ • Avoiding the quick fix│    │ • Deal with disenchantment│
        └────────────────────────┘    └──────────────────────────┘

                    ┌─────────────────────────────────────────┐
                    │  Act II: Creating a New Vision           │
                    └─────────────────────────────────────────┘

        Organizational dynamics              Individual dynamics
        ┌────────────────────────┐    ┌──────────────────────────┐
        │ A motivating vision     │    │ Transitions              │
        │ • Create a vision       │    │ • Death and rebirth process│
        │ • Mobilize commitment   │    │ • Perspective on both endings│
        │                         │    │   and new beginnings     │
        └────────────────────────┘    └──────────────────────────┘

                    ┌─────────────────────────────────────────┐
                    │  Act III: Institutionalizing Change      │
                    └─────────────────────────────────────────┘

        Organizational dynamics              Individual dynamics
        ┌────────────────────────┐    ┌──────────────────────────┐
        │ Social architecture     │    │ New beginnings           │
        │ • Creative destruction  │    │ • Inner realignment      │
        │ • Reweaving the social  │    │ • New scripts            │
        │   fabric                │    │ • New energy             │
        │ • Motivating people     │    │                          │
        └────────────────────────┘    └──────────────────────────┘

                    ┌─────────────────────────────────────────┐
                    │  Epilogue: History Repeats Itself        │
                    └─────────────────────────────────────────┘
```

Figure 9.1. Transformational leadership: a three-act drama. (From Tichy and Devanna: The transformational leader. *Training and Development Journal,* July, 1986b; 27–32. Copyright © 1986, *Training & Development Journal,* American Society for Training and Development. Reprinted with permission. All rights reserved.

organizational practices are founded. A second method is the building external networks through professional associations such as the American College of Healthcare Executives. These groups can provide feedback to individuals about major changes occurring in the hospital's environment.

Arranging for administrative staff to visit other provider organizations can provide insights into the operational practices of other health services delivery institutions. The inclusion of varied referents into the management evaluation process can assure that multiple viewpoints on organizational performance are considered. Finally, rotation or changing of management personnel can provide fresh viewpoints for the organization.

Each of these activities suggests that health care executives must be willing to open their institutions to external influences on a routine though controlled basis.

Having systematically informed the organization of significant problems and the need for change, transformational leaders must be prepared to handle three types of change. These are technical change, political change, and cultural change (Tichy, 1983). There are obstacles associated with each of the areas.

Problems resulting from technical changes have several sources in health care organizations, including disruption of interaction patterns or interpersonal relationships, parochial concerns for "our way of doing things," sunk costs in current medical technology, low tolerance for ambiguity among professionals with specialized skills, and lack of skill to cope with change.

Political forces create resistance to change because modifications of a health services organization redistribute key resources such as power, personnel, prestige, and money. These shifts in resources threaten existing coalitions, call into question prior support by superiors, and create personal self-doubt. Major concerns among medical staff today are associated with the possibility of declining resources available to pay for medical services.

Finally, the culture of an organization can be the source of resistance to change. The culture provides the current understanding of the expectations for behavior, core values, stories and myths, and language of the organization. Through reference to these elements of culture, changes can be viewed in ways ranging from inappropriate to irrelevant to immoral.

Many, if not all, of the forms of resistance resulting from these sources occur because individuals in an organization perceive that negative consequences will occur, regardless of whether they ever do. The perceptual nature of resistance suggests that management concerns during such periods should be more with symbolic than substantive issues (Pfeffer, 1979). Based on his own work (Cameron, 1984) and that of others (Bennis et al., 1969), Cameron (1986) lists seven approaches to overcoming resistance, which follow from a symbolic approach to management.

1. Involve those affected by the change in all stages of diagnosing the need and planning the response.
2. Provide those involved with a feeling of control of at least how and when, if not what, changes are to occur.
3. Acquire the support of key formal and informal leaders.
4. Establish trust and support of key groups in the organization.
5. Demonstrate an understanding of the perspective of resistors and the advantages to them of change.
6. Assist those involved in seeing the changes as an opportunity (i.e., internally initiated and reducing burdens) rather than a threat.
7. Tie changes to the existing core values of the organization or, if need be, large cultural groups.

Once a performance gap has been identified and the need for change recognized, the transformational leader must create a vision to carry the organization into the future. A vision motivates individuals to accept and accomplish the challenge that is articulated. It also provides guiding principles that individuals can internalize and reflect on in making specific decisions. In mature organizations, a vision is typically not the creation of one individual but rather reflects the commitment of a larger constituency group. Through this statement, the organization considers the future and provides some insight as to how the organization will function technically, politically, and culturally. This vision provides the framework for the opinions that senior managers will form about the changes that will be occurring.

The process of creating a vision requires both right-brain and left-brain thinking. Right-brain thinking provides the organization's conception of the future manifested as a whole, without specifying detail and emotional content. The vision so conceived results from intuition, not analysis. Group process techniques such as designing an ideal competitor or projecting an organization 5 years into the future have been used to facilitate right-brain thinking.

Alternatively, left-brain thinking is a rational analysis of an organization's future using the techniques traditionally associated with strategic planning. As Cameron (1986) suggests, this type of thinking should provide answers to questions such as the following: What business are we in? What major obstacles do we face? What information do we require? What are our resources? How do we communicate our vision?

The next step in the transformational process, the creation of commitment to the vision, is analogous to the forming of attitudes that occurs during the initiation stage of the innovation process. At this point the leader interested in fostering change must influence the attitudes that developing within the organization. Leaders must use symbolic means to communicate the significance of the vision for the organization and its members. For example, a public commitment to change by key managers or health care professionals increases the likelihood that their behavior will be congruent with these statements. In addition, encouraging involvement in the change process will lead to identification with it. Leaders must assist in connecting the organization's projected future to its past while at the same time justifying the change and putting the past to rest. Cameron (1986) indicates that commitment also can be enhanced by setting effective goals, selecting and socializing key people, and starting with simple successes.

With the vision in place, the transformational leader can turn to institutionalizing the change, which is equivalent to the implementation stage of innovation. The most fundamental aspect of implementing the vision consists of changing the technical, political, and cultural networks of the organization. To manage these networks, the transformational leader must first dissect and understand the existing networks. In general, the technical networks can be analyzed by examining information flows in the organization; political networks can be analyzed through formal and informal influence exchanges; and culture networks can be understood through friendship relationships in which norms and values are shared. Because of the complexity of these network analyses and their subsequent management, this phase of the transformational process is far more difficult to accomplish than vision creation.

The task of the transformational leader, thus, is to weaken old networks and establish new ones. An issue during this process is typically the alleviation of political uncertainty. As structures change, it becomes unclear who controls resources, and the significance of decisions grows more ambiguous. Until such issues are resolved, technical and cultural problems cannot be affected. For example, the restructuring of many religiously oriented, not-for-profit health systems has resulted in elaborate efforts to renew philosophy and mission. These activities continue in many systems as they seek to define "mission effectiveness" in an environment in which the distribution of influence between lay and religious leaders has not been fully resolved. To the frustration of those who have experienced a time of clear authority and direct action, executive time appears to be spent excessively on process rather than action.

The essential question for health care organizations as well as others is not whether bureaucratic structures will remain, but how these structures can be used to reduce uncertainty for the organization. According to Tichy and Devanna 1986a, two options are available to control technical, political, and cultural uncertainty. One can leave the area creating the uncertainty, or one can try to design the organization to respond to it. In the second instance, there are two alternatives: (a) a mechanistic approach that attempts to increase predictability of action, and (b) an organic approach that attempts to make the organization more flexible, hence more responsive to diverse problems. Health care organizations tend to focus on mechanistic solutions. Competitive pressures, however, emphasize the need to respond to rapid environmental changes and to create pressures for organic structures.

Finally, transformational leaders institutionalize the vision through the management of people, and thus the human resources system is key to the process. Each aspect of a human resources system (selection, performance appraisal, compensation, and development) can play a role in the transformational process. The appraisal system can provide information on the availability of appropriate personnel to accomplish a change, as well as offering a measure of whether strategic objectives are being accomplished. The reward structure can encourage innovation and the development of a supportive culture. Through selection, succession, and development, transformational leadership can be expanded. Selection and succession planning can be used to screen individuals with regard to both their skills and their values. A balance, however, between homogeneity of characteristics and the potential for innovation must be maintained.

The remainder of this chapter focuses on leadership development. Although this area is perhaps no more important to the transformational process than the other aspects of human resources management, its relationship to change has received little systematic attention in the literature.

LEADERSHIP DEVELOPMENT

Some writers begin by asking not "How does one develop transformational leaders?" but rather "Can one develop such leadership?" In general, the answer provided by most contemporary writers is yes; however, the extent to which nurture is viewed as more significant than nature varies. After carefully analyzing the potential for development, Zaleznik (1977) concluded that leaders, as opposed to managers, can be assisted in reaching positions of authority but that a distinct personality structure is essential. Leaders have "twice-born" personalities, resulting from weak parental attachments that produced "inner directedness" and a sense of separateness. Unlike managers, who are more socially oriented, leaders seek risk rather than security, focus on the meaning of action rather than policy and procedures, and are able to empathize with others rather than merely understanding their roles in a structure.

Stronger advocacy of the ability to develop transactional and transformational leadership can be found. This view has been articulated by proponents of professional graduate education in health administration (Kurz, 1986) and business (Tarr, 1986). In general, the skills emphasized in such programs, such as quantitative decision making, accounting, or finance, appear to be those associated with transactional leadership. In addition, transformational leadership can be learned and managed. Having identified the behavioral characteristics of such leaders, programs that develop those qualities can be created.

Although the extent to which leadership can be developed has not been resolved, it is clear that some degree of training can occur. Hall (1984) provides a conceptual framework for the outcomes of human resources development based on the relationship of time frame to locus of concern. The topology resulting from the cross-tabulation of these factors is presented in Table 9.2.

Performance refers to the accomplishment of current work objectives, and *attitudes* to the individual's knowledge, feelings, and beliefs about work or ones career. *Adaptability* is the extent to which an individual is preparing for future roles, and *identity* the degree to which the individual is integrating his response to these changes over time.

If individuals are to be trained to cope with the transformational leadership process, each of these outcomes must be addressed. Performance and attitudes as outcomes are the core of traditional management development programs, as well as graduate health administration programs and business schools. These efforts have focused on providing the skills and attitudes needed by future line or staff managers to accomplish their tasks. In fact, such programs have been quite successful in developing what Zaleznik (1977) terms managers or transactional leaders.

Analyses of transformational leadership suggest that emphasis must also be placed on adaptability and identity as outcomes of the development process. As illustrated in Figure 9.1, Tichy and Devanna (1986b) indicate that the process of transformational leadership includes both organizational and individual dynamics. From the perspective of task accomplishment, individuals must be selected and developed with regard to future roles rather than merely present ones. If transformation is to be a continuing process, and in health care this is a reasonable assumption for the foreseeable future, then leaders must be prepared to achieve the long-range objectives of the organization.

Transformational leadership is also a psychodynamic process consisting of endings, transitions, and new beginnings. Each of these stages has significant impact on the identity of both individual leaders and followers. After a change event has occurred, participants must begin to separate themselves from former expectations and the satisfactions that came from these. This phase will undoubtedly create some personal confusion and self-doubt. If handled correctly, one can move to a new role in which some values of the past are maintained while at the same time a base is created for new

TABLE 9.2. Outcome Measures of Career Effectivenss

Time Frame	Locus of Concern	
	Task	*Self*
Short term	Performance	Attitudes
Long term	Adaptability	Identity

SOURCE: Hall, D. T. Human Resource Development and Organizational Effectiveness. In *Strategic Human Resource Management*, C. J. Fombrun, N. M. Tichy and M. A. Devanna (Eds.). Copyright © 1984, John Wiley & Sons, Inc. Reprinted by permission of John Wiley & Sons, Inc.

opportunities. As health care organizations break old networks and establish new ones, the potential for revitalizing the self occurs.

If leadership development is possible and is a more complex process than traditional management development, it is important to specify capabilities to foster for transformational leadership. Although these skills may not be mutually exclusive or exhaustive, six competencies requiring development are suggested.

1. The first competency is *planful opportunism* (Tichy and Devanna, 1986a). This capability consists of the systematic and accidental collection of data both inside and outside the organization, to be used in diagnosing the problem(s) at hand. In addition, it requires the development of alternative futures for the organization based on the facts that were gathered.

2. A second skill to be acquired is the ability to *empathize* with others, to understand their feelings about a situation and to respond to their needs (Zaleznik, 1977). The transformational leader can create successful endings and beginnings only through an awareness of what events and decisions mean to participants.

3. The transformational leader must also develop his or her skill in *visioning* (Tichy and Devanna, 1986a). This ability involves both right-brain activities, which unleash individual creativity, and left-brain activities, which provide guidelines for implementing the vision. As noted earlier, creativity may be of greater significance for executive than for supervisory decision making (Rusmore, 1984). In any event, it is a difficult skill to teach, although some systematic approaches have been developed (Center for Creative Leadership, 1986; Rice, 1985).

4. *Symbolic leadership* is the ability of management to explain, rationalize, and legitimize actions taken by an organization (Pfeffer, 1979). The actions themselves are likely to be determined by resource interdependencies and other environmental imperatives or power disparities. The task of management is to make action meaningful to participants through the development of a social consensus about the activities being undertaken.

5. An additional competence is the ability to perform *network analysis* (Tichy and Devanna, 1986a). This process consists of work-flow analysis of the technical activities of the organization, analysis of key influences in the political area, and an understanding of the organization's norms, values, and beliefs as well as their custodians in the culture. The technical, political, and cultural networks can then be modified—perhaps simultaneously—through management of symbols (symbolic leadership) or chance events (planful opportunism).

6. Finally, leaders must be schooled more fully in the elements and opportunities of an integrated *human resources system*. The focus in most health care organizations as well as elsewhere in industry has been on technical systems of appraisal and compensation for employees, with little consideration of the strategic significance of human resources activities. Increased regulation and competition in health care has sharpened the need for management to assess systematically the human resources that will be needed to respond effectively to a complex and uncertain environment (Fombrun, 1984).

How can these competencies be developed to enhance transformational leadership? The acquisition of the six abilities will require an instructional process that involves the presentation of new information and the practice of these competencies. Three strategies are needed to modify an individual's performance, attitudes, adaptability, and identity, namely: *cognitive* strategies, *behavioral* strategies, and *environmental* strategies (Hall, 1984).

A cognitive strategy consists of communicating information in a didactic or interactive fashion and is the most widely used method at present. For health care managers, this method includes seminars sponsored by professional societies such as the American College of Healthcare Executives and executive development programs on university campuses, as well as on-site, in-service programs. Development based on cognitive strategies remains the core of educational efforts of managers at all levels, and many companies (e.g., IBM, General Electric, Whirlpool) have ongoing programs that involve virtually all managers at some point during the year.

A behavioral strategy is intended to change an individual's behavior within a specific environmental context. Unlike the other strategies, the outcome sought is typically more concrete and limited, even if the consequences of the change are substantial. An environmental strategy, on the other hand, alters the situation to produce opportunities for substantial personal development or modifies the structure to affect the culture of the employment setting. Although neither behavioral nor environmental strategies have been used as extensively as cognitive strategies, both hold great potential for development of transformational leadership.

Mentoring, a behavioral strategy, is a fundamental method in the development of leaders. Only through one-to-one relationships of senior and junior executives can aggressiveness and individual initiative be fostered; peer relations are constraining and will stifle potential leaders. Similarly, job rotation, an environmental strategy, may be essential to the development of executives for vertically integrated health care systems. In the past, hospital executives began their careers at the entry level and advanced through line positions to the executive level. In a highly diversified health care organization, the technical, cultural, and political networks of alternative delivery systems, long-term care units, and hospitals will differ and might best be understood through management experience in them.

Finally, where in the organization should the development of transformational leadership be directed? Although transformational leadership is often described at the executive level (Tichy and Devanna, 1986a; Cameron, 1986), such training must occur at all management levels of the health care organization. The expansion of the role of department head has begun, and attempts are being made to respond to the growing need for development opportunities (Henderson et al., 1985). Directors of pharmacies and physical therapy departments have been encouraged to market their services outside the hospital, and medical records supervisors are asked to manage information systems that integrate clinical and financial data. As critical decision-making moves down in the organization, these individuals will need to understand the process of transformational leadership to cope successfully with the demand for change.

SUMMARY

This chapter has described the relationship between organizational change, leadership, and leadership development. Technical change and transition can be accomplished employing traditional management skills—those associated with the transactional leadership model. The development of these capabilities has been a part of contemporary management development and graduate education in health administration and business.

Transformation—changes in organizational means and ends—requires a different model of leadership, transformational leadership. This type of leadership is most effectively undertaken by those who have already achieved substantial competence as transactional leaders. Transformational leadership, however, demands greater adaptability from participants, as well as changes in identity as shifts occur in the fundamental nature of the organization.

The new competencies needed by transformational leaders are typically more qualitative and include planful opportunism, empathy, visioning, symbolic leadership, network analysis, and effective management of human resources. In addition, the development of these skills will require strategies directed at work behavior and environments as well as cognitive understanding. Greater development of transformational leadership is essential if health care organizations are to cope with the demands of a continually changing environment.

REFERENCES

Bass, B. *Leadership and Performance Beyond Expectations.* New York: Free Press, 1985.

Bennis, W. G., K. Benne, and R. Chin. *The Planning of Change.* New York: Holt, Rinehart & Winston, 1969.

Blanchard, K., and S. Johnson. *The One Minute Manager.* New York: Morrow, 1982.

Bowers, D. G., and S. E. Seashore. "Predicting organizational effectiveness with a four-factor theory of leadership." *Administrative Science Quarterly,* 11:238–263 (September 1966).

Burns, J. M. *Leadership.* New York: Harper & Row, 1978.

Burns, L. R., and S. W. Becker. "Leadership and decision making." In S. M. Shortell and A. D. Kaluzny, (Eds.), *Health Care Management: A Text in Organization Theory and Behavior,* Second Ed., New York: Wiley, 1988.

Cameron, K. S. "The effectiveness of ineffectiveness." In B. M. Staw and L. L. Cummings, (Eds.), *Research in Organizational Behavior,* Vol. 6. Greenwich, CT: JAI Press, 1984.

Cameron, K. S., and D. O. Ulrich. "Transformational leadership in colleges and universities." In J. R. Smart, (Ed.), *Higher Education: A Handbook of Theory and Research.* New York: Agathon, 1986.

Center for Creative Leadership. *Creativity Development at the Center for Creative Leadership.* CCL: Greensboro, NC, 1986.

Child, J. "Organization structure, environment and performance: The role of strategic choice." *Sociology,* 6:1–22 (1972).

Downs, A. *Inside Bureaucracy.* Boston: Little, Brown, 1966.

Evans, M. G. "Extensions of a path-goal theory of motivation." *Journal of Applied Psychology,* 59:172–178 (1974).

Fayol, H. *General and Industrial Management.* London: Pitman, 1949.

Fombrun, C. L. "The external context of human resource management." In C. J.

Fombrun, N. M. Tichy, and M. A. Devanna, Eds., *Strategic Human Resource Management*. New York: Wiley, 1984.

Georgopoulos, B. S., G. M. Mahoney, and N. W. Jones. "A path-goal approach to productivity." *Journal of Applied Psychology*, 41:345–353 (1957).

Hall, D. T. "Human resource development and organizational effectiveness." In C. J. Fombrun, N. M. Tichy, and M. A. Devanna, (Eds.), *Strategic Human Resource Management*. New York: Wiley, 1984.

Henderson, A. C., C. C. Haddock, and T. C. Dolan. "An assessment of hospital middle managers' continuing education needs." *Journal of Health Administration Education*, 3:415–425 (Fall 1985).

House, R. J. "A path-goal theory of leadership effectiveness." *Administrative Science Quarterly*, 16:321–338 (1971).

House, R. J. "A 1976 theory of charismatic leadership." In J. G. Hunt and L. L. Larson, (Eds.), *Leadership: The Cutting Edge*. Carbondale: Southern Illinois University Press, 1977.

House, R. J., and M. L. Baetz, "Leadership: Some empirical generalizations and new research directions." In B. M. Staw (Ed.) *Research in Organizational Behavior*, Vol. 1. Greenwich, CT: JAI Press, 1979.

House, R. J., and T. R. Mitchell. "Path-goal theory of leadership." *Journal of Contemporary Business*, 5:81–97 (1974).

Kaluzny, A. D., and S. R. Hernandez. "Organizational change and innovation." In S. M. Shortell and A. D. Kaluzny, (Eds.), *Health Care Management: A Text in Organization Theory and Behavior*. New York: Wiley, 1987.

Kaluzny, A. D., and J. E. Veney, "Types of change and hospital planning strategies." *American Journal of Health Planning*, 1:13–19 (1977).

Kanter, R. M. *The Change Masters: Innovation for Productivity in the American Corporation*. New York: Simon Schuster, 1983.

Kurz, R. "Health administration education: Assumptions, guidelines, and future directions." *Journal of Health Administration Education*, 3(3):382–392 (1986).

Lieberson, S., and J. F. O'Connor. "Leadership and organizational performance: A study of large corporations." *American Sociological Review*, 37:117–129 (April 1972).

Lorange, P. *Corporate Planning: An Executive Viewpoint*. Englewood Cliffs, NJ: Prentice-Hall, 1980.

March, J. and H. Simon, *Organizations*. New York: Wiley, 1958.

Peters, T. J., and R. H. Waterman, Jr. *In Search of Excellence*. New York: Harper & Row, 1982.

Pfeffer, J. "Management as symbolic action: The creation and maintenance of organizational paradigms." In L. L. Cummings and B. Staw, (Eds.), *Research in Organizational Behavior*, Vol. 3. Greenwich, CT: JAI Press, 1979.

Pfeffer, J., and G. P. Salancik, *The External Control of Organizations*, New York: Harper & Row, 1978.

Quinn, J. B. *Strategies for Change: Logical Incrementalism*. Homewood, IL: Irwin, 1980.

Rice, Berkeley. "Teaching managers the art of creativity." *USAir*, VII (September, 1985).

Rusmore, J. T. "Executive performance and intellectual ability in organizational levels." Advanced Human Systems Institute, San Jose State University, 1984.

Salancik, G. P., and J. Pfeffer. "Constraints on administrator description: The limited influence of mayors on city budgets." *Urban Affairs Quarterly,* 12:475–498 (June 1977).

Scott, W. R. *Organizations: Rational, Natural, and Open Systems.* Englewood Cliffs, NJ: Prentice-Hall, 1981.

Smith, J. E., K. P. Carson, and R. A. Alexander. "Leadership: It can make a difference." *Academy of Management Journal,* 27:765–776 (1984).

Tarr, Curtis. "Can leadership be taught?" *Cornell Enterprise,* 2(2):23–25, (Spring 1986).

Taylor, F. *The Principles of Scientific Management.* New York: Harper & Row, 1911.

Tichy, N. M. *Managing Strategic Change: Technical, Political, and Cultural Dynamics.* New York: Wiley, 1983.

Tichy, N. M., and M. A. Devanna. *The Transformational Leader.* New York: Wiley, 1986(a).

Tichy, N. M., and M. A. Devanna. "The transformational leader." *Training and Development Journal,* pp. 27–32 (July 1986b).

Weber, M. *The Theory of Social and Economic Organization* translated and edited by A. M. Henderson and Talcott Parsons. New York: Oxford University Press, 1947.

Weiner, N., and T. A. Mahoney. "A model of corporate performance as a function of environmental, organizational and leadership influences." *Academy of Management Journal,* 24:453–470 (1981).

Zaleznik, A. Managers and leaders: Are they different?" *Harvard Business Review,* 55:67–80 (1977).

Zaltman, G., R. Duncan, and J. Holbek. *Innovations and Organizations.* New York: Wiley, 1973.

PART FOUR
Human Resources Process Systems

CHAPTER 10

Recruitment

Jacqueline Landau
Geoffrey A. Hoare

NEW RECRUITING NEEDS

Recruiting involves seeking and screening out qualified applicants and forming a pool from which job candidates are selected. As such, it is an integral part of human resources management. Health care organizations cannot function if they fail to attract quality applicants to fulfill their manpower requirements. This is particularly important in times of low unemployment when fewer people are available to work. Service organizations such as hospitals and clinics are labor intensive, hence vulnerable to shortages. Selection strategies, training programs, compensation systems, and career development programs can have only limited success if organizations cannot attract the work force they need.

There are few industries that employ personnel with such a wide range of expertise as health care. Physicians, nurses, technicians, and semiskilled employees work interdependently to care for a patient. All these care givers and support staff are becoming increasingly dependent on administrators who know about the characteristics of the service area, ways to generate demand for service through marketing, and the intricacies of reimbursement (Kaluzny and Shortell, 1983). Several trends are making this already complex situation more complicated.

The health care industry is dramatically changing the mix of types of services offered as the burden of disease shifts. The United States is moving from an era of treating cases of acute, infectious diseases to managing patients with chronic diseases that periodically experience acute episodes (Miller and Miller, 1981). At the same time as third-party payers have demanded reductions in the cost of care, user-friendly technologies have been perfected that allow a number of treatments to be given outside the hospital. The proliferation of clinical settings, including ambulatory surgery, skilled nursing and long-term care facilities, and home care, has created demands for types of personnel different from those found in a traditional community hospital. Furthermore, the linking of hospitals and alternative care facilities into multiunit diversified health care corporations has created a number of new management tasks, and hence manpower requirements (Hoare, 1987). These positions in the corporate infrastructure, in headquarters and regional offices, for example, may entail planning, productivity analysis, or performance auditing. All the while, tertiary care facilities continue to need highly trained staff to work with increasingly complex technology and very sick patients.

Because of all these factors, the range and mix of skills required by health care personnel is expanding. This will complicate the work of a health care corporation's personnel department. For example, a large community hospital recently built a 12-bed hyperbaric chamber to treat burn and trauma patients but found few staff qualified to work there. The hospital had to either train nurses to work in those conditions or recruit deep-sea divers familiar with the hyperbaric technology and train them in nursing. The recruitment challenge created by this proliferation in requirements for skills and knowledge is further complicated by the location of the facilities. There are real differences in labor markets both between urban and rural settings, and among regions of the country (Rosenblatt and Moscovice, 1982). Different recruiting strategies are necessary, then, for different types of staff and facilities of various types.

The discussion in this chapter can be simplified by dividing health care personnel into four categories:

Unskilled.   Housekeeping, dietary, security
Technical workers.   Nurses, technicians, therapists
Managers.   CEO, department heads, controller, personnel director
Physicians.   Contract, salaried, or merely medical staff members

This division will allow us to describe the common issues in recruiting as well as the distinctive problems that arise within each category. Although every issue will need to be addressed for each type of employee, as Table 10.1 indicates, different recruitment strategies may be needed depending on the type of health care position to be filled: unskilled, technical, managerial, or physician. The asterisks indicate that for specific employees there may be special problems that will affect strategies in terms of, for example, the amount of planning needed, the time frame, or the geographic scope of the effort.

THE RECRUITMENT PROCESS

The recruiting process consists of three basic steps: planning, implementation, and evaluation.

To develop an effective recruiting program, six familiar questions must be answered: who, what, why, where, when, and how? The organization's plan explains *why* a position is required. An organization must then decide *what* types of positions must be filled and *what* types of people are needed to fill those positions, *who* will recruit each job category, *where* candidates will be recruited from, and *when* the recruiting process must begin to ensure that the positions are filled by a specified time. Once these decisions have been made, the plan can be *implemented* and a pool of

TABLE 10.1. Recruitment Problems with Different Categories of Personnel

Recruiting Issues	Unskilled Workers	Technical/Professionals	Managers	Physicians
New needs			X	
Planning				
Why?		X	X	X
Who?				
What and when?			X	X
Where?		X	X	
Implementation				
Attracting		X		X
Screening				X
The law	X			X
Evaluation		X		
Future needs		X	X	X

candidates selected. To do this one needs to ask *how* the organization will attract qualified candidates and disqualify those who are clearly unsuitable. Then the program is *evaluated*, and the planning process begins once more. The steps in the recruitment process are listed in Figure 10.1.

RECRUITMENT PLANNING

Why? Demands for New Strategies in Human Resources

Recruitment, selection, and placement are closely linked functions of human resources management and can be leverage points for bringing about strategic change in an organization's goals (Tichy et al., 1982) or culture (Schein, 1985). In professionally dominated organizations, practitioners have deeply ingrained work patterns as a result of their professional training and socialization. Recruiting and placing individuals who use different treatment protocols or have different work norms is one of the only ways to change these organizations (Mintzberg, 1979).

Because of dramatic changes in the type of work in health care organizations, human resources planning and its recruitment component need to be closely tied to the overall planning process. Planning can be divided into three time frames (Ackoff, 1970):

- Strategic planning (3–10 year horizon)
- Tactical planning (1–3 year horizon)
- Operational planning (1 year horizon)

Why?

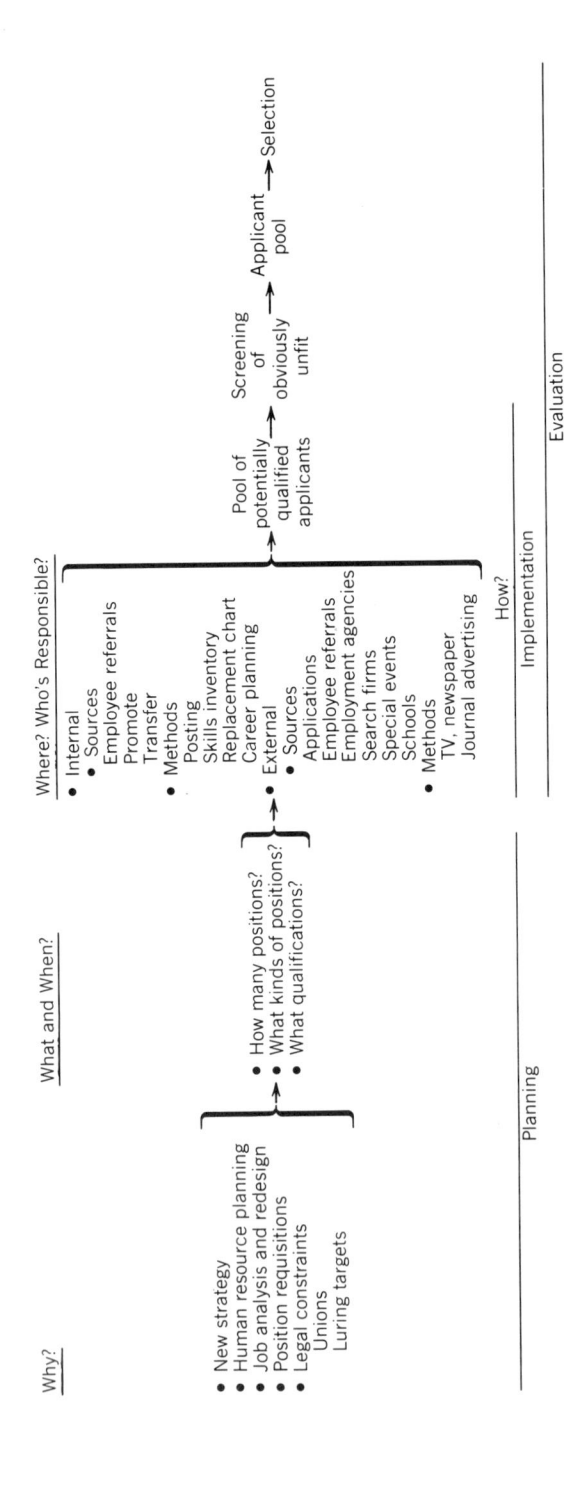

- New strategy
- Human resource planning
- Job analysis and redesign
- Position requisitions
- Legal constraints
- Unions
- Luring targets

What and When?

- How many positions?
- What kinds of positions?
- What qualifications?

Where? Who's Responsible?

- Internal
  - Sources
    - Employee referrals
    - Promote
    - Transfer
  - Methods
    - Posting
    - Skills inventory
    - Replacement chart
    - Career planning
- External
  - Sources
    - Applications
    - Employee referrals
    - Employment agencies
    - Search firms
    - Special events
    - Schools
  - Methods
    - TV, newspaper
    - Journal advertising

Pool of potentially qualified applicants → Screening of obviously unfit → Applicant pool → Selection

Planning	Implementation	Evaluation
	How?

Figure 10.1. Steps in the recruitment process. Reprinted by permission from *Personnel and Human Resource Management*, Second Ed., by Randall S. Schuler; Copyright © 1984 by West Publishing Company. All rights reserved.

Plans focusing on each of these time frames should be updated periodically. Information generated by this ongoing planning process could change the focus or the scope of the recruiting effort. Table 10.2 outlines the linkages between organization-wide planning and recruitment.

Changes in an organization's strategic plans, such as diversification and vertical integration of services (Goldsmith, 1981), could require dramatic changes in human resources, which in turn would call for new recruitment plans. For example, if a hospital starts an HMO, some physicians may need to be recruited. Their availability will need to be determined before the feasibility of this strategy can be assessed. Over a longer period, the case mix of an institution may change as a result of demographic changes in the service area from migration or from aging of the population. This could create incremental annual changes in staffing patterns, which would accumulate over time into significantly different staffing patterns, hence modified manpower requirements and recruiting needs.

At the tactical level, as hospitals "downsize" and shift more services into satellite offices, internal recruiting strategies may need to change. There may be a number of employees who would rather transfer to a satellite clinic than be laid off. Testing and training may be necessary, though, to prepare these individuals for new jobs. Such changes would affect annual recruiting quotas and the size of the recruiting budget.

The reasons *why* an organization recruits different kinds of employees or physicians will continue to change. As these evolutionary changes occur, an effective recruiting strategy will help to prevent:

- A shortage of labor

TABLE 10.2. Linkages Between Organization-Wide and Recruitment Planning

Strategic Planning	Tactical Planning	Operational Planning
Organization-wide Tasks	*Organization-wide Tasks*	*Organization-wide Tasks*
Define mission	Identify new programs and ventures	Develop budgets
Compare strengths and weaknesses with environment's opportunities and threats	Specify objectives	Quantify performance goals
	Specify manpower needs: amount and type	
Articulate strategy		
Set long-range goals		
Recruiting Tasks	*Related Recruiting Tasks*	*Related Recruiting Tasks*

Assess feasibility of attracting different categories of needed staff	Project recruiting needs based on:
	  New positions
	  Attrition
	  Layoffs
	  Productivity changes
		Set recruiting goals by job category based on:
		  Projected surplus or deficits
		  Staffing authorizations
		  Succession planning
		  Promotions and transfers

Analyze
  Labor force composition
  Labor force supply
  Demographic changes
  Internal labor force
	Develop recruiting strategies for each staff category	

SOURCE: Cherrington, D. J. *Personnel Management: The Management of Human Resources*, Second Ed. Dubuque, Iowa: William C. Brown, 1987.

- Excessive use of overtime or excessive use of contract labor, such as pool nurses
- Excesses of full-time equivalent (FTE) employees compared to some standard such as occupied beds or patient visits
- High training costs of new hires
- High turnover of new hires, suggesting a mismatch between the applicants' skills and actual job demands
- Physician or patient dissatisfaction with the staff's performance
- A poor image in the eyes of potential employees, suggesting that the organization has not been favorably represented in recruiting activities

What and When? Number and Type of Positions and Timing of the Search

Before any recruiting takes place, the management must determine how many positions need to be filled, and what kinds. For physicians, this is a strategic decision under the purview of the board, medical staff, and top administrator that affects the service mix and, in the long run, the viability of the organization. Decisions about nurses, department managers, and hourly workers have been the responsibility of the personnel function in the organization. In small organizations these decisions are usually made on an ad hoc basis when line management informs personnel through employee requisition forms that a position has been vacated or a new one created. However, large organizations, particularly organizations that need highly skilled employees, must forecast their needs far in advance to ensure adequate staffing. Their need for employees in the future is determined by the strategic plans and the goals of the organization. Equal Employment Opportunity Commission (EEOC) regulations and affirmative action goals must also be taken into consideration. Organizations may be required to recruit specific minorities to redress past inequities. These needs must be compared to projected availability of human resources, and demands must be reconciled with supply. The results of this process will indicate to personnel staff the number and kinds of positions that need to be filled within a specified time frame.

The personnel staff must then determine how long it will take to successfully fill positions. Factors to be considered include the state of the economy, the technical and educational requirements of the positions to be filled, and lead time for posting and advertising positions, and receiving and screening responses. Unskilled labor may be recruited in a week, but it may take 18 months to recruit a physician specialist. Nursing has presented a chronic recruiting problem (Filoromo and Ziff, 1980; Moscovice, 1984). For a variety of reasons relating to the unequal distribution of nurses (Goldsmith, 1981; Bureau of Health Professionals, 1980; Hanft, 1981; Moscovice, 1984), including external factors such as family obligations (Price, 1981) and undesirable working conditions in some facilities (Rosenblatt and Moscovice, 1982), it will continue to be difficult to recruit nurses. Special efforts such as recruiting trips to professional conventions and job fairs will have to be planned far in advance.

The economy affects recruitment in several ways. First, the amount of unemployment will determine the size of the available labor pool. This will affect various employee categories differently, but the amount of unemployment should be considered when timing and deciding the scope of a recruitment effort. Regional migration patterns that are economically motivated can also change the labor pool, hence recruitment plans. The technical and educational requirements of positions can tax the local labor pool. In these instances a wider search is needed, lengthening the amount of time required. Institutions have helped set up schools with specialized training programs for LPNs, RNs, or various allied health positions, to compensate for local labor shortages.

Decisions also have to be made on how many individuals to recruit for each job opening, since not all candidates will be acceptable, and some acceptable candidates will decline job offers. Historical data can be used to calculate yield ratios, comparing the number of candidates at each stage of the selection process to the number of candidates who make it to the next stage. Figure 10.2 shows an example of what yield ratios might look like when recruiting a nurse anesthetist.

In the past, physician recruitment either has not been done at all or has been done by the medical staff, but now, because of the strategic importance of recruiting the right kinds of subspecialists, hospital administration has become involved. An institution cannot implement a plan to diversify into certain services unless it has medical staff members willing to admit patients of those kinds. Also, as the number of staff model HMOs grows, the number of salaried physicians will increase. Recruiting and selecting these physicians will become an increasingly important responsibility of management (Berger and Schoen, 1981).

Deciding what types of physician are needed on the hospital's medical staff, hence, what types need to be recruited, is one area in which recruitment planning is inextricably tied to strategic and operational planning. This is always an executive function, but it is a support service that could be centralized in larger multiunit corporations.

Analyzing your medical staff for replacement purposes entails several steps.

- Categorize staff by specialty and prioritize based on importance to current plans.
- Note which physicians contribute more to the facility's profits. For cost-plus patients, indicators include number of admissions, total patient days, and number of ancillary procedures. As prospective payment becomes the norm, though, more sophisticated measures will need to be developed. Someone who admits many patients for long stays that are not totally reimbursed could jeopardize the hospital's financial position.
- Note whether these physicians are located in parts of the service area where the facility captures a strong market share. Also note whether they are in areas where the hospital hopes to improve its position.
- The physicians' ages should also be listed. In this way, one can predict when they will

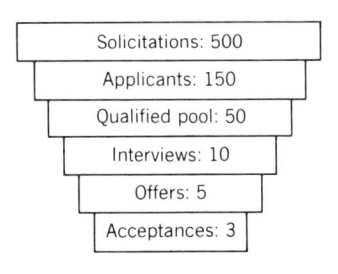

Solicitations: 500
Applicants: 150
Qualified pool: 50
Interviews: 10
Offers: 5
Acceptances: 3

Figure 10.2. Yield ratios for different recruiting steps. Adapted from Calhoun, R. P. *Managing Personnel.* New York: Harper & Row, 1966, pp. 147–148 and Cherrington, D. J. *Personnel Management: The Management of Human Resources,* 2nd ed. Dubuque, Iowa: William C. Brown, 1987.

be reducing their practice or retiring. Knowing when medical staff in key subspecialties are retiring will indicate which type of staff to recruit and when.

Table 10.3 shows how this information can be displayed for easier analysis. Such an array quickly reveals any gaps in the current complement of physicians on the staff (Kropf and Greenberg, 1984) that may be due to the unavailability of certain key support staff (and facilities). This would indicate further recruitment needs. Also, future gaps can be identified as key physicians retire or reduce their practice. Any plans should also compare medical staff needs to the projected array of specialists at competing institutions. A similar analysis can be done of the age structure of senior management personnel, who are often costly to replace (Ray, 1977).

As shown in Table 10.3, efforts should soon be under way to replace some of the chief admitters at this hypothetical facility. National Standards indicate shortages of chief admitters (see Goldsmith, 1981, p. 208; Jacobsen and Rimm, 1986; Steinwachs et al., 1986), and several key admitters will be retiring soon. To recruit a physician and for him or her in turn to build up a practice takes several years, so efforts should be started now. Possibly each of the senior key admitters could be encouraged to take on a partner who will eventually take over the practice.

Once a position has been targeted, it can take 18 months and cost up to $15,000 to recruit a physician (Cejka and Taylor, 1986; Garofolo, 1984). The board or medical

TABLE 10.3. Physician Analysis for Recruitment

Specialties	Percentage of Profits	Ages	Years to Retirement	Additional Physicians Needed Based on Rational Norms (GEMENAC)	Office Location Important?
I Subspecialty (prioritized by strategic importance)				13	
Chief admitters					
1. (Prioritized	5%	59	6		X
2. by	3%	35	30		
3. productivity	3%	42	23		
4. or	5%	58	7		X
5. contribution					
to					X
n. profits)					
II. Subspecialty				4	
Chief admitters					
1.	4%	56	9		
2.	1%	51	14		X
⋮					
n.					

SOURCE: Adapted from Kropf, R., and J. A. Greenberg. Strategic Analysis for Hospitals. Germantown, MD: Aspen, 1983.

staff may be reluctant to accelerate this process by incurring the cost of identifying the need and using a search firm. They should consider, however, what revenue could be generated if the needed physician were admitting 6 months sooner (Garofolo, 1984).

Who Is Responsible for Recruiting?

In most large organizations, recruiting is a function of the personnel department. Depending on the size of the department, recruiting may be the responsibility of the director, a staffing unit, or a recruiting unit. The personnel staff recommends policy and strategy to top management, seeks applicants from various sources, including the organization itself, screens applicants, and forms an applicant pool (Cherrington, 1987). When recruiting managers, physicians, or some technical staff, line managers, such as department heads or clinical chiefs, will take a more active role in the recruitment process. The complexity of federal regulations often requires that a personnel department member, such as an affirmative action officer, be in a position of functional authority (Koontz et al., 1984) over line staff, perhaps with authority to veto certain applicants or to reopen the process if correct procedures were not followed. This protects the organization from litigation from applicants who felt they were not treated fairly. A recruitment plan, then, should state who should be involved and how each person will participate in the relevant phases of the process.

As care-giving facilities coalesce or merge into multiunit alliances and systems, many of the technical aspects of recruitment and other personnel functions are being centralized. Indeed, recruiting on a wider geographic scope is espoused as one of the benefits of system affiliation (Zuckerman, 1979). The recruitment functions that lend themselves to centralization include:

- Evaluating a facility's needs
- Developing a computerized skills bank that can be adapted for use by facilities
- Advertising and screening for needed technicians, nurses, managers, or physicians
- Tracking of laws and regulations, such as affirmative action guidelines
- Training facility staff to comply with regulations, for example, when screening applicants
- Evaluating the cost effectiveness of facilities' recruitment systems

Where? Recruiting Sources and Methods

Internal Recruiting

One of the best *sources* of job applicants consists of the present employees of the organization. They are familiar with the culture and operations of the organization, and therefore, may be promoted or transferred into open positions with confidence that they will learn quickly. Also, promotions are one of the best motivators organizations have at their disposal. Career opportunity at all hierarchical levels can account for the way people involve themselves in their work. Studies have shown that dramatic increases in opportunity can lead to increased career aspirations, higher work commitment, and a sense of organizational responsibility (Kanter, 1977; Landau and

Hammer, 1986; Vardi and Hammer, 1977). Sometimes union contracts require management to consider internal candidates before looking outside the organization. The contract may also specify criteria to be used in developing an internal applicant pool (e.g., those with the greatest seniority must be considered first).

There are several *methods* an organization can use to facilitate internal recruiting. First, job vacancies can be posted in public access areas and in company newsletters or bulletins. This gives the employees the opportunity to apply for positions that fit their needs. Second, the organization can maintain a computerized skills inventory. This inventory may contain information on technical skills, supervisory experience, degrees obtained, language skills, and even career aspirations. When a position becomes vacant, or a new position is created, the organization will be able to quickly locate any potential candidates.

A third method of identifying candidates for management slots is succession planning and the use of a replacement chart (see Figure 10.3). It is a convenient way to show who is qualified to fill the different managerial positions in the facility. This approach is applicable only internally in large facilities, but it is a useful way to keep track of managers working in various units in the system. By identifying transfers that best match the needs of the positions with the career aspirations of the employees, there should be a reduction in turnover as a result of increasing employees' satisfaction.

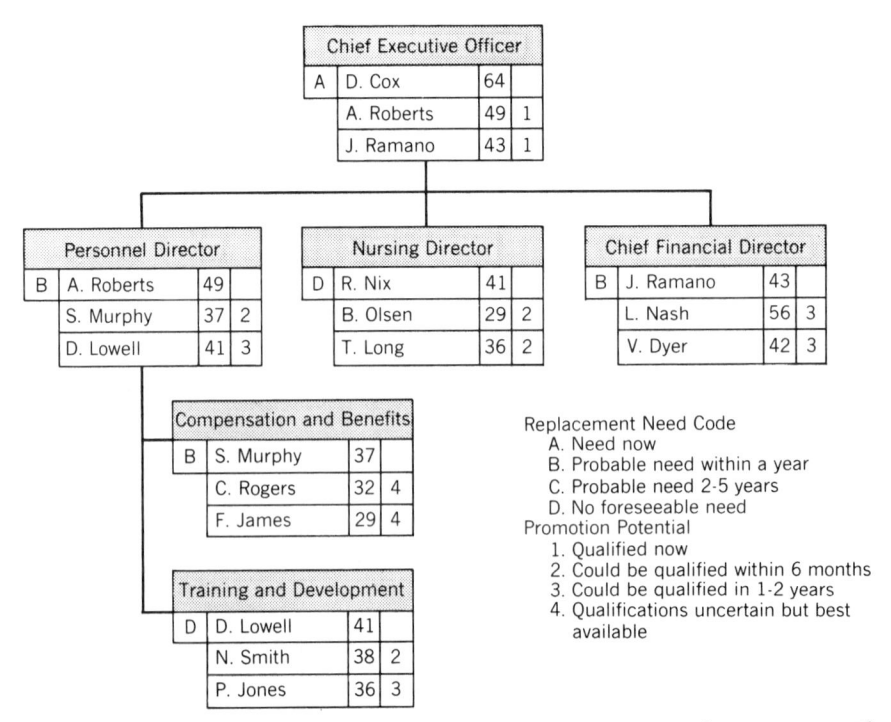

	Chief Executive Officer		
A	D. Cox	64	
	A. Roberts	49	1
	J. Ramano	43	1

	Personnel Director		
B	A. Roberts	49	
	S. Murphy	37	2
	D. Lowell	41	3

	Nursing Director		
D	R. Nix	41	
	B. Olsen	29	2
	T. Long	36	2

	Chief Financial Director		
B	J. Ramano	43	
	L. Nash	56	3
	V. Dyer	42	3

	Compensation and Benefits		
B	S. Murphy	37	
	C. Rogers	32	4
	F. James	29	4

	Training and Development		
D	D. Lowell	41	
	N. Smith	38	2
	P. Jones	36	3

Replacement Need Code
A. Need now
B. Probable need within a year
C. Probable need 2-5 years
D. No foreseeable need
Promotion Potential
1. Qualified now
2. Could be qualified within 6 months
3. Could be qualified in 1-2 years
4. Qualifications uncertain but best available

Figure 10.3. Replacement chart. From Cherrington, David J. *Personnel Management: The Management of Human Resources*, Second Ed. Copyright © 1983, 1987, William C. Brown Publishers, Dubuque, Iowa. All Rights Reserved. Reprinted by permission.

External Recruiting

Although there are several advantages to recruiting internally, there are also some disadvantages. When organizations need to shift their direction quickly, bringing in new people with new ideas and new ways of doing things may be easier than trying to change all current employees. Also, the manpower an organization needs for the future may not always be available internally. This is particularly true in health care, organizations are expanding rapidly and technological developments constantly create a demand for new skills. Then, the organization must depend on the external labor market to fulfill its manpower requirements.

External recruitment becomes necessary as rapid diversification creates the need for new knowledge and skills not found in a traditional health care setting. For example, insurance sales persons are needed to market prepaid health service packages to employers. There has been some movement to recruit managers from outside industries just to infuse more "business sense" into hospitals or systems. It is not clear that this will be cost effective because of the unique characteristics of service industries, in general (Sasser et al., 1978), and hospitals, in particular (Shortell and Kaluzny, 1983). The advantages and disadvantages of internal and external recruitment are summarized in Table 10.4.

The scope of the external search will depend on the types of position to be filled. The organization can generally rely on the local labor market to fill lower level unskilled and semiskilled positions. Nurses or highly trained technicians may need to be recruited from a region-wide labor market. This may require strong affiliations with allied health schools in the area. Managers may need to be recruited nationally from

TABLE 10.4. Advantages and Disadvantages of Internal and External Recruiting

Promotion from Within	
Advantages	Disadvantages
Greater motivation for good performance	Creates a narrowing of thinking and stale ideas
Greater promotion opportunities for present employees	Creates political infighting and pressures to compete
Better opportunity to assess abilities	
Improves morale and organizational loyalty	Requires a strong management development program
Able to perform the job with little lost time	

External Hiring	
Advantages	Disadvantages
New ideas and new insight	Loss of time due to adjustment
Able to make changes without needing to please constituent groups	Destroys incentive of present employees to strive for promotions
Does not change the present organizational hierarchy as much	No information is available about the individual's ability to fit with the rest of the organization

SOURCE: From Cherrington, David J., *Personnel Management: The Management of Human Resources*, Second Ed. Copyright 1983, 1987 William C. Brown Publishers, Dubuque, Iowa. All Rights Reserved. Reprinted by permission.

health care and related industries. Physicians may need to be recruited nationally for certain subspecialties. Residency programs have traditionally been used to attract new physicians to an area in the hope that they will remain after completing the residency.

Whether the scope of the search is narrow or wide, an organization can utilize many different *sources* to attract and identify applicants. These include direct applications; employee, medical staff, and other referrals; private and public employment agencies; executive search firms; special events; and high schools, colleges, and universities. Television, radio, journal, and newspaper advertisements are the chief *methods* used to attract recruits.

Direct Applications. If the organization has a good reputation and convenient location, direct applications (e.g., walk-ins and unsolicited write-in applications) may provide a large enough pool to fill all semiskilled and some technical positions at very little cost.

Employee, Medical Staff and Other Referrals. Many organizations rely on current employees to refer qualified applicants. Some organizations give cash bonuses to employees who encourage a friend or relative to apply for a position, if the applicant is hired. Additional cash bonuses may be paid if the referral remains employed by the organization for a specified time period.

The medical staff is probably not an adequate recruitment source for physicians because of its geographic limits and the changing needs of the hospital. However, the hospital exists in a complex network of relationships with suppliers, salespeople, licensing agents, and the like. These contacts are used to generate potential applicants. For example, sales representatives of suppliers to physicians' offices often can identify experienced but unhappy physicians.

Private and Public Employment Agencies. Private employment agencies charge fees to perform many of the prescreening functions an organization would usually perform, such as finding qualified applicants, doing preliminary interviewing, and obtaining application blank information. Many of them specialize in, for example, nursing or technical personnel. The quality of private employment agencies varies tremendously, so the organization may have to do some searching before finding one that adequately meets its needs.

Public employment agencies are operated by the states under the auspices of the U.S. Training and Employment Service. All men and women who collect unemployment insurance are required to use this service, which is free to both seekers and organizations. Public employment agencies now utilize computer systems to match applicants with jobs. Few organizations, however, take advantage of this service, since employers complain that the referrals often are poorly qualified and not always interested in accepting employment (Heneman et al., 1986).

Executive Search Firms. These highly specialized private employment agencies usually seek candidates currently earning $40,000 or more. They are more expensive and more aggressive than other private agencies, generally charging half an executive's annual salary as a fee. Whereas most private employment agencies rely on candidates who are actively looking for a position, executive search firms, often called "headhunters," will approach highly qualified professionals who are currently employed and had not been looking. Industry norms forbid hospitals to recruit employees directly from one another. Thus, headhunters may serve a useful purpose by broadening the pool of applicants for organizations that need upper level executives.

Care should be taken, however, to contract with a reputable firm. Table 10.5 lists some characteristics that distinguish reputable from disreputable executive search firms.

Special Events. Job fairs and meetings of professional societies provide a convenient setting for an organization's recruiter to meet with potential employees or search firms. Job fairs are often regional events in which employers in certain industries can meet with interested applicants and promote their organization's job openings.

High Schools, Colleges, and Universities. Educational institutions are good sources of applicants for positions that require very little experience. High school and vocational schools are good applicant sources for low skilled and trade positions. In some communities, businesses develop and partially fund training programs in conjunction with educational institutions. This arrangement provides the businesses with a readily available pool of qualified applicants.

Nursing schools have often been developed by larger hospitals. The professionalization of the field, with a concomitant raising of educational requirements, has forced many of the schools not associated with colleges or universities to close in recent years. There continues to be a shortage of specialized nurses, in operating rooms and intensive care units, for example, suggesting a need for more continuing education programs.

Colleges and universities are good applicant sources for nursing and managerial positions as well as physicians. Many universities now have specialized programs in health care administration. The Association of University Programs in Health Administration (AUPHA) publishes a directory of all member graduate and undergraduate health administration programs in the United States and Canada. A number of these require students to participate in either summer internships or longer residencies before graduating. Also, there are a growing number of competitive fellowships being offered. These usually require a 2-year commitment and often are based in the corporate offices of multiunit systems.

Finally, hospitals' affiliation with medical schools in order to employ residents has long been a means of recruiting physicians in local communities (Cohen, 1979; Yanish, 1979). Search firms recruit directly at medical schools or residency sites.

Other than working with the *sources* mentioned above, the major *method* of recruiting is through advertising.

Advertisements. Organizations can advertise position openings in newspapers and journals, on television and radio, and on even billboards. Organizations often consult with advertising agencies to determine the most effective medium and content of a recruitment campaign.

Advertisements should convey the qualifications and major responsibilities of the position, as well as the working conditions. If the organization is a government contractor, the advertisement must indicate that the organization is an equal opportunity employer. An advertisement implying that a candidate of a particular race, sex, religion, age, and/or nationality is preferred over others is illegal, unless the

TABLE 10.5. Characteristics of Reputable Search Firms

Service	Reputable Agencies	Questionable Agencies
Recruiting	Engage in selective recruiting occasionally, and then only to fill the urgent needs of client companies at their specific requests. Most applicants obtained through newspaper advertising.	Do all their recruiting through a large, heavily indoctrinated staff of "counselors." Prospects are called at home between 6 and 8 P.M. and Saturdays. This builds up reservoir of applicants. No selective recruiting for specific jobs.
Applicant motivation	Encourage applicants to seek positions commensurate with their highest skills, to set salary requirements at high but realistic levels.	Promise salary advantage to raise recruitment volume. Later, urge applicants to reduce salary demands to assure faster placements.
Résumés	Always provide personnel managers with résumés whether requested or not. Encourage applicants to prepare detailed résumés in a professional manner. Help them revise their résumés when necessary.	Do not provide personnel managers with résumés if it can be avoided. Insist on blind interviews. Standard pretext: "We are selling you the man, not an employment record."
Screening applicants	Interview in depth, to be certain that applicant is qualified in every particular for job he is seeking. Give approved tests when need is indicated. Question applicant to determine whether he is emotionally qualified for job and is likely to have a rapport with his prospective employer.	Spend minimum of time to get basic information. No testing by "counselors." Start here to maintain control of applicant and condition him to accept first job offer.
Screening companies	Study employers' record for fair employment practice. Do not send applicants to companies known to have poor reputations for job security, salaries or benefits.	None.
Placement follow-up	After a reasonable interval, contact both applicant and his new employer to determine whether both are satisfied.	None.
Advertising	Advertise only the existing job, not an imaginary opening so dramatized as to attract applicants.	None, or very little.
Counselors	Employ only experienced, professional counselors with proven records for placing the right people in the right job.	No previous experience in personnel work is required. Persons with background in selling goods or services are preferred. Persuasive telephone voice is an asset.

TABLE 10.5. (*Continued*)

Service	Reputable Agencies	Questionable Agencies
Job orders	Never send an applicant out for an interview unless a specific job order requesting a man with his qualifications has been received.	Rarely match applicants with job orders, except for general job category. Theory is that companies will be satisfied with applicants who approximate job descriptions. Frequently try to "pump in" applicants.
Personnel managers	Try to understand their problems. Don't demean them by attempts to reach department or division managers.	Regard company personnel managers as inefficient clerks. Try to bypass them whenever possible.
Fees	Either make firm agreements with client companies so that the latter pay fees, or inform applicants that fees will be chargeable to them.	Tell applicants virtually all fees are company-paid, but have no firm, contractual agreements with companies. Applicants must sign contracts requiring them to pay the fees if they accept jobs and the companies refuse to pay.

SOURCE: From Metzger, N. *Handbook of Health Care Human Resources Management.* Reprinted with permission of Aspen Publishers, Inc., Copyright 1981.

organization can prove that these characteristics are bona fide occupational requirements. According to Section 703(e) of Title VII of the Civil Rights Act of 1964, an organization may discriminate on the basis of race, sex, religion, age, or national origin if the attribute is a bona fide occupational qualification (BFOQ), reasonably necessary to the normal operation of that particular business or enterprise.

In developing an advertising campaign, an organization must also take into consideration the message it wants to convey to the public, and the audience it wants to target. Lower level, unskilled, and semiskilled workers can best be reached through want ads in local newspapers, or radio advertisements. Professional and technical workers are more likely to respond to advertisements in trade publications. Placing advertisements in publications of these types is particularly worthwhile if the level of expertise required for the position is not available through the local labor market. Trade and professional journals are distributed nationally and sometimes internationally, so they reach a wider and more specialized audience. A good list of journals and mailing lists for advertising job openings to the nursing labor market is given by Filoromo and Ziff (1980).

Evaluating Recruitment Sources

Organizations need to periodically evaluate the effectiveness of various recruiting sources. Questions that should be asked about each source include the following.

1. What is the ratio of costs to benefits in terms of number of applicants referred, interviewed, selected, and hired?

2. How effective are applicants hired from various sources in terms of job performance and absenteeism?
3. How long do hires who were obtained from a particular source remain employed by the organization?

Only a few studies have investigated the effectiveness of different recruitment sources. This research has found that employee referrals are the best source in terms of employee tenure, while newspaper advertisements and employment agencies are the worst (Decker and Cornelius, 1979; Gannon, 1971; Heneman and Schwab, 1986). A study of research scientists showed that employees recruited through college placement offices and newspapers did not perform as well as those who responded to a professional journal or convention advertisement or made contact based on their own initiative (Breaugh, 1981). Those who were recruited through newspaper ads were also absent twice as frequently as others (Breaugh, 1981). Clearly, however, the effectiveness of a recruitment source is going to depend on the type of position that needs to be filled. Table 10.6 shows the best recruiting methods for different job categories.

HOW? IMPLEMENTING RECRUITING PLANS

Attracting Recruits

A well-developed plan is essential for attracting job applicants. This is particularly important when the unemployment rate is low, or the organization needs to fill highly specialized positions for which few candidates are available. Organizations can convey information about themselves through the sources described above including tele-

TABLE 10.6. Relevant Recruiting Sources by Personnel Category

Source	Unskilled	Technical/ Professional	Managers	Physicians
Employee referrals	×	×		
Promotion	×	×	×	
Transfers	×	×	×	
Job posting	×	×		
Direct applications	×	×		
Employment agencies	×	×	×	
Search firms			×	×
Special events		×		
Schools		×	×	×
Temporary help	×	×		
Interns, residents			×	×
Newspaper advertisements	×	×	×	
Journal advertisements		×	×	×

vision, trade journals, community events, and brochures; however, the recruiter is most important in conveying a positive image of the organization to the applicant during screening interviews.

Research by Dean and Wanous (1984) has shown that turnover in organizations could be decreased by as much as 28 percent and satisfaction increased as well if candidates were given more realistic job previews. They argue that candidates should be told both the positive and negative aspects of a job and career path within an organization so that they might make informed choices. Breaugh (1983), however, states that this conclusion may be premature. Realistic job preveiws may be effective in reducing turnover and increasing satisfaction only when candidates can be selective about accepting job offers, when realistic information is not readily available about the job from other sources, and when the candidate might have difficulty in coping with certain job demands.

The effectiveness of any recruiting message depends not only on its content, but also on the credibility of its source. Organizations should select their recruiters carefully and make sure they are trained for the position. Studies have shown that candidates prefer recruiters who are middle-aged, have an important position within the organization, are pleasant and enthusiastic, and are knowledgeable about the candidate and the job vacancy (Rynes and Miller, 1983). Unfortunately, many organizations give the job of recruiting to new employees, who have little credibility with applicants and may not know very much about the organization. They may be effective in screening applicants, but they cannot provide candidates with the information needed.

Recruitment of Nurses and Physicians

Extraordinary efforts may be needed to attract nurses or physicians to a facility. Although the oversupply of physicians or the cutbacks in hospital staff may create a surplus of physicians and nurses in some areas, this could better be described as maldistribution. Rural and inner-city hospitals will continue to experience shortages. Shortages will have a relatively larger impact on the smaller staff of rural hospitals, affecting their ability to provide adequate coverage.

The centrality of nurses and physicians to hospitals justifies the sometimes extraordinary efforts made to attract them. The factors mentioned above—realistic job previews and credible interviewers—are important when enticing nurses to accept a position. In an attempt to attract nurses, organizations have tried emphasizing the quality of working life at the hospital, the interesting variety of patients and treatment modalities, the educational opportunities, flexible working scheduling with differential pay for evening or weekend shifts, and/or help with housing. Free vacations have even been offered.

Physicians are concerned about the same kinds of issues as nurses or other professional staff, and because of the physicians' centrality to the functioning of the hospital, these issues have to be carefully addressed to successfully recruit a productive, balanced medical staff. Physicians are hard to recruit for several reasons (Garofolo, 1984).

- They do not place a high priority on career development, but rather focus on professional development.

- They are constantly being recruited.
- They have little time for interviewing.
- They may be indecisive with nonclinical decisions.

One needs to look at the practice opportunity through the eyes of the candidate to see if all of his or her concerns can be addressed. One then needs to use several information sources (letters, phone interviews, visits, and negotiations) to convey the message that the physician's needs can and will be met by the hospital and the community. Physicians will want to have their professional interests fulfilled, their financial risks eliminated, and their families made happy. Table 10.7 lists the factors physicians and nurses look for in a practice opportunity and some methods by which health care organizations can address these factors.

It is clear from Table 10.7 that attracting physicians and nurses is a multifaceted task that involves not only the administration and the candidate, but key members of the hospital and community. Because of the conflicts of interest of those involved in physician and nurse recruitment, search firms advocate use of a neutral facilitator to:

- Keep the process moving and attend to the numerous details
- Manage the appropriate participation of key individuals
- Help the organization establish screening criteria and the ground rules of the process
- Screen all leads
- Follow up on inquiries and commitments
- Attend to the needs of the successful candidate's family

One should be ready to make an offer during the site visit if the candidate is appropriate for the job. Be prepared to negotiate!

Screening Recruits

In addition to attracting good candidates the recruitment process serves to disqualify unsuitable candidates. Reviewing applications or résumés and interviewing are the main ways to perform initial screening. It is important that this process be done consistently. Consistent screening is possible only if clear criteria are applied to each applicant by trained screeners. Choosing appropriate screening and selection criteria is discussed in Chapter 11. For screening purposes it is only necessary that the criteria be legal and define the prerequisite knowledge and skill for the job.

The recruitment interview, besides providing an opportunity to impart information about the organization, is a time to screen out unsuitable job applicants. Whereas the organization wants to use the interview to attract good candidates and eliminate unsuitable candidates, from the candidate's perspective the purposes are to receive an invitation to be seriously considered for a position and to find out as much as possible about the job and the organization. In both cases the first item usually takes top priority, so both the candidate and the organization spend their time trying to look good. As a result, neither the candidate nor the organization obtains the information necessary to make an informed decision. The organization may select an employee based on

TABLE 10.7. Incentives for Physicians and Nurses

Practice Opportunity Factors	Incentives
Financial	
Income	Income guarantees during the transition

Insurance
  Malpractice
  Health
  Disability
  Life
	Complete coverage or at group rates; professional help at designing and managing a coverage plan

Pension program	Complete coverage or at group rates; help at designing and managing a coverage plan
Productivity	Performance incentives, bonuses
Vacation	Costs paid
Continuing education	Leaves and allowances
Practice environment	
Hospital	A case mix that fits the physician's interests; special staff, equipment, and facilities; facility's future plans
Office	Assistance in locating, setting up, staffing, managing an office; low-cost rent or low-interest loans
Professional relationships	
Peers	Professional membership; well-orchestrated interviews; group practice membership; or help in recruiting a partner
Referrals	Clear information on utilization patterns; introductions to primary care physicians or prepaid plans; a physician referral service
Coverage	Help in joining or forming a group
Medical school or teaching opportunities	A plus for some physicians
Community	
Housing	Help in locating and financing a home; moving costs
Quality of school systems	Clear information on both public and private schools through college
Job opportunities for spouse	Information and professional contacts and help in job search
Social and cultural opportunities	Club memberships; personal contacts; a well-planned visit
Recreational opportunities	Club membership; visit
Location; climate, geography population	Visit

inaccurate information, and the employee may accept a position without really knowing what the job or organization is like. Careful use of screening criteria in a well-delineated interviewing procedure can maximize the amount of information gleaned from an interview.

During the screening process it is very important to keep accurate records of applicants and to respond promptly to all contacts. If good records are not kept, the selection process cannot proceed smoothly and qualified candidates may be lost because of slow organizational response. Recruiting entails a significant contact with the community; therefore, all applicants should be treated politely. A person who is not suitable for the current position may be an ideal candidate for a future position.

Recruitment and the Law

Recruitment is the first step in the selection process. There can be no selection process if an organization does not have a pool of candidates. Since choice of recruitment policies and procedures influences the number and kinds of individuals who will make up the applicant pool, the entire recruitment process must meet EEOC guidelines. Title VII of the Civil Rights Act of 1964 prohibits discrimination on the basis of sex, race, religion, or national origin. Organizations must evaluate whether their choice of recruiting source is disproportionately eliminating minority groups from the applicant pool. They must also make sure that they screen candidates on the basis of bona fide occupational qualifications only.

According to the provisions of the EEOC uniform guidelines, the following questions may not be asked of candidates either in screening interviews or on application forms, unless the organization can prove that the qualification is a business necessity (Lowell and DeLoach, 1982):

1. Names used previously. Names could indicate a candidate's marital status, sex, or national origin.
2. Height and weight, unless these are BFOQs.
3. Age. The Age Discrimination in Employment Act of 1967 protects employees from 40 to 70 years of age.
4. Religion.
5. Race/color.
6. Citizenship. Asking whether candidates are citizens of the United States is legal, but asking whether are naturalized or native-born citizens or asking for the date of citizenship is illegal.
7. National origin.
8. Education requirements, unless these are a BFOQ. Asking for dates of attendance and graduation may be construed as illegal since these dates may indicate an applicant's age.
9. Military status. An organization may ask if the candidate is a veteran but may not ask questions regarding type of discharge or branch of military served in.
10. Arrest records. Applicants cannot be disqualified because of arrest records, since an arrest does not mean the individual was guilty, and minorities are arrested more frequently than nonminorities.
11. Name and/or address of a relative, unless the applicant is a minor.

12. List of physical handicaps, defects, or past illnesses. An organization may ask whether the applicant has any physical handicaps or illnesses that might interfere with job performance.
13. Marital status and/or names and ages of children, since these questions adversely impact women.
14. Sex. This question cannot be included on an application blank unless it is a BFOQ.
15. Housing. Questions cannot be asked regarding whether an applicant owns his or her own home, rents, or lives in a house or apartment.

Several of these questions can be asked after an applicant has been hired; for example, number and ages of children is information needed for insurance purposes. Also, questions such as race and/or national origin can be asked if they are to be utilized for research purposes only (filling out EEOC reports), and this is clearly stated on the application blank.

Recruiting physicians creates a number of different legal problems relating to not-for-profit tax status (Mulhausen and Tracy, 1986) and Medicare fraud and abuse laws. Some of the incentives needed to attract physicians, such as subsidized office space or income guarantees, may constitute "private inurement." This is forbidden if a not-for-profit status is to be maintained with the IRS. If those or similar inducements can be construed as the quid pro quo for referrals of Medicare patients, the hospital is at risk of violating fraud and abuse laws. Therefore, thought must be given to what incentives will be offered when recruiting physicians.

EVALUATING THE RECRUITMENT PROCESS

Evaluating the recruitment process is extremely important both to ensure the success of the institution's strategy and to allow the efficient functioning of other personnel functions such as selection or training. Strategically important indicators, although not solely measures of poor recruiting, would focus on physicians or professional support staff. These include:

- Loss of market share to competitors from lack of subspecialists in high demand areas
- Low quality of care, hence an unacceptable incidence of errors or even malpractice claims because of insufficient or unqualified staff
- Physician dissatisfaction with the quality of the nursing and support staff

Other personnel activities can depend on the effectiveness of recruiting. For example, any selection activity will have limited utility if quality applicants cannot be located within a specified time frame. For this reason all aspects of the recruitment process, including sources (internal or external), methods, and administration, should be evaluated in terms of the costs and benefits to the organization. A variety of measures can be used at different steps in the recruiting process to evaluate the effectiveness of attracting qualified applicants. For each recruiting source and method, data can be collected on the number of applicants, costs per applicant, and time required to locate applicants (Milkovich and Glueck, 1985). Once candidates have been screened and selected, data can also be collected on the costs per hire and time lapsed per hire. Also, yield ratios can be calculated comparing visits offered to the total number of applicants, offers extended to either the number of visits accepted or number of qualified applicants, and offers accepted compared to the number of offers extended (Milkovich and Glueck, 1985). Data should also be kept on the number of minority and female applicants and hires compared to the total number of applicants and hires, to ensure that the organization is meeting its equal employment opportunity and affirmative action goals.

The determination that a recruiting system is attracting applicants in a timely manner is not sufficient to judge its effectiveness. The quality of applicants should also be assessed by tracking the progress of new hires in terms of job performance and turnover. These behavioral outcomes, however, may be indicators of a poor work environment, an inefficient selection system, a poor training program, or other problems not related to the recruitment process. These other possibilities should be investigated before any changes are made in the recruitment process.

Evaluation of the costs of recruitment should include both direct and indirect costs. Direct costs include fees paid to agencies, telephone, travel expenses, entertainment, salaries of recruiting staff, and advertising fees. Indirect costs include the time involvement of operating managers, physicians, and board members (Milkovich and Glueck, 1985). Because these costs are usually underestimated (Cejka and Taylor, 1986) when initial plans, if any, are made, the search often seems to be not cost effective.

Finally, the administrative components of the recruitment process should be evaluated. Are data on candidates stored so that information is easily accessible? Is the organization responding to applicants in a timely manner? If the response time is too long, good candidates may be lost to other organizations. Also a lack of response to unqualified candidates may promote a poor public image.

SUMMARY

Recruiting is and will continue to be an important function of human resources in health care. As illustrated in Table 10.8, the relevance of recruitment varies by job category. The difficulty of answering the questions of what, where, why, and when also varies by job category. All levels of the organization are involved in an effective recruiting campaign.

The continuation of current trends in health care means that recruiting will remain a strategically important function. Diversification into alternative service delivery settings will continue to require staff with new sets of skills. Because of vertical integration into larger, multiunit, comprehensive health care organizations, skilled managers will continue to be in demand. Complicated clinical management decisions revolving around the issues of quality and quantity of care will increasingly create a need for physician administrators.

These new staffing needs will mean that health care organizations will be competing for personnel with other industries as never before. This will require sophisticated recruiting to assure an adequate supply of applicants. At the same time, cost containment pressures may decrease the attractiveness of health care facilities as places to work. An effective recruiting system could provide a real competitive advantage to an organization.

TABLE 10.8. Summary of Key Issues by Personnel Category

	Staff Categories			
Recruiting Issues	Unskilled Workers	Technical/ Professionals	Managers	Physicians
New needs?		Yes	Yes	Yes, primary care gatekeepers
Planning				
Why?		New strategy	New strategy	More control
Who's responsible?	Personnel	Department heads; personnel	Top management	Management and medical staff
What positions?	Same	New settings— alternative delivery system (ADS)	New setting— ADS	Salaried, gatekeepers
When?	Short lead time	Moderate lead time	Moderate lead time	Moderate to long lead time
Where available?	Local	Local, regional; lateral placement	Regional, national; other industries	National; residency programs
Implementation: How?				
Attracting	Ads, walk-in, agencies, etc.	Ads, agencies, etc.	Search firms, national trade journals	Search firms; joint ventures; service incentives
Screening		Professional registry	Important	Important
The law[a]	EEOC/AA; unions	EEOC/AA, unions		Tax, Medicare, antitrust questions
Evaluation	Short-range; operational measures	Short to mid-range impact	Strategic impact and measures	Strategic impact and measures
Future needs		Changing	Increased importance	

[a] EEOC = Equal Employment Opportunity Commission; AA = affirmative action.

ACKNOWLEDGMENTS

The authors acknowledge the help of a number of human resources executives who provided information on their organizations' practices, particularly Michael Howe, Senior Vice-President, Human Resources, HealthOne; Frank St. Denis, Vice-President of Human Resources, HealthWest; Larry Smally, Vice-President of Human Resources,

Hospital Group, National Medical Enterprises; and Robin Walker, Vice-President, Tyler & Company.

Geoffrey Hoare was at the Department of Health Systems Management, Tulane University School of Public Health, while this chapter was written and is grateful for the support he received there.

REFERENCES

Ackoff, R. L. *A Concept of Corporate Planning,* New York: Wiley, 1970.

Berger, J. E., and S. Schoen. "Future of physician recruiting." *Group Practice Journal,* pp. 7–10 (July 1981).

Breaugh, J. A. "Relationships between recruiting sources and employee performance, absenteeism, and work attitudes." *Academy of Management Journal,* 24:142–147 (1981).

Breaugh, J. A. "Realistic job previews: A Critical appraisal and future research directions." *Academy of Management Review,* 8:612–619 (1983).

Bureau of Health Professions. "The recruitment shortage of registered nurses: A new look at the issue." Washington, DC: U.S. Department of Health and Human Services, 1980.

Calhoun, R. P. *Managing Personnel.* New York: Harper & Row, 1966, pp. 147–148.

Cejka, S. A., and M. W. Taylor. "When is the right time to add a physician?" *Medical Group Management,* 33(5):18 (September–October 1986).

Cherrington, D. J. *Personnel Management, The Management of Human Resources,* 2nd ed. Dubuque: W. C. Brown, 1987.

Cohen, P. D. "Medical school and hospital affiliation relationships: An interorganizational perspective." *Health Care Management Review,* 6(1):43–50 (Winter 1979).

Dean, R. A., and J. P. Wanous. "Effects of realistic job previews on hiring bank tellers." *Journal of Applied Psychology,* 69:61–68 (1984).

Decker, P. J., and G. T. Cornelius. "A note on recruiting sources and job survival rates." *Journal of Applied Psychology,* 64:463–464 (1979).

Filoromo, T., and D. Ziff. *Nurse Recruitment: Strategies for Success.* Rockville, MD: Aspen Systems, 1980.

Gannon, M. J. "Source of referral and employee turnover." *Journal of Applied Psychology,* 226–228 (1971).

Garofolo, F. "What medical staffs need to know about recruiting physicians." *Hospital Medical Staff,* pp. 18–24 (September 1984).

Goldsmith, J. *Can Hospitals Survive?* Homewood, IL: Dow Jones-Irwin, 1981.

Hanft, R. S. "Health manpower." In S. T. Jonas, Ed., *Health Care Delivery in the United States.* New York: Springer, 1981.

Heneman, H. G., D. P. Schwab, J. A. Fossum, and L. D. Dyer. *Personnel and Human Resource Management.* Homewood, IL: Dow Jones-Irwin, 1986.

Hoare, G. A. "New managerial roles in multi-organizational systems: Implications for health administration." *Journal of Health Administration Education,* 5(3) (Summer 1987).

Jacobsen, S. J., and A. A. Rimm. "Primary care physicians: HMOs versus the GMENAC." *New England Journal of Medicine*, 315(5):324 (1986).

Kaluzny, A. D., and S. M. Shortell. "Challenges for the future." In S. M. Shortell and A. D. Kaluzny, Eds., *Health Care Management*. 1st ed., New York: Wiley, 1983.

Kanter, R. M. *Men and Women of the Corporation*. New York: Basic Books, 1977.

Koontz, H., and C. O'Donnell, H. Weihrich. *Management*, 8th ed. New York: McGraw-Hill, 1984.

Kropf, R., and J. Greenberg. *Strategic Analysis for Hospitals*. Rockville, MD: Aspen Systems, 1984.

Landau, J., and T. H. Hammer. "Clerical employees' perceptions of intraorganizational career opportunities." *Academy of Management Journal*, 29:385–404 (1986).

Lowell, R. S., and J. A. DeLoach. "Equal employment opportunity: Are you overlooking the application form?" *Personnel*, pp. 49–55 (July–August 1982).

Milkovich, G. T., and W. F. Glueck. *Personnel/Human Resource Management: A Diagnostic Approach*, 4th ed. Plano, TX: Business Publications, 1985.

Miller, A. E., and M. G. Miller. *Options for Health and Health Care: The Coming of Postclinical Medicine*. New York: Wiley, 1981.

Mintzberg, H. *The Structuring of Organizations*. Englewood Cliffs, NJ: Prentice-Hall, 1979.

Moscovice, I. "Health care personnel." In S. J. Williams and P. R. Torrens, Eds., *Introduction to Health Services*. 2nd ed. New York: Wiley, 1984.

Mulhausen, M. R., and K. L. Tracy, "What are the risks of physician recruitment programs?" *Healthcare Financial Management*, pp. 42–56 (August 1986).

Price, J. L. *Professional Turnover: the Case of Nurses*. New York: SP Medical and Scientific Books, 1981.

Ray, K. "Managerial manpower planning—A systematic approach." *Long Range Planning*, 10(2):21–30 (1977).

Rosenblatt, R. A., and I. S. Moscovice. *Rural Health Care*. New York: Wiley, 1982.

Rynes, S. L., and H. G. Miller. "Recruiter and job influences on candidates for employment." *Journal of Applied Psychology*, 68:147–154 (1983).

Sasser, W. E., R. P. Olsen, and D. D. Wyckoff. *Management of Service Operations*. Boston: Allyn & Bacon, 1978.

Schein, E. *Organizational Culture and Leadership*. San Francisco: Jossey Bass, 1985.

Schuler, R. S. *Personnel and Human Resources Management*, 2nd ed. St. Paul, MN: West Publishing, 1984.

Shortell, S. M. and A. D. Kaluzny, "Organization Theory and Health Care Management." In S. M. Shortell and A. D. Kaluzny, Eds., *Health Care Management*, 1st ed. New York: Wiley, 1983.

Steinwachs, D. M., J. P. Weiner, S. Shapiro, P. Batalden, K. Coltin, and F. Wasserman. "A comparison of the requirements for primary care physicians in HMOs with projections made by the GMENAC." *New England Journal of Medicine*, 314:217–222 (1986).

Tichy, N., C. J. Fombrun, and M. A. Devanna. "Strategic human resource management." *Sloan Management Review*, pp. 47–61 (Winter 1982).

Vardi, Y., and T. H. Hammer. "Intraorganizational career mobility and career perceptions among rank and file employees in different technologies. *Academy of Management Journal,* 20:622–634 (1977).

Yanish, D. L. "Small hospitals, recruit family as well as M.D." *Modern Health Care,* p. 64 (November 1984).

Zuckerman, H. S. "Multihospital systems: Promise or performance." *Inquiry,* 291–314 (Winter 1979).

CHAPTER 11

Selection and Placement

Jacqueline Landau
Dan Fogel
Lisa Frey

As health care organizations grow larger and adapt more quickly due to changing environments, the selection of personnel becomes increasingly important. Health care is a service industry, with care levels ranging from clinic/outpatient, emergency, and ambulatory surgery, to inpatient, long-term/skilled nursing. Hospitals are in the process of diversifying. In the past few years, hundreds of nonprofit hospitals have been restructured into holding companies so that they could move into some of the other health care areas being exploited by the for-profits. Diversification is forcing hospitals to strengthen their marketing and long-range planning departments, and to find staffs to handle such issues as cost accounting, labor productivity, and financial modeling (Coddington et al., 1985). Also, the advent of Medicare's prospective payment system (PPS) requires hospitals to become increasingly competitive by building a market niche, creating product differentiation, and more efficiently using resources.

As mentioned in Chapter 10, these changes create needs for individuals with qualifications different from those required in the past. In addition to having the appropriate credentials and fitting in with the organization's culture, employees need to be aware of the economic ramifications of their use of resources and delivery of patient care. For example, previously most health care administrators received formal training in clinical, public health, or health care/public administration. Now, increasing numbers of administrators have MBAs and private sector business experience. These administrators provide health care organizations with the entrepreneurial spirit and understanding of business/economic issues needed for survival today.

At the same time that the need for employees with flexibility and diverse skills and abilities has increased, the number of hospital employees has decreased (Coddington et al., 1985). Health care organizations must learn to do more with less, which means that matching the right person to the right job has become increasingly important. Organizations can ill afford a nonproductive, dissatisfied employee. The skills, abilities, motivations, and career goals of individuals must match the requirements of the job. Therefore, the selection process must be planned and implemented with care.

DESIGNING THE SELECTION PROGRAM

Designing a selection program involves four primary steps: conducting a thorough job analysis; defining employee effectiveness and deciding how it will be measured (choosing criteria); choosing predictors of employee effectiveness and deciding what the selection instruments should be; and collecting scores on both the criteria and predictors and examining the relationship between criteria and predictors (testing the validity of the selection instruments) (Heneman et al., 1986).

The objective of these steps is to produce a reliable and valid selection process that is in compliance with federal regulations. The organization must then decide, on the basis of a cost-benefit analysis, what should be the cutoff scores on selection instruments and how and when candidates should be evaluated throughout the hiring process (see Table 11.1).

Job Analysis

Job analysis consists of two components: describing the tasks, responsibilities, activities, and working conditions of a job (job description), and identifying the qualifications needed for adequate job performance, including skills, abilities, and experience (job specifications). Without this information, the organization will not know the type of people it needs to hire, what kinds of instruments need to be developed to predict job performance, or how to evaluate the effectiveness of the selection process. *The Uniform Guidelines on Employee Selection Procedures*, which

TABLE 11.1. Steps in Designing a Selection Program

Step	Key Tasks to Accomplish Program
Job analysis	Write job description; identify qualifications for adequate job performance
Choose criteria	Choose selection criteria; choose ways to measure these criteria
Choose predictors	Select information that will predict criteria; identify selection tools to obtain information
Collect data	Collect data on criteria and predictors
Study relationship between criteria and predictors	Establish reliability, validity, and utility of instruments
Federal law compliance	Study process in light of federal laws
Establish cutoff scores	Establish minimal levels acceptable
Establish how and when candidates should be evaluated throughout process	Establish applicant pool; establish hiring process in company; set up records

provides a framework for determining the proper use of tests and other selection procedures to ensure that organizations will be in compliance with federal law prohibiting discrimination, specifically state that a thorough job analysis must be the basis for the design of a selection procedure (Equal Employment Opportunity Commission, 1978).

The first step in the job analysis process is deciding what kinds of data to collect. Relevant information may include tasks performed and percentage of time spent on each task, working conditions, supervisory responsibilities, reporting relationships, job hazards, physical requirements, machines and equipment used to perform the job, and materials or services produced. The type of data collected depends on (a) how the data are going to be used, (b) the accessibility and accuracy of the data, and (c) the relevance of the data to the job.

This information may be gathered using several different methods, such as observation, interviews, diaries, recording instruments, and questionnaires. Factors to consider when choosing a method include geography, physical environment, technology, complexity and routinization of the job, background of the job holders, and social factors (Henderson, 1979). For example, if the environment is hazardous or the job complex, direct observation would be a very difficult undertaking. For hazardous jobs, unobtrusive observational techniques, such as recording devices, could safely provide information. For complex jobs, interviews or questionnaires might be the best choice for gathering data. Diaries could be useful if job incumbents were committed and motivated to record their daily activities.

Choice of method will also depend on the job analysis procedure selected by the organization: conventional or quantitative. The conventional procedure begins with a job analyst interviewing supervisors and job incumbents and administering a questionnaire. On the basis of the subjective interpretation of the analyst, the data are summarized into a standard format. Then supervisors and incumbents are given an opportunity to change and approve the description. The final description usually contains three sections. In the first section the job is identified by title, location, and number of incumbents (Milkovich and Newman, 1984). In the second section the job is defined in terms of its purpose, its relationship to other jobs and overall organizational objectives, and the end results (Milkovich and Newman, 1984). In the third section the job is described in terms of its major duties, degree of discretion, supervisory responsibility, and perhaps training and experience required (Milkovich and Newman, 1984). Table 11.2 shows an example of a conventional job description for a registered nurse.

Another example of a conventional approach to job analysis is the functional procedure used by the U.S. Training and Employment Service. Information on this procedure is contained in the *Handbook for Analyzing Jobs*, (U.S. Department of Labor, 1972) and descriptions for approximately 20,000 jobs are listed in the *Dictionary of Occupational Titles* (U.S. Department of Labor, 1977). The descriptions include five categories of information: worker functions; work fields; machines, tools, and equipment; materials, products, subject matter, and services; and worker traits. Jobs are described according to their relationship to data, people, and things, and identified by the function the worker performs. Within each of these three categories workers may perform several functions ranging from simple to complex. (For a modified version of this approach see Fine and Wiley 1971).

Some critics believe that conventional job analysis procedures are too subjective because they depend heavily on the opinion of the job analyst. Quantitative methods rely on computer analysis of tasks and worker traits. Job incumbents are asked to

TABLE 11.2. Conventional Job Description

Title:	Registered Nurse
STAFF NURSE:	Less than 24-hour care units: Labor/Delivery, Dialysis, Emergency Room, Operating Room, Recovery Room, Clinic
SUMMARY:	To function as a member of the total nursing team on their assigned unit. The staff nurse is expected to utilize the nursing process in planning care with the patient, significant others, and the health care team. Through assessment and planning, she or he should formulate and document and continuously evaluate the nursing care plans, nurses' notes, accurate patient care conferences.
RELATIONSHIPS:	Responsible to head nurse and/or charge nurse Workers supervised: Ancillary nursing personnel of unit Interrelationships: Works closely with all members of the health care team, patients, and their families The staff nurse is responsible and accountable for quality nursing care of assigned patients. The staff nurse is also responsible for duties assigned by the charge nurse.
ASSESSMENT:	1. Initial assessment of patient is complete according to unit time guide. 2. Basic head-to-toe assessment at beginning and conclusion of procedure/stay.
PLANNING:	1. Documentation will reflect standard of care specific to the given care unit. 2. Documentation of nursing care will reflect compliance with the medical plan of care. 3. Documentation of patient care planning will be reflected in the nursing documentation. 4. Patient teaching/D/C planning will be reflected in nursing documentation as appropriate to the needs of the patient. 5. Psychosocial needs of patients will be identified.
INTERVENTION:	1. Transcribes routine physician's orders (includes proper utilization of forms, requisitions) according to policy. 2. Administers medications all routes according to policy; assess outcomes for effectiveness, toxicity, side effects; is aware of usual dosage. 3. Initiates and monitors IV therapy including fluids, blood, blood products, TPN according to policy. 4. Performs IV assessments and care (assessment of rate, site; dressings and tubing changes). 5. Handles controlled substances according to policy. 6. Monitors and records vital parameters (TPRs, BPs, weights, I & Os, head circumferences, abdominal girths). 7. Provides comfort, hygiene of patients (A.M. care, P.M. care, baths, mouth care, skin care, shampoos, positioning, T & D hose, etc.). 8. Collects specimens: ensures correct disposition, tests as appropriate (urine, blood, sputum, stools; tests specimens for S & A, hematest, reducing substances, pH's, etc.). 9. Uses equipment correctly and appropriately to patient's needs, (Med. pumps, Gomso suction, beds, scales, Emerson suction, Pluravacs, oxygen equipment).

TABLE 11.2. (*Continued*)

	10. Complies with all infection control policy and procedures (isolation, hand washing, etc.).
	11. Utilizes appropriate resources/supplies when assisting with/performing procedures (dressings, wound care, drains and drain care, urethral catheterizations; assisting with LPs, paracentesis, thorencentesis, bone marrows, kidney biopsies, insertions of central lines, cutdowns).
	12. Recognizes and handles emergency situations in a prudent manner.
	13. Notifies charge nurse/physician of change in patient status or condition.
	14. Completes documentation of all patient care activities.
	15. Provides nutritional support (PO feedings, calorie counts, gastrostomy feedings, N.G. feedings).
	16. Provides for safety of patient within environment (side rails, call lights; correct use: restraints, wheelchairs, stretchers, etc.).
	17. Provides post mortem care including completion of death certificate according to policy.
EVALUATION:	1. Reassesses patient as situation warrants and documents accordingly.
	2. Completes transfer stamp or discharge assessment, as appropriate.
	3. Contributes information to the discharge planning form as indicated.
MISCELLANEOUS:	1. Correctly utilizes incident reports/first report of injury.
	2. Knows visiting policies and enforces as appropriate.
	3. Participates in change of shift report, transfer report, etc.
	4. Correctly uses phone system, paging system, on-call schedules, Executone system.
	5. Conforms with dress code.
	6. Aware of and able to use correctly inter- and intradepartmental resources (consultations, social service, diabetic teaching program, biomedical engineering, etc.).
	7. Maintains a professional and therapeutic relationships with patient, family, health care team.
	8. Checks crash carts as assigned.
	9. Handles patient classification (demonstrates ability to classify patients by TMC H/C acuity system).
	10. Utilizes manual, knows location, content.
	11. Certified to initiate IVs and/or venipunctures.

complete a structured questionnaire that includes items on all aspects of their job. They are usually asked to rate the listed tasks in terms of their importance, and the time it takes to perform and learn them. Knowledge, skills, and abilities are measured in terms of degree of importance and prior experience required (Milkovich and Glueck, 1985). Since questionnaires take considerable time, expense, and effort to develop, they are usually purchased from consultants who have spent years validating and refining the instruments. Well-known examples of these questionnaires include the Position

Analysis Questionnaire and Control Data Corporation's Position Description Questionnaire.* Once the surveys have been completed, the data are entered onto a computer, statistically analyzed, and summarized. The costs of analyzing the data should be considered before choosing a quantitative procedure.

Choosing Criteria

The second step in the selection process is choosing which criteria to use, and how to measure the criteria. Questions to be considered include:

Does the organization want to select people who will be successful filling current job vacancies, or future positions?

Does the organization want to hire people who are likely to leave after a couple of years, or is it looking for stable, long-tenured employees?

Does the organization want employees who already have specific skills and abilities, or does it want employees who are able to learn quickly on the job?

The global criterion for almost every organization is job performance, but the aspects of job performance that are important will vary from organization to organization and from job to job. Once the global criterion has been chosen, more specific criteria are selected. If the global criterion is job performance, job descriptions can be used for determining more specific criteria. If the global criterion is the ability to move rapidly up the hierarchical ladder, job descriptions of future positions rather than current vacancies need to be referred to.

Next, methods for measuring the specific criteria determined must be chosen. A frequently used measure of job performance is the score on a performance appraisal instrument; however, for low-skilled, highly mechanized jobs, a count of how many items are produced within a certain time frame without wastage may be more indicative.

Choosing Predictors

The next step in the selection process is choosing the predictors and selection instruments. Based on knowledge of job specifications, it is hypothesized that certain individual characteristics are related to job performance. For example, a relationship might be hypothesized between the ability to cope with organizational change and successful job performance as a hospital administrator. Next, it must be decided how to measure these characteristics. Perhaps the ability to cope with change could be demonstrated by previous management experience in a turbulent organizational environment. This information could be obtained through a résumé, interview, and reference check. Today organizations have a wide choice of selection instruments, including ability and personality tests, job simulations, interviews, reference checks,

---

*Information on the PAQ and the PDQ, respectively, is available from PAQ Services, P. O. Box 337, Logan, Utah 84321, and from Control Data Corporation, 8200 34th Avenue South, Minneapolis, Minnesota.

application blanks, and assessment centers. Choice of method will depend on the characteristic being measured, the type of position being filled, and the costs to the organization. Each method has advantages and disadvantages, which are discussed later in the chapter. Table 11.3 summarizes the selection instruments frequently used by health care institutions.

The Relationship Between Predictors and Criteria

The relationship between criteria and predictors can be determined through statistical or nonstatistical methods. But to determine whether the use of the predictors improves the organization's chances of selecting the best people for the job, the reliability and the validity of the chosen predictors must be ascertained. *Reliability* refers to the accuracy of the measuring instrument. *Validity* refers to the extent to which a score on the predictor is related to success within the organization.

Reliability

Reliability, as mentioned earlier, refers to the consistency of the measuring instruments. If two interviewers interview the same candidate using the same questions and format, will their conclusions about the candidate be similar? If one candidate takes the same intelligence test twice in one week, will he or she receive approximately the same score on the test each time? Both these questions address the issue of reliability.

The reliability of a selection instrument can be estimated using several different methods, including test-retest reliability, alternate forms reliability, split-halves reliability, and interrater reliability. Test-retest reliability is determined by measuring candidates more than once on the same instrument. If the scores are approximately the

TABLE 11.3. Selection Instruments in Health Care Institutions

Instrument	Benefits	Problem Areas
Tests	Many are available	Aptitude tests are controversial; difficult to validate
Interview	Easy to use; can help obtain information on "intangibles"; have credibility in court cases	Unreliable and probably not valid in most cases; time-consuming
Reference checks	Verify application	Reference givers not willing to state true feelings
Job simulation	Verify application	Costly
Credentialing	Rigorous	Costly; leads to in-breeding
Licensing	External validation of skills	May eliminate effective employees
Application blanks	Focuses information in one location; very good for initial screening	Tendency to analyze data subjectively
Assessment center	Job related; reliable and valid	Very costly
Controversial methods	Unique data	Reliability/validity uncertain; invasion of privacy

same each time, the instrument is reliable. To determine alternate forms reliability, candidates are measured on two different versions of an instrument that are essentially intended to measure the same characteristics. Again, if the scores are approximately the same, the instrument can be judged to be reliable. To estimate reliability using the split-halves method, a candidate is measured on one instrument. Then the items on the first half of the test are correlated with the items on the second half of the test. If the correlation is high, the instrument is reliable. This method can only be used, however, if the instrument purports to measure only one characteristic.

Whereas the foregoing methods of determining reliability measure the consistency of the instrument, interrater reliability measures the consistency of *use* of the instrument. For example, two interviewers who use identical instruments to evaluate a candidate may reach different conclusions depending on how each one processes the information. In this case, interrater reliability would be very low.

Reliability is a very important concept. An unreliable selection instrument cannot be valid, although a reliable instrument will not necessarily give valid results. An inconsistent measuring instrument cannot predict job performance. However, a measuring instrument such as an intelligence test may be highly reliable, yet not valid, if intelligence is not a bona fide occupational qualification for a particular position. Organizations must determine the reliability of their selection instruments before they check the validity of their selection process.

Validity

The issue of validity has to do with whether you are measuring what you think you are measuring. In other words, are your selection instruments really measuring job performance? Validity can be determined through either data-based methods or nonempirical, logical methods.

There are two data-based or empirical methods: the predictive validity model and the concurrent validity model. Predictive validity is determined by collecting predictor data, such as scores on a test, from job applicants. Applicants are then selected for hire on the basis of some other predictor—for example, how well they perform in an interview. Several months later, performance data are collected on these new hires and correlated with the original assessments. If the correlation is high, the test can be said to predict job performance, hence to be valid.

Although this strategy of determining validity is methodologically sound, it has some serious drawbacks. First, the test scores cannot be validated until several months have elapsed, and most organizations cannot wait that long. Second, to conclude anything from statistical analyses, an organization needs a relatively large sample of new hires at a particular point in time.

The concurrent validity model is a less time-consuming, more practical method. Information on the performance of current employees is collected, and scores are obtained from the current employees on the predictor in question. These scores are then correlated with the performance data. Although this method is much quicker to use, it does have a few disadvantages. First, current employees have already been screened by the organization, so they are likely to receive higher scores on the predictor than a sample of applicants would receive. There will be less variance among scores of current employees, which will limit the maximum possible correlation coefficient between the predictor and the criterion, a problem referred to as restriction of range. Second, employees who were not able to perform the job probably will have left the organization either voluntarily or involuntarily, decreasing the variance of the criterion. Third, current employees will have had the opportunity to learn on the job, which could increase their scores on the predictors. Despite these problems, the *Uniform Guidelines* (Equal Employment Opportunity Commission, 1978), recommends the use of this procedure when predictive validity is not feasible.

Under certain conditions, empirical, data-based validity testing is neither practical nor necessary. For example, a small nursing home is not going to have enough employees in any one job category to be able to conduct an empirical study. When statistical data-based validity is not practical, validity may be determined through nonempirical, logical methods. The organization may have to rely on results from studies in other organizations, or, as recommended by the *Uniform Guidelines*, may use a content validity strategy to make a rational judgment about whether there is a similarity between the predictors and job performance, and whether the predictors tap the entire domain of job performance. This strategy can easily be defended when the selection instrument is a job simulation, in which the applicant is asked to perform the actual tasks required by the position.

A controversy has developed in recent years over whether it is necessary to validate each component of the selection process or just the overall results. The *Uniform Guidelines* state that as long as the total selection process does not have an adverse impact, the organization does not have to validate each component of the process. This is called the "bottom line" concept. However, in 1982 the Supreme Court, in *Connecticut v. Teal* [29 FEP 1 (1982)], ruled five to four that the results of one step of the selection process were enough to establish a case of disparate impact. Although the the *Teal* decision has been severely criticized (Blumrosen, 1984; Thompson and Christiansen, 1984), organizations must be ready to prove that no portion of their selection process causes adverse impact.

THE SELECTION PROCESS AND FEDERAL REGULATIONS

The legislative acts that have the greatest impact on the selection process are the Civil Rights Act of 1964 (Title VII, amended in 1972), the Age Discrimination in Employment Act of 1967 (amended in 1978), and the Vocational Rehabilitation Act of 1973. As briefly mentioned in Chapter 10, Title VII declares it illegal for an organization to:

1. Fail, refuse to hire, discharge any individual, or otherwise to discriminate against any individual with respect to his compensation, terms, conditions, or privileges of employment because of the individual's race, color, religion, sex, or national origin; or
2. To limit, segregate, or classify employees or applicants for employment in any way which would deprive, or tend to deprive, any individual of employment opportunities or otherwise adversely affect his stature as an employee because of such individual's race, color, religion, sex, or national origin. (Arvey, 1979b).

The Age Discrimination Act and the Vocational Rehabilitation Act extend protection to employees between the ages of 40 and 70 years, and to handicapped individuals, respectively. The 1978 *Uniform Guidelines on Employee Selection Procedures* and the *Principles for the Validation and Use of Personnel Selection Procedures* (American Psychological Association, 1975) provide a framework for determining the proper use of selection procedures. Both publications encourage organizations to use selection procedures that are valid (predictive of job performance). An organization does not have to conduct validity studies unless the selection process adversely impacts a particular group of individuals. However, the courts have not agreed on the exact definition of adverse impact.

The two methods that have been used most frequently to determine adverse impact are disparate rejection rates and population comparisons. The disparate rejection rate method involves comparing the ratio of minority hires to minority applicants and the ratio of nonminority hires to nonminority applicants. The *Uniform Guidelines* state that a selection rate for any racial, ethnic, or sex subgroup that is less that four-fifths (i.e., 80 percent) of the rate for the group with the highest rate will generally be regarded as evidence of adverse impact.

For certain types of position, for example, heart surgeon, there may be fewer minority than nonminority individuals with the requisite job qualifications. In such cases, according to the *Uniform Guidelines* case, a large discrepancy between the ratio of the number of minorities in the relevant labor market with the appropriate qualifications to the number of minorities in the relevant labor market and the ratio of the number of nonminorities in the relevant labor market with the appropriate qualifications to the total number of nonminorities in the relevant market, can be used to establish proof that a particular minority group is less likely to have the requisite qualifications.

The *Uniform Guidelines* population comparison method compares the ratio of the number of minorities employed to the total number of employees with the ratio of the number of nonminorities in the relevant geographic area to the total number of people in the relevant geographic area. The meaning of the term "relevant geographic area," however, is open to interpretation.

If adverse impact is demonstrated, the organization must prove that its selection procedures are not only valid and reliable, but a business necessity or a bona fide occupational qualification (BFOQ) as well. Validity and reliability can be determined through experimental design and statistical analysis, but proving that a procedure is a business necessity or BFOQ is largely a judgment call. An example of the use of a criterion based on business necessity would be hiring applicants of a certain race to demonstrate cosmetics formulated for members of that race. An example of a BFOQ would be hiring only females to be attendants in a ladies' room.

To further comply with regulations, organizations must make certain that job-related questions are asked on application blanks and during interviews. Chapter 10 outlines the questions that are forbidden by federal law, and certain states have additional regulations. For example, in Massachusetts applicants cannot be asked if they have been convicted of a first-degree misdemeanor, but only if they have ever been convicted of a felony, since studies have shown that blacks are disproportionately convicted for misdemeanors (Massachusetts General Laws, 1982).

SELECTION INSTRUMENTS

Tests

Tests have been developed to tap numerous individual characteristics, including intelligence, dexterity, aptitudes, interests, and personality, and many organizations use tests as part of the selection process. Health care organizations typically include testing as part of the selection procedure for lower level positions, and for some professional positions. Medical transcriptionist, pharmalogical, frequency computation, and clerical tests, are some of the more frequently used instruments. Considerable controversy exists over the use of tests as selection devices, however. Much of the discrimination litigation has focused on the adverse impact of tests on minority groups. In a landmark case, *Griggs v. Duke Power* [3 FEP 175 (1971)], the Supreme Court declared that scores on an intelligence test could not be used as a basis for choosing employees for transfers because the test had an adverse impact on minority candidates. Duke Power lost the case because it was unable to prove that intelligence was related to job performance. If a test is used for selection purposes, the organization must be ready to prove that the test is both reliable and valid. Since this process can be very expensive and time-consuming, some organizations have abandoned using tests altogether. However, some tests have been proven to be both reliable and valid predictors of job performance in occupations of certain types. The *Standards for Educational and Psychological Tests* of the American Psychological Association (1974) provide guidelines for evaluating psychological tests. Information on reliability, validity, and group norms for various tests is provided in the *Mental Measurements Yearbooks* (see, e.g., Buros, 1978).

A widespread controversy has developed over the use of aptitude tests in personnel selection. Schmidt and Hunter claim that professionally developed aptitude tests "are valid predictors of performance on the job and in training for all jobs in all settings, and are equally valid for minority and majority applicants" (1984, p. 200). They also claim that the use of cognitive ability tests to select employees can produce considerable labor cost savings for the organization. Their view contradicts the findings of many studies conducted before 1980, and the theories incorporated into the *Uniform Guidelines*. According to Schmidt and Hunter, (1984) most of the conflicting results among studies regarding the adverse impact and validity of aptitude tests have been caused by statistical error. Therefore, they argue that organizations should not be required to validate tests for particular subgroups of employees and occupations if information on the reliability and validity of these tests for other organizations and groups is already available.

The Interview

The interview is one of the most frequently used selection devices. There are several interview formats, including nondirective, semistructured, and structured. In the nondirective interview, the interviewee determines the course of the interview; few, if any, questions are planned in advance. In the semistructured interview basic questions are planned in advance to guide the flow of the interview. However, the interviewer is not restricted to the use of these questions, but may probe into other areas when appropriate. The structured interview resembles a questionnaire to which the interviewee responds orally. Both the question and the response formats are planned in advance of the interview. This method of interviewing is usually the most reliable, but not very practical for selecting employees. Generally, the semistructured interview is preferred. The nondirective interview is too unreliable for selection purposes, although this method is often used by untrained interviewers.

Interviews can be conducted on an individual or group basis. The individual interview is the typical one-on-one situation. The interrater reliability of interviews given on an individual basis is impossible to determine, unless each interviewer asks each candidate all the same questions. However, such collecting of redundant information is a waste of time. In group interviews, either one or more interviewers may collect information from a group of candidates simultaneously, or multiple interviewers may question one candidate at the same time. The former type of interview is particularly stressful for the candidates, who are placed in a highly competitive situation. The latter type of interview can save time when the decision on who to hire is to be made by more than one person.

Little evidence exists, however, to indicate that the interview is either reliable or valid (Arvey and Campion, 1982). Interviews are subject to a variety of biases and prejudices. In a review of studies on interviews, Schmitt concluded that interviewers tended to rate more favorably candidates who were similar to themselves in attitudes and race (Schmitt, 1976). Also, interviewers tended to rate male candidates more favorably than female applicants, but the outcome depended partially on the type of position the candidate was applying for. Females were rated lower when applying for typically "masculine" jobs, whereas males were rated lower when applying for typically "feminine" jobs (Arvey, 1979a). Some evidence indicates that white applicants are sometimes rated more favorably than black applicants (McIntyre et al., 1980), but in one study of large organizations the reverse was found to be true, probably due to awareness of the importance of complying with equal employment opportunity legislation (Newman and Kryzstofiak, 1979).

Studies have also indicated other biases in the employment interview. Negative information tends to be weighted more heavily than positive information (Schmitt, 1976). Therefore, candidates are taught to respond to the often asked question "What are your weaknesses?" by suggesting a strength (e.g., "I work too hard."). Some studies have shown that the outcome of the interview may depend on the strength of the preceding interviewees (Wesley et al., 1972). College recruitment offices suggest that average students not schedule interviews to follow the session of someone who is at the top of the class. First impressions have also been found to have a significant impact on the outcome of the interview (Farr, 1973). However, the interview is not the first piece of information an interviewer has about a candidate. An interviewer almost always sees a résumé and sometimes a grade point average, references, and/or an application blank for each candidate. On the basis of this information the interviewer forms a hypothesis about the candidate and seeks to confirm it throughout the interview (Dipboye, 1982). One study showed that 88 percent of the time the postinterview decision is made on the basis of preinterview information (Springbett, 1958).

Since the interview is a time-consuming method fraught with biases, why is it used so frequently to select candidates? First, the interview serves purposes other than selection. The interview is an opportunity for the organization to sell itself and to provide information to the candidate. Also, the interview provides a vehicle for candidate and employer to get to know each other and to decide whether they will be able to work well together. Unfortunately, this decision is often based on faulty information, since both the interviewer and candidate, in attempting to make favorable impressions, often suppress important and realistic information. Finally, very few cases dealing with interviews have been litigated. Therefore, interviews are safer selection instruments to use than tests, which must be validated and checked for reliability. Discrimination in the interview is extremely difficult to prove, since witnesses are seldom present, and the candidate often has no way of knowing how the interview information is combined with other data to reach a decision.

Organizations can increase the reliability and validity of their selection interview by following a few basic guidelines. First, interviewers should have adequate job specifications; they should know exactly what type of position needs to be filled and the necessary qualifications for that position. Second, interviewers should be trained to be aware of and make efforts to reduce any biases they may have or stereotypes they may rely on. Third, interviewers should have a definite interview plan, with specific questions to be asked of each candidate. Fourth, a technique should be developed for recording and interpreting findings from the interview.

Reference Checks

Reference checks have become increasingly important in recent years as more and more people have been caught providing fraudulent information on application blanks. Through the use of reference or background checks, an organization can verify an applicant's educational background and work experience. Backgrounds are usually checked through phone interviews and by asking applicants for letters of recommendation. For lower level employees, the former approach is used most frequently, whereas written references are frequently requested when the applicant is a professional.

Many organizations now prefer telephone checks, even though they are very time-consuming, because few people are willing to write an unfavorable reference. Individuals making selection decisions based on written recommendations have difficulty discriminating between honestly and politely positive statements. This is a highly subjective judgment, open to all sorts of biases. Since the passage of the 1974 Freedom of Information Act and various state regulations regarding privacy, people have been even more reluctant to write unfavorable references, because applicants may have access to the records, and authors of such letters could be subject to litigation. For fear of litigation, some organizations will not even allow their employees to give references over the phone. They are permitted only to verify that an applicant was employed by the organization and to give the dates of employment and job title.

Job Simulation

Job simulations, methods of duplicating actual job tasks before full-time employment, are used in a unique way in the health care industry. Employees who have been hired may be given provisional status until they can demonstrate specific job-related skills. Nurses, for example, are asked to pass a series of tests to demonstrate their clinical skills and ability to use particular equipment and procedures, such as starting IVs. Nurses who fail these tests during the probationary period are dismissed or trained. Probationary periods range from 2 weeks to 6 months, although 3 months is the average.

Credentialing

A unique selection process in hospitals is the credentialing of physicians for appointment to the medical staff. The objectives of this process are (a) to ensure that care is rendered by appropriately qualified individuals, (b) to ensure that each eligible applicant is afforded equal opportunity to be appointed to the medical staff, and (c) to assure that adequate information pertaining to selection criteria is reviewed by appropriate individuals and committees before a final recommendation is made to the board of directors. A typical procedure for credentialing is presented in Table 11.4. This complex example involves the board of directors and the medical staff. The medical staff does not usually get involved in selection, however, unless the applicant is a chief administrator or a physician.

TABLE 11.4. Sample Credentialing Policy and Procedure

Qualifications for Appointment

1. Appointment to the medical staff is a privilege that shall be extended only to professionally competent individuals who continuously meet the qualifications, standards, and requirements set forth in these bylaws and in such policies as are adopted from time to time by the board.

2. Only physicians, dentists, oral surgeons, and podiatrists who (a) are currently licensed to practice in this state; (b) possess current, valid professional liability coverage in amounts satisfactory to the hospital; and (c) can document their background, experience, training, and demonstrated competence, their adherence to the ethics of their profession, their good reputation and character, and their ability to work harmoniously with others sufficiently to convince the hospital that all patients treated by them in the hospital will receive quality care and that the hospital and its medical staff will be able to operate in an orderly manner shall be qualified for appointment to the medical staff. The word "character" is intended to include the applicant's mental and emotional stability.

3. No individual shall be entitled to appointment to the medical staff or to the exercise of clinical privileges in the hospital merely by virtue of the fact that (a) he is licensed to practice any profession in this or any other state, (b) he is a member of any particular professional organization, or (c) he had in the past, or currently has, medical staff appointment or privileges in other hospitals.

4. No individual shall be denied appointment on the basis of sex, race, creed, color, or national origin.

Conditions of Appointment

Duration of initial provisional appointment. All initial appointments to the medical staff regardless of the category of the staff to which the appointment is made and all initial clinical privileges shall be provisional for a period of 12 months from the date of the appointment or longer if recommended by the executive committee. During the term of this provisional appointment he shall be evaluated by the chairman of the department or departments in which he has clinical privileges, and by the relevant committees of the medical staff and the hospital as to his clinical competence and as to his general behavior and conduct in the hospital. Provisional clinical privileges shall be adjusted to reflect clinical competence at the end of the provisional period, or sooner if warranted. Continued appointment after the provisional period shall be conditioned on an evaluation of the factors to be considered for reappointment set forth in these bylaws.

Rights and duties of appointees. Appointment to the medical staff shall confer on the appointee only such clinical privileges as have been granted by the board and shall require that each appointee assume such reasonable duties and responsibilities as the board or the medical staff shall require.

Time requirements for promotion. The period of time and qualification requirements stated in these bylaws for promotion from provisional status may be altered as to specific applicants by the board on its own motion or as recommended to the board by the executive committee.

TABLE 11.4. (*Continued*)

Application for Initial Appointment and Clinical Privileges

Information. Applications shall be in writing, and shall be submitted on forms prescribed by the board. Application information should include:

- a. the names and addresses of at least three physicians, oral surgeons, podiatrists, or other practitioners, as appropriate, who have had recent extensive experience in observing and working with the applicant and who can provide adequate information pertaining to the applicant and who can provide adequate information pertaining to the applicant's present professional competence and character, and the name and complete address of at least one appointee to the active staff not professionally associated in practice with the applicant who will attest to this;
- b. the names and addresses of the chairmen of each department of any and all hospitals or other institutions at which the applicant has worked or trained (i.e., the individuals who served as chairmen at the time the applicant worked in the particular department). If the number of hospitals the applicant has worked in is great or if a number of years have passed since the applicant worked at a particular hospital, the credentials committee and the board may take into consideration the applicant's good faith effort to produce this information;
- c. whether the applicant's medical staff appointment or clinical privileges have ever been resigned, denied, revoked, suspended, reduced, or not renewed at any other hospital or health care facility;
- d. whether the applicant has ever withdrawn his application for appointment, reappointment, and clinical privileges, or his medical staff appointment before final decision by a hospital's or health care facility's governing board;
- e. whether the applicant's membership in local, state, or national professional societies or his license to practice any profession in any state, or his narcotic license has ever been suspended, modified, or terminated. The submitted application shall include a copy of all the applicant's current licenses to practice, as well as a copy of his narcotics licenses, medical, dental, or podiatry school diploma, and certificates from all postgraduate training programs completed;
- f. whether the applicant has currently in force professional liability coverage or qualifies under the state's Malpractice Act, the name of the Insurance company and the amount and classification of such coverage;
- g. applicant's malpractice claims;
- h. a consent to the release of information from the applicant's present and past malpractice insurance carriers;
- i. applicant's physical and mental health;
- j. convictions of a felony and details about any such instance;
- k. citizenship and visa status of the applicant;
- l. such other information as the board may require.

Undertakings. Every application for staff appointment or reappointment shall be signed by the applicant and shall contain:

- a. the applicant's specific acknowledgment of his obligation upon appointment to the medical staff to provide continuous care and supervision to all patients within the hospital for whom he has responsibility;
- b. an agreement to abide by all bylaws and policies of the hospital, including all bylaws, rules, and regulations of the medical staff as shall be in force from time to time during the time he is appointed to the medical staff;

(*Continued*)

TABLE 11.4. (*Continued*)

    c. an agreement to accept committee assignments and such other reasonable duties and responsibilities as shall be assigned to him by the board and the medical staff;

    d. an agreement to provide the hospital current information regarding all questions on the application form at any time;

    e. a statement that the applicant has received and had an opportunity to read a copy of the bylaws of the hospital and bylaws, rules, and regulations of the medical staff as are in force at the time of his application and that he has agreed to be bound by the terms thereof in all matters relating to consideration of his application without regard to whether or not he is granted appointment to the medical staff or clinical privileges;

    f. a statement of willingness to appear for personal interviews;

    g. a statement that any misrepresentation or misstatement in, or omission from the application whether intentional or not, shall constitute cause for automatic and immediate rejection of the application resulting in denial of appointment and clinical privileges. In the event that an appointment had been granted prior to the discovery of such misrepresentation, misstatement or omission, such discovery may result in summary dismissal from the medical staff;

    h. a statement that the applicant will: (1) refrain from fee splitting or other inducements relating to patient referral; (2) refrain from delegating responsibility for diagnoses or care of hospitalized patients to any individual who is not qualified to undertake this responsibility or who is not adequately supervised; (3) refrain from deceiving patients as to the identity of an operating surgeon or any other individual providing treatment or services; (4) seek consultation whenever necessary; and (5) abide by generally recognized ethical principles applicable to his profession.

Burden of providing information. The applicant shall have the burden of producing adequate information for a proper evaluation of his competence, character, ethics, and other qualifications, and of resolving any doubts about such qualifications.

SOURCE: Adapted from the bylaws of a 500-bed, denominational hospital in the south-central part of the United States.

Licensing

The need for selection testing in the health care environment is minimized by strict licensing and certification requirements. Any health care facility approved by the Joint Commission for the Accreditation of Hospitals (JCAH) is required to use licensing as a selection criteria for most clinical and professional/technical positions. Registered nurses can have a variety of educational backgrounds ranging from diploma programs to either 2 or 4 years of college. However, anyone who is certified as a registered nurse must take and pass the state board examination. Specialty certification on either the state or national level furthers the role of certification in the selection process.

Application Blanks

Almost all organizations require applicants to complete an application blank as the first step in the selection process. The types of questions typically asked are illustrated in Figure 11.1. The application blank serves several purposes. First, the instrument can be used to disqualify applicants who are obviously unsuitable for the position. Second, the application blank can serve as a reliability check, easily verified through references checks, and compared to information obtained through interviewing. Third, data needed for personnel files, such as address and social security number, can be collected. Fourth, application blanks may contain information needed for reports to the Equal Employment Opportunity Commission on sex, race, and national origin, but only if these items are blocked off in a special section stating that these items will be used for research purposes only. To be on the safe side, many organizations have eliminated these potentially problematic questions. Instead, they have each applicant fill out an additional card, which is kept in a separate file.

Critics of application blanks claim that the information they contain is analyzed much too subjectively, often by employees who have no idea what they should be looking for. Researchers believe that the application blank could be a much more effective selection tool if items were weighted on the basis of statistical analysis in terms of their importance. Then the screening staff could simply score the information according to a predetermined format. These instruments, called weighted application blanks, have been found to be valid predictors of job performance (Reilly and Chao, 1982).

Assessment Centers

The assessment center is used primarily to identify candidates who would be successful in managerial positions. Candidates are evaluated by a group of assessors as they perform a series of individual and group tests, usually over 2 or 3 days. Although assessment centers were developed during World War II to evaluate candidates for the Office of Strategic Services, the first industrial use of such a facility is attributed to AT&T (Howard, 1984). The purpose of AT&T's first programs was to decide which of their current employees were promotable to higher levels of management.

Assessment center participants are usually evaluated on the following dimensions: leadership, organizing and planning, decision making, oral and written communication skills, initiative, energy, analytical ability, resistance to stress, use of delegation, behavior flexibility, human relations competence, originality, controlling, self-direction, and overall potential (Howard, 1984). These dimensions are assessed by evaluating behavior in a host of situational tests, including in baskets, leaderless group discussions, and fact-finding exercises. In addition to these situational tests, participants are interviewed and asked to take projective and paper-and-pencil aptitude tests.

In the in-basket exercise the candidate plays the role of a manager who has just returned to the office after a week-long trip and finds a huge stack of letters, memos, reports, and messages (the contents of the in basket). The manager will be leaving town again the next day and must respond in some way to each item. After the exercise, the candidate is frequently asked to explain his or her actions and decision processes.

In the leaderless group discussion, candidates are given a problem such as deciding who to promote, or what disciplinary action to take against an employee. They are required to come to a group decision about the problem. As the name implies, no leader is designated.

In the fact-finding exercise, individual candidates are given information about an organization, its problem, and possible alternative solutions. Group members then select, present, and justify the best alternative from the limited information available, both orally and in writing.

NAME LAST	FIRST	MIDDLE	SOCIAL SECURITY NO.
PHONE NO. ADDRESS NUMBER	STREET	CITY	STATE ZIP

EDUCATION

HIGH SCHOOL

Name of School(s) _____
Address _____
City, State _____
When Did You Attend? _____
Did You Graduate? _____ GED _____

COLLEGE

Name of School(s) _____
Address _____
City, State _____
Where Did You Attend   From _____ mo./ _____ yr. to _____ mo./ _____ yr.
Did You Graduate? _____ mo./ _____ yr. If No, How Many Years Credit Do You Have? 1 2 3 4
Your Degree _____
Major Subject(s) Studies _____

OTHER: Graduate School, Business School, Vocational Training and/or Military Service Information Below:

Name _____
Address _____   Dates of Attendance _____ mo./ _____ yr. to _____ mo./ _____ yr.
City, State _____

PROFESSIONAL REGISTRATION NUMBER (If Applicable) _____

US Citizen _____ Yes _____ No. If No, Type of Visa. _____
Have You Ever Been Convicted of a Felony: Yes _____ No _____
If Yes, Give Date and Details: _____
What Prompted You to Apply at the Medical Center _____
Have you ever been employed by the Center? _____ If yes, specify location and dates of employment. _____

LIST POSITION FOR WHICH YOU ARE APPLYING:
1. _____ 2. _____ 3. _____   SALARY DESIRED _____

WHAT SHIFT CAN YOU WORK?
Day: From _____ to _____
Evening: From _____ to _____
Night: From _____ to _____
Willing To Work Weekends? Yes _____ No _____
Date Available For Work _____

TYPE OF EMPLOYMENT DESIRED: **LENGTH OF EMPLOYMENT DESIRED:**
Full Time _____   Temporary _____
Part Time _____   Staff _____

SKILLS:
Typing _____ WPM   Medical Terminology _____
Dictaphone _____   Foreign Language _____
Shorthand _____   10 Key Adding Machine _____
Word Processing Machines _____   Calculator _____ WPM
OTHER: _____   Types _____

Below This Line For Personnel Use Only.

Interviews Scheduled:
Date/Time _____ With _____ Position _____

Hire Authorization: (Circle)   Full Time   Part Time   Staff   Temporary
8 Hours   10 Hours   12 Hours   Flexpool
Replacing _____
Position Control Number _____
Title _____   Date to Begin: _____
Hours: _____   Dept. No. to be Charged _____
Salary: _____   Signed: _____
Experience: _____

284

LIST ALL EMPLOYERS. IF ADDITIONAL SPACE IS NEEDED COMPLETE ADDITIONAL FORM.

MOST RECENT OR PRESENT EMPLOYER: May we contact your present employer? ________ Yes ________ No

Company Name: ______________________
Address ______________________
City ______________________
State, Zip ______________________
Supervisor's Name ______________________
Supervisor's Title ______________________
Supervisor's Phone No. ________ ________
AREA CODE NUMBER

Your Title ____________ Unit ____________
Dates of Employment (Month & Year)
From ________ 19 ____ to ________ 19 ____
Name Under Which Employed ______________
Wages/Salary Circle One: PART TIME FULL TIME
Starting $ ____________ per: hr / mo / yr
Final $ ____________ per: hr / mo / yr

Why Did You Leave? ______________________
____
____
Description of Duties ______________________
____
____
____

Company Name: ______________________
Address ______________________
City ______________________
State, Zip ______________________
Supervisor's Name ______________________
Supervisor's Title ______________________
Supervisor's Phone No. ________ ________
AREA CODE NUMBER

Your Title ____________ Unit ____________
Dates of Employment (Month & Year)
From ________ 19 ____ to ________ 19 ____
Name Under Which Employed ______________
Wages/Salary Circle One: PART TIME FULL TIME
Starting $ ____________ per: hr / mo / yr
Final $ ____________ per: hr / mo / yr

Why Did You Leave? ______________________
____
____
Description of Duties ______________________
____
____
____

Company Name: ______________________
Address ______________________
City ______________________
State, Zip ______________________
Supervisor's Name ______________________
Supervisor's Title ______________________
Supervisor's Phone No. ________ ________
AREA CODE NUMBER

Your Title ____________ Unit ____________
Dates of Employment (Month & Year)
From ________ 19 ____ to ________ 19 ____
Name Under Which Employed ______________
Wages/Salary Circle One: PART TIME FULL TIME
Starting $ ____________ per: hr / mo / yr
Final $ ____________ per: hr / mo / yr

Why Did You Leave? ______________________
____
____
Description of Duties ______________________
____
____
____

PERSONAL REFERENCES:

1. ____
2. ____
3. ____

NAME	ADDRESS	OCCUPATION	TELEPHONE

I certify that this information, to the best of my knowledge, is true and complete. I understand that any misrepresentation or willful omission of facts is cause for immediate discharge. I hereby authorize the hospital to write my former employers. I understand that as a condition of initial and continued employment, I may be required to take a physical examination. I also agree, upon termination of such employment to take a physical examination. I also agree, upon termination of such employment to return any hospital property issued to me or allow value of same to be deducted from my wages.

It will be necessary, in order for us to fully evaluate your application for employment to check references with your previous employers and the education institutions you have attended. Your signature below acknowledges that you are aware of our intent to check these references and you approve of our doing so. Any information we receive will be kept entirely confidential. In accepting employment at the hospital I understand that my employment and/or compensation can be terminated at anytime with or without notice or cause by the hospital.

SIGNED: ______________ DATE: ______________

Figure 11.1. Application blank.

After the exercises have been completed, the assessors review candidate performance and complete a written evaluation. Each candidate is usually evaluated by more than one assessor. The assessors may be members of management who themselves have gone through the assessment center, or they may be psychologists. The assessors must be extensively trained in evaluation.

Studies have shown that the interrater reliabilities for the final assessment evaluations and for several components of the assessment process are high. The validity of the overall evaluation has also been shown to be high compared to the validity of alternative selection procedures; however, some of the components of the assessment center are not valid across all organizations (Howard, 1984). The main drawback to the assessment center is that it is extremely costly and time-consuming—it is a feasible selection alternative only for large organizations or when selection may be critical to the survival of the firm. To date, assessment centers have not been used extensively as selection tools in the health care industry. However, in a sampling of 10 local hospitals, one reported using an assessment center to determine the promotability potential of department heads to higher level administrative positions. The assessment included a role-playing exercise, an in-basket test, an interview, and applicants were evaluated on 15 different criteria.

Physical Examinations

Many organizations require job applicants to undergo a physical exam before a hiring decision is made. The results may be used to screen out candidates with physical or mental handicaps that would prevent them from performing the job, or to protect the organization from workers' compensation claims based on preemployment conditions (Cherrington, 1987). Additionally, health care organizations use physical exams to screen out applicants with contagious diseases. Whether all applicants for health care positions should be tested for HTLV-III antibodies is currently one of the most controversial issues facing the health care industry today. The HTLV-III blood test indicates whether a person has been infected by the AIDS virus, but not whether that person has or will develop the acquired immunodeficiency syndrome (Kadzielski, 1986). Although state regulations require most health care providers to screen employees for contagious diseases, AIDS cannot be transmitted through casual contact, and therefore these regulations may not apply (Harris, 1987). Also, a number of states and communities have passed laws prohibiting discrimination against people with AIDS (Levine, 1986). Health care providers can, however, discriminate against employees with HTLV-III infections if the condition would prevent the employee from performing the job or poses a danger to the health and safety of others (Kadzielski, 1986).

Controversial Selection Methods

Organizations are continually searching for new methods of selecting employees. Several selection methods that have become increasingly popular in recent years have generated considerable controversy. These include honesty testing through the use of polygraphs, graphology, and drug screening.

Although the use of polygraphs by employers has been prohibited in 22 states, several organizations use these devices to identify possible security risks, particularly when low-level employees are being hired to handle large sums of money. Critics of polygraphs (including recently the American Medical Association) claim that the validity of the technology has not been proven, and the dangers of a false-positive outcome (data indicating that the applicant is lying, when in fact he or she is telling the truth) are too great. Furthermore, critics argue that polygraph testing is an invasion of privacy because questions are often asked regarding such topics as arrest records and marital history (Kaler, 1985). Supporters argue, however, that polygraphs do not have to be 100 percent accurate to be useful, since other selection devices, such as tests, which explain as little as 25 percent of the variance of performance, are often accepted by the courts (Horvath, 1985). Proponents also argue that selection decisions are never based solely on polygraph test results, and the information these tests provide is not currently obtainable by any other means (Horvath, 1985). The courts, however, have usually sided with the critics (*Newsweek*, 1986). Some health care organizations use paper-and-pencil tests, rather than polygraphs, to assess honesty.

Graphologists believe that a person's suitability for a job can be determined by handwriting analysis. They claim that graphology is a very accurate assessment device because it is next to impossible for applicants to disguise their handwriting. One company, the pioneer in computerized handwriting analysis, reports that its clients have ranged from construction companies and security firms to entertainment companies (*Newsweek*, 1986). The validity of graphology, however, has not been determined, and an organization would have a difficult time defending a graphology-based decision in court.

The most controversial screening device at the current time is drug testing. Nearly 25 percent of the Fortune 500 corporations now do routine urinalysis on employees and job applicants to try to detect disease or illegal drug use (*Fortune*, 1985). Just 3 years ago 10 percent of the same corporations were participating in these screening tests (*Fortune*, 1985). This testing is sometimes combined with blood tests to detect drug abuse, genetic predisposition to disease, and the presence of the AIDS virus. Detecting drug abuse is particularly relevant for health care organizations, since employees have easy access to drugs.

The Centers for Disease Control (CDC) estimate that it costs less than $5 to do an initial screening and $20–$50 for confirmation of the results (*Fortune*, 1985). However, the CDC found that some laboratories testing for drugs had a false-positive error rate of up to 66 percent (*Fortune*, 1985). Some facilities gave false-negative results (indicating that a person is not taking a drug when he or she really is) 100 percent of the time when testing for certain drugs (*Fortune*, 1985). In other words, tossing a coin would have predicted the sample results with about the same reliability. Human resources professionals can expect, however, to see an increase in this type of testing because of the magnitude of the problem. The Alcohol, Drug Abuse and Mental Health Administration estimated that reduced productivity due to alcohol and drug abuse cost the United States more than $99 billion in 1983 (Dogoloff et al., 1985). On the other hand, until drug and blood screening procedures are perfected, organizations that use these procedures to disqualify applicants may find themselves in court fighting a downhill battle. One large laboratory that does drug screenings claims that the most important step is to determine whether the use of drugs being screened for impairs performance. The technology of effective drug testing will quickly surpass the ability of organizations to set and use effective policies and procedures related to drug screening (personal communication from Dr. Thomas F. Puckett Laboratory, P.O. Box 1549

Hattiesburg, Mississippi 39402). Some health care organizations use paper-and-pencil tests to assess tendencies toward violence and drug abuse, rather than urinalysis tests. These tests, however, are usually required only for security personnel.

DETERMINING THE UTILITY OF THE SELECTION PROCESS

The selection process does not end once the selection instruments have been demonstrated to be reliable and valid. It must be decided how the individual components of the process are to be combined and what the cutoff scores on the instruments should be. Three procedures can be used to select employees: the multiple-hurdles approach, the compensatory approach, and the hybrid approach (Milkovich and Glueck, 1985). If the organization chooses the multiple-hurdles approach, the applicant must pass one component of the process before proceeding to the next. A cutoff point is chosen for each predictor. If the organization uses a compensatory approach, each candidate completes all components and a hiring decision is based on a weighted composite of all the predictors. The premise of this approach is that strengths in certain areas can compensate for weaknesses in others. With the hybrid approach, candidates must demonstrate some minimum requirement (e.g., college degree or certain number of years of experience) before continuing, so that those who are clearly unqualified can be quickly eliminated. However, once the candidates have passed the minimum qualifications, they proceed through the rest of the process. This is the most common selection approach.

Regardless of the type of approach used by the organization, the usefulness of the predictors needs to be determined. The square of the regression coefficient (coefficient of determination) indicates the amount of variance in job performance explained by the predictors (the validity of the predictors), but it does not account for situational factors that might influence the utility of the procedure. Taylor and Russell (1939) suggested considering three factors when determining the utility of a selection device: the validity coefficient, the base rate, and the selection ratio. As noted in Chapter 10, the selection ratio is the number of applicants hired compared to the total number of applicants. If the selection ratio is high, spending time and money developing selection instruments is a futile endeavor, since almost everyone who applies for the job needs to be hired. This, of course, indicates that the organization has an ineffective recruitment strategy. If the selection ratio is low, the organization can afford to be very selective, and thus predictors are needed that can discriminate between qualified and unqualified candidates.

The base rate of success is the percentage of the current work force considered to be effective job performers. If the base rate is already high, a new predictor would need to have an extremely high validity coefficient to be of any use. If the base rate is low, even a predictor with marginal validity could save the organization money.

Using the base rate, selection ratio, and validity coefficients, Taylor and Russell (1939) developed tables that indicate the improvement in successful job performance that could be expected using a new predictor. Table 11.5 shows that with a base rate of 50 percent and a selection ratio of .20, if a new predictor has a validity coefficient of .40, the base rate of successful performance could increase from 50 to 73 percent.

No selection procedure, however, is 100 percent valid. Errors will always occur, and an organization must decide, before a cutoff point is chosen for a particular predictor, which type of error is least costly: false negative or false positive. Figure 11.2 shows two sets of the four possible situations resulting from selection. In each case, cell 2 includes applicants who received high scores on the predictors, were hired, and performed well on the job. Cell 3 includes applicants who were not selected because they received low scores on the predictors and would not have been successful. In both cases, no errors have been made. Cell 1, however, includes applicants who were rejected because they received low scores on the predictors but would have been satisfactory employees if hired. This is called a false-negative error. Cell 4 contains applicants who were hired because they scored high on the predictors, but performed unsatisfactorily. This is called a false-positive error. The costs of both false positives and false negatives have to be weighed before cutoff points for predictors are established. A high cutoff point will increase the possibility of false negatives as indicated by the proportionately larger size of cell 1 in Figure 11.2a, and a low cutoff point will increase the possibility of false positives, as seen in cell 4 of Figure 11.2b. The cost of making a false-positive error could be extremely high when filling a physician's position or chief administrator position. However, when a file clerk position needs to be filled, the organization may be more willing to risk a false-positive than a false-negative decision. The costs of training or replacing an unsuccessful candidate would not be high, and recruiting costs could be reduced by lowering the cutoff point.

Recently, methods have been developed that enable the organization to assess the value of its selection procedure in dollar terms. The method of Schmidt et al. (1979) is based on asking supervisors to estimate the dollar value of a job at different performance levels. They then estimate the value added by a selection device, based on its ability to improve performance. Cascio (1982), instead of asking supervisors to estimate the dollar value, assumes that the average salary paid for a job reflects its value. Jobs are broken down into major activities, and supervisors assign a portion of the total salary to each activity. Then the supervisors rate their subordinates' performance on each activity and calculate how much the job is currently worth, and how much it would be worth if the selection instrument improved performance. These methods are currently being refined and hold promise for the future.

TABLE 11.5. Example of Taylor-Russell Tables[a]

Correlation Coefficient	Selection Ratio			
	.20	.40	.60	.80
.20	.61	.58	.55	.53
.30	.67	.62	.58	.54
.40	.73	.66	.61	.56

SOURCE: Adapted from Taylor, H. C. and Russell, J. T. The Relationship of Validity coefficients to the practical effectiveness of tests in selection: Discussion and tables. *Journal of Applied Psychology*, vol. 28, 1939, 565–578. Copyright 1939 by the American Psychological Association. Adapted by permission of the publisher and author.
[a] Proportion of employees considered satisfactory (base rate) = .50.

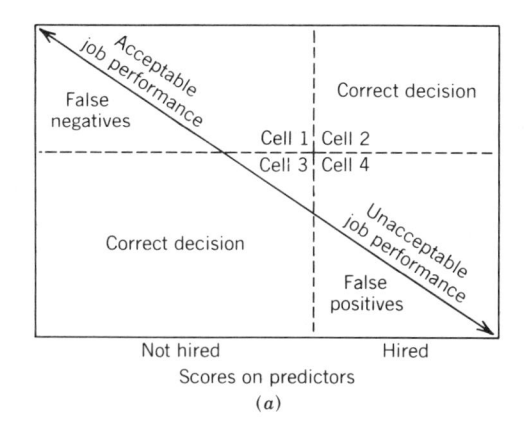

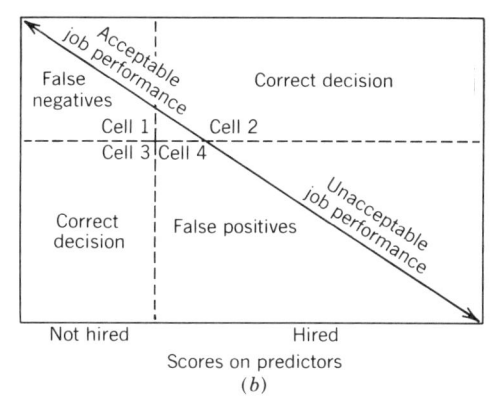

Figure 11.2. Types of selection errors.

SUMMARY

Selection of employees for health care organizations is more important than ever before. Organizations are growing and changing more quickly because of unstable environments. Moreover, while the need for employees with flexibility and diverse skills and abilities has increased, the number of employees in health care has decreased. The manager's challenge is to select the employee who can offer the widest variety of skills to the organization.

This chapter outlines the critical steps in selecting employees for a health care environment. The process begins with a job analysis, to establish the parameters for selecting employees in a particular job category. After job analysis, the human resources professional proceeds through a series of technical steps to ensure the reliability and validity of the selection process. Finally, determining the utility of the selection process helps to personalize and further validate an organization's approach to selecting employees.

The future of health care will include a renewed interest in human resource selection. Human resources professionals must become involved in strategic planning by helping an organization both to define its need for employees and to obtain an effective work force. Selection could very well be one of the critical dimensions that defines the competitive edge of a particular health care institution.

REFERENCES

American Psychological Association. *Standards for Educational and Psychological Tests*. Washington, DC: American Psychological Association, Inc., 1974.

American Psychological Association, Division of Industrial-Organizational Psychology. *Principles for the Validation and Use of Personnel Selection Procedures*. Dayton, OH: The Industrial-Organizational Psychologist, 1975.

Arvey, R. D. *Fairness in Selecting Employees*. Reading, MA: Addison-Wesley, 1979a.

Arvey, R. D. "Unfair discrimination in the employment interview: Legal and psychological aspects." *Psychological Bulletin*, 86:736–765 (1979b).

Arvey, R. D., and J. E. Campion. "The employment interview: A summary and review of recent research." *Personnel Psychology*, 35:281–322 (1982).

Blumrosen, A. W. "The bottom line after *Connecticut v. Teal*." In R. S. Schuler and S. A. Youngblood, Eds., *Readings in Personnel and Human Resource Management*, 2nd ed. St. Paul, MN: West Publishing Company, 1984.

Buros, O. K. *The Ninth Mental Measurement Yearbook*. Highland Park, NJ: Gryphon Press, 1978.

Cascio, W. F. *Costing Human Resources: The Financial Impact of Behavior in Organizations*. Boston: Kent, 1982.

Coddington, D. C., L. E. Palmquist, and W. B. Trollinger. "Strategies for survival in the hospital industry." *Harvard Business Review*, p. 137 (May–June 1985).

Dipboye, R. L. "Self-fulfilling prophecies in the selection-recruitment interview." *Academy of Management Review*, 7:579–586 (1982).

Dogoloff, L., R. Angarola, and S. Price. *Urine Testing in the Workplace*. Washington, DC: American Council for Drug Education, 1985.

Equal Employment Opportunity Commission, "Uniform guidelines on employee selection procedures," *Federal Register*. Vol. 43, No. 166, Friday, August 25, 1978.

Farr, J. L. "Response requirements and primacy effects in a simulated selection interview." *Journal of Applied Psychology*, 57:228–233 (1973).

Fine, S. A., and W. W. Wiley. *An Introduction to Functional Job Analysis. A Scaling of Selected Tasks from the Social Welfare Field, Methods for Manpower Analysis*, Vol. 4. Kalamazoo, MI: W. E. Upjohn Institute for Employment Research, 1971.

Fortune. "Medical testing," 112(4):58 (Aug. 19, 1985).

Harris, S. "AIDS poses employer dilemma." *American Health Care Association Journal*, 12:43–44 (1987).

Henderson, R. I. *Compensation Management: Rewarding Performance*. Reston, VA: Reston Publishing Company, 1979.

Heneman, H. G., D. P. Schwab, J. A. Fossum, and L. D. Dyer. *Personnel/Human Resource Management*. Homewood, IL: Irwin, 1986.

Horvath, F. "Job screening." *Society*, 22(6):43–46 (October 1985).

Howard, A. "An assessment of assessment centers." In R. S. Schuler and S. A. Youngblood, Eds., *Readings in Personnel and Human Resource Management*, 2nd ed. St. Paul, MN: West Publishing Company, 1984, pp. 183–199.

Kadzielski, M. A. "Legal implications for health care providers." *Health Progress*, pp. 48–52 (May 1986).

Kaler, I. K. "A mole among the gerbils?" *Newsweek*, pp. 14–15 (March 11, 1985).

Levine, H. Z. "AIDS in the work place." *Personnel*, 63(3):56–64 (March 1986).

Massachusetts General Laws Annotated, Vol. 22A. St. Paul, MN: West Publishing Company, 1982, p. 427.

McIntyre, S. D. J. Moberg, and B. Z. Posner. "Preferential treatment in preselection decisions according to sex and race." *Academy of Management Journal*, 23:738–749 (1980).

Milkovich, G. T., and W. F. Glueck. *Personnel/Human Resource Management: A Diagnostic Approach*. Plano, TX: Business Publications, 1985.

Milkovich, G. T., and J. M. Newman. *Compensation*. Plano, TX: Business Publications, Inc., 1984.

Newman, J. M., and F. Kryzstofiak. "Self reports versus unobstructive measures: Balancing method variables and ethical concerns in employment discrimination research." *Journal of Applied Psychology*, 64:82–85 (1979).

Newsweek. "Can you pass the job test?" May 5, pp. 46–53 (1986).

Reilly, R. R., and G. T. Chao. "Validity and fairness of some alternative employee selection procedures." *Personnel Psychology*, 35:1–62 (1982).

Schmidt, F. L., and J. E. Hunter. "Employment testing: Old theories and new research findings." In R. S. Schuler and S. A. Youngblood, Eds., *Readings in Personnel and Human Resource Management*, 2nd ed. St. Paul, MN: West Publishing, 1984, pp. 200–212.

Schmidt, F. L., J. E. Hunter, R. C. McKenzie, and T. W. Muldrow. "Impact of valid selection procedures on work-force productivity." *Journal of Applied Psychology*, 64:609–626 (1979).

Schmitt, N. "Social and situational determinants of interview decisions: Implications for the employment interview." *Personnel Psychology*, 29:79–101 (1976).

Springbett, B. M. "Factors affecting the final decision in the employment interview." *Canadian Journal of Psychology*, 12:13–22 (1958).

Taylor, H. C., and J. T. Russell. "The relationship of validity coefficients to the practical effectiveness of tests in selection: Discussion and tables." *Journal of Applied Psychology*, 23:565–578 (1939).

Thompson, D. E., and P. S. Christiansen. "Court acceptance of Uniform Guidelines provisions. The bottom line and the search for alternatives. In R. S. Schuler and S. A. Youngblood, Eds., *Readings in Personnel and Human Resource Management*, 2nd ed. St. Paul, MN: West Publishing, 1984.

U.S. Department of Labor, Manpower Administration. *Handbook for Analyzing Jobs*. Washington, DC: Government Printing Office, 1972.

U.S. Department of Labor. *Dictionary of Occupational Titles.* Washington, DC: Government Printing Office, 1977.

Wesley, K. N., G. A. Yukl, G. A. Kovacs, and R. E. Sanders. "Importance of contrast effects in employment interviews." *Journal of Applied Psychology,* 56:45–48 (1972).

CHAPTER 12

Training and Development

Howard L. Smith
Myron D. Fottler

In most organizations, including those in the health services sector, employee training and staff development programs are an apparent enigma (Levine, 1981). On the one hand, such programs are prevalent in organizations. Few executives would admit that they lack a formal orientation, training, or staff development program. This is especially true in the health care field, where continuing education is a requirement for many professional licenses. On the other hand, most executives would also admit that they do not believe their program is a primary strategic factor that is instrumental in achieving exceptional organizational performance. Health care executives need to fully contemplate this apparent paradox. If employee training and staff development programs are so prevalent in organizations, why are they not more highly valued for their contribution to organizational performance?

The answers to this question are varied, but they inevitably distill to one main idea. Employee training and staff development programs are undervalued because traditionally executives have not expected them to dramatically improve or otherwise affect organizational performance. As a result, most staff development programs have lived up (or down) to the expectations that have been set for them. Without an active incentive to perform otherwise, staff development programs have sought and attained mediocre performance. Naturally, there are exceptions to this generalization, but they are difficult to identify. As a rule, there has been little motivation for staff development programs either in health care or in business corporations to produce measurable improvements in employee or organizational performance. Without a proper incentive, many staff development programs have become complacent and content with less than extraordinary standards.

Any health care executive who is interested in the strategic management of people must at some point address the paradox presented by employee training and staff development programs. In short, such programs can be either a liability, a minimal expectation, or an asset. Staff development programs are a liability when they excessively consume resources and do not offer an adequate return on investment. They represent a minimal expectation when they help health care professionals retain credentials or provide an adequate orientation to the new employee. However, these outcomes provide less than a health care executive should tolerate. At the very least, employee training and staff development programs should contribute extensively to the strategic management of an organization. They should establish a foundation for excellence in personal, work unit, program, departmental, and organizational performance.

This chapter explores the nature and scope of employee training and staff development programs in health care organizations. An overview of staff development programs provides a complete understanding of the different dimensions such programs can assume within health care organizations. A rationale is presented for linking employee development to improved performance. This orientation underscores the discussion above in the sense that health care executives have expected too little in the way of meaningful results for too long from staff development programs. Yet, there are new forces in the health care organization environment that are instrumental in rectifying these past deficiencies.

This chapter also defines the primary components of a staff development program through a visual model. This model is primarily a tool to better understand the basic ingredients underlying a staff development program. The staff development needs of specific personnel in health care organizations—physicians, nurses, administrators—are addressed in the context of fine-tuning development programs to meet the special needs of key professionals. Additionally, programmatic planning for staff development is elucidated in terms of setting objectives and assessing alternatives. Finally, the importance of assessing the performance of employee development programs is discussed, along with an exploration of feasible methodologies, anticipated outcomes, and managerial implications.

DEFINITIONS, GOALS, AND THE IMPORTANCE OF EMPLOYEE DEVELOPMENT

Employee development should be an integral aspect of the strategic management of organizations. By deriving the highest quality output attainable from each and every staff member, an organization is prepared to meet competition, respond to regulatory constraints, overcome reimbursement limitations, or address the key factors that eventually separate successful organizations from unsuccessful organizations. This conclusion is true for virtually every organization, but there is added significance for health care organizations in view of their labor intensity.

Before staff development can begin to attain its proper prominence in health care organizations, however, it is essential that executives understand the many different definitions and concepts of staff development (McGehee and Thayer, 1961; Snow and Grant, 1980; Swansburg, 1968). Without question, the term "staff development" means different things to different people. To some it may merely represents the introduction an employee receives upon entering an organization. To another it may be training and skill development acquired throughout the job experience. To still another, employee development may imply a methodical effort to facilitate employee growth over the entire career span. Staff development consists of all these efforts and more.

The challenge before health care executives is to understand the broader meaning of staff development not only as a concept, but also as a method for formulating and implementing the strategic direction of an organization. This can be accomplished only when executives are well grounded in the continuum of concepts representing staff development. The goal is to envision the organization as a unique ensemble of human resources. As these resources develop and grow in depth, the beneficiary is ultimately the organization. Its capacities are much more robust, its set of skills more diverse and better prepared for utilization. In short, the organization becomes a logical extension of its human components. By developing the individuals who comprise the organization, staff development programs concomitantly nurture the growth of the parent organization.

The implications of this philosophical view for staff development are far-reaching. By viewing staff development as a strategic prerequisite for organizational development, a more realistic management approach can be conceived. Staff development is then recognized as one of the best means available to executives to actually shape the capability and identity of their organizations. This is consistent with the research on organization culture (e.g., Japanese management, searches for excellence), which suggests that exceptional organizational performance is not an artifact of human effort but is a distinct end result from human activity (Pascale and Athos, 1981; Peters and Waterman, 1982). Thus the organization can achieve excellence in low-cost, high-quality health care delivery (or other service and production efforts) only when the staff members comprising the organization are supported and encouraged in the pursuit of the limits of their abilities. By constantly raising these limits, staff development becomes a primary strategy for long-run excellence in organizational performance.

The Continuum of Staff Development

Just what constitutes staff or employee development? The terms need to be clarified before we progress further in the analysis of this strategic concept. The map for negotiating the terminology presented in Table 12.1 is not presumed to exhaust all possibilities, but it indicates a wide variety of employee training and staff development programs. In fact, there are so many programs that we offer some general categories that are most representative of the options normally available in organizations. It is our belief that Table 12.1 is a convenient beginning from which to create a staff development program: One or all of its options may ultimately provide the foundation for programmatic activity by an organization.

The implications from Table 12.1 for the strategic management of employee development programs are numerous, but one point should be underscored. There is no single best way to organize a staff development program. A contingency approach is preferred (Hofer, 1975). In other words, executives need to examine the specific setting of an organization to determine what type of staff development program is appropriate. Relevant variables in that setting such as task technology, type of service or product, number of employees, prior staff development program efforts, budgeted resources, extent of external constraints, organizational goals, and experience of the work force ultimately determine the nature of a staff development program for any

TABLE 12.1. Differentiating Employee Development Concepts

Concept	Objectives	Scope of Skill Diversity	Emphasis on Personal/Career Growth	Training Site	Frequency
Orientation training	To introduce staff to the mores, behaviors, and expectations of an organization	Narrow	Narrow	Internal	Single instance
Training	To teach staff specific skills, concepts, or attitudes	↕	↕	Internal	Sporadic
In-service education	To teach staff members about skills, facts, attitudes, and behaviors largely through internal programs	↕	↕	Internal	Continuous
Continuing education	To facilitate the efforts of staff members to remain current in the knowledge base of their trade or profession through external programs designed to achieve external standards	↕	↕	External	Continuous
Career development	To expand the capabilities of staff beyond a narrow range of skills toward a more holistically prepared person	Broad	Broad	Internal and external	Continuous

specific organization. In some cases only an employee orientation program is needed. In other cases, a comprehensive program is appropriate. The program content, objectives, and scope of effort are contingent on the situation.

For example, a private group practice consisting of three pediatricians with two secretarial and two nursing staff members has decidedly unique staff development needs compared to a 30-physician, multispecialty group practice with 90 supporting staff members. Among the factors characterizing the pediatric group practice are the following: there are fewer resources available to fund staff development; no organizational administrative infrastructure is present to plan for or acquire staff development resources; there is much closer supervision between the physicians and the support staff (thereby allowing more opportunities for on-the-job training); the physicians are positioned better to interpret performance variances or motivational difficulties and to react accordingly; and the set of skills needed to run the clinic is less diverse. Precisely the opposite can be said for the multispecialty group practice. It is a larger scale organization, and its staff development needs, correspondingly, are much more varied and complex. Hence, it must adopt an entirely different model if it expects to manage the strategic aspects of staff development.

Table 12.1 also implies that a number of terms are used interchangeably when referring to staff development. Employee development, staff development training, orientation, management development, in-service education, continuing education, and other terms are often used synonymously. In many cases a person will use different terms to describe similar staff development efforts. The value of Table 12.1 is that it clarifies the diversity of terms and concepts. Because staff development programs are quite variable, it is vital to use the correct term with the appropriate concept. However, it is also possible to refer to the generic activity of orienting, training, teaching, or otherwise developing employees as staff development or employee development. We will use these terms interchangeably throughout this chapter to imply effort directed toward developing human resources.

As indicated in Table 12.1, at least five major staff development efforts should be identified on the continuum of an employee training and development program. These efforts are defined according to their respective objectives, scope of skill diversity, emphasis on personal/career growth, training site, and frequency. It is useful to repeat that a staff development program should be viewed as a continuum of programmatic effort. There is no single best way. In other words, training should not be evaluated as somehow less important for organizational performance than continuing education. In the right organizational context, a training program may indeed be even more valuable in contributing to organizational performance than a continuing education program.

Objectives

There is a wide range of objectives underlying staff development programs. Orientation training attempts to introduce staff to the values, mores, beliefs, behaviors, standard operating procedures, policies, and expectations of an organization. When a small nursing home expands its capacity by 30 beds, for example, it must hire additional employees. All new employees may meet as a group to learn about the mission and goals of the nursing home, emphasis on patient satisfaction, as well as standard operating procedures for reporting to and leaving work, and other prevalent policies needed to complete work each day. The new employees may be divided into smaller groups according to department to receive specific information needed for successful integration in their respective departments.

In contrast, career development is designed to continually expand the capabilities of staff beyond a narrow range of skills and toward a more comprehensive set of capabilities. Each employee is viewed as requiring continual nurturing in terms of education, training, mentoring, and experience in order to develop into a valuable asset to the organization. For example, the nursing home administrators in a large chain of nursing homes may receive periodic training and in-service education on specific managerial skills from the corporate office. They are also encouraged through a system of promotional incentives and reimbursement to continue their education at local universities. Additionally, a plan is developed for them to progress beyond the facility level to the corporate level. As should be apparent in Table 12.1, orientation training and career development are at opposite ends of a continuum because of the end results each strives to accomplish.

Training is generally focused on teaching staff members specific skills, concepts, or attitudes. For example, if the head nurse in the urology ward observes that some patients' catheters are becoming clogged after surgery, nurses may receive a 20-minute refresher course on the proper irrigation of catheters. Or, the admission clerks may receive 4 days of training about guest relations and insurance counseling. In these cases the training serves to provide the employee with a specific skill or to reinforce previously learned behavior.

In-service education is concerned with teaching staff members skills, facts, attitudes, behaviors, or concepts through largely internally designed programs. In-service education is often confused with continuing education. Continuing education attempts to achieve broader professional or vocational requirements—usually according to externally established standards—than in-service education. In contrast, an in-service program addresses more specific skills, facts, attitudes, or behaviors through internally generated efforts. Continuing education usually relies on external training or education resources (e.g., college or university courses) to accomplish its objectives. In this sense, in-service education is more nearly aligned with training; that is, emphasis is on learning specific facts or skills. Continuing education more closely approximates career development; that is, in the course of continuous learning of relevant skills, facts, attitudes, concepts, or other knowledge, the staff member evolves as a person or a professional.

Skill Diversity and Personal Growth

Table 12.1 provides an overview of the variability among staff development concepts on the extent of skill diversity and emphasis on personal/career growth. Although not every staff development program can be neatly characterized in this manner, and there is much variance in how programs compare to the interpretations offered in Table 12.1, it is generally true that orientation training programs are narrow with respect to skill diversity and personal/career growth. By contrast, career development programs address a very broad scope of skill diversity. They also are directed toward a wide spectrum of personal or career growth subjects. The orientation program emphasizes one specific event in an employee's career—acquaintance with the employing organization. Career development emphasizes the evolution of employees and their careers. Granted, the initial orientation program is invaluable for getting off to a proper start, but it cannot replace the successive acquisition of a wide variety of skills that provide a foundation for personal growth.

Training Site and Frequency

Staff development programs can generally be offered through two basic methodologies. The training site may be either internal or external. Internally generated programs rely extensively on in-house resources for education or training, and external programs mainly incorporate external resources, but the two forms are not necessarily mutually exclusive. Table 12.1, however, suggests the main emphasis for single programs of different types. Orientation training almost always incorporates on-site training, with internal personnel or staff presenting the content of the program. The same is typically true for training and in-service education programs. Continuing education programs, on the other hand, rely mainly on an external experience from professional societies or from teaching institutions to accomplish their goals. Career development combines both an extensive internal focus (i.e., heavy reliance on mentoring and job rotation) and an external focus (i.e., participation in professional societies and academic training).

The frequency of staff development programs also varies as indicated by Table 12.1. Orientation training usually happens once or is limited to a short series of courses over a relatively brief period. Career development, by way of comparison, begins when an employee joins an organization and continues until retirement. The frequency of application of the other staff development concepts varies from sporadic to continuous. The point is that the staff development program objective is the criterion that determines the frequency; there should be no preconceptions on this score.

Importance of Employee Development

Although the importance of staff development programs has already been alluded to, it is vital to underscore their strategic relevance. Such programs should provide a distinctive ability of organizations to perform better. In this sense, staff development is valuable because it helps organizations in the following areas:

- Accomplishment of work unit, departmental, and organizational goals
- Accomplishment of individual goals
- Integration of individual and organizational goals
- Implementation of a philosophy of human development within the organization culture

These factors make staff development invaluable to the health care organization facing a competitive market in which third-party reimbursement is increasingly constrained and large-scale corporations are actively involved.

Figure 12.1 captures the essence of the rising importance of staff development to organizational performance, namely that staff development programs should lead to better performance. It is this causal link, which heretofore has been neglected by staff development programs, their directors, and health care executives, that provides the strategic relevance of employee training and staff development programs. Quite simply, staff development programs have been permitted to languish in a state of

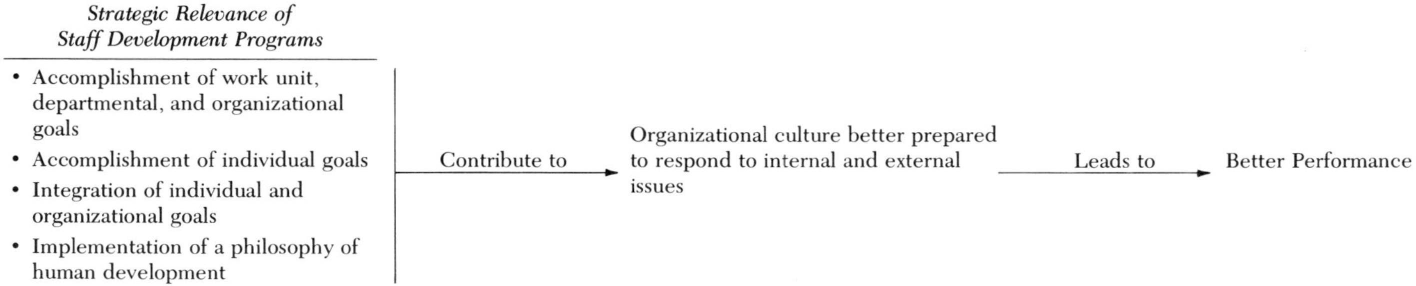

Strategic Relevance of
Staff Development Programs

- Accomplishment of work unit, departmental, and organizational goals
- Accomplishment of individual goals
- Integration of individual and organizational goals
- Implementation of a philosophy of human development

Contribute to → Organizational culture better prepared to respond to internal and external issues

Leads to → Better Performance

Figure 12.1. The causal link of staff development's strategic relevance to better performance.

mediocrity. As a result they have not promoted the type of organizational culture that is essential for attaining the highest levels of performance.

The strategic relevance of staff development programs must be founded on their ability to attain goals—this is the acid test of most organizational investments. However, staff development efforts can achieve the highest level of success when they also address the individuals comprising organizations. Consequently, these programs must primarily accomplish organizational goals while attending to the aspirations of individuals. Really effective employee training will help to integrate the individual and the organization. This is a very powerful combination. When employee and organizational interests coincide, there are phenomenal opportunities for instilling a corporate culture committed to excellence in performance, however those goals are defined.

With the proper combination of individual and organizational emphasis, it is apparent that the organization is interested in implementing a philosophy of human development. Health care executives need to take such a possibility into serious consideration. The creation of an organizational culture that supports the growth of its employees is more likely to be one that accomplishes goals while remaining flexible enough to address the challenges of the future. By investing in the development of its human resources, an organization automatically invests in its own future.

Figure 12.1 conveys these points graphically. The strategic relevance of staff development programs contributes to an organizational culture that is better prepared to respond to internal and external issues. In turn, this leads to better performance. This is the causal link. A focus on goal attainment nurtures an organizational culture that can meet the challenges of the organizational environment. In the final analysis it is this culture that comes to represent the high standards of performance.

Organizational Goal Accomplishment

As far as organizations and their managers are concerned, the primary strategic value of staff development is that it facilitates goal accomplishment (Morrow-Winn, 1983). Specifically, *any* staff development program should produce such tangible end results as the following:

- Higher employee productivity
- Understanding and achievement of cost control
- Activation of employee responsibility for improving organizational services and performance
- Understanding and achievement of quality control
- Adaptability to change, resulting in constructive responses that build capacities for goal attainment
- Integration of personnel within a corporate culture that promotes organizational identity while underscoring service
- Knowledge of the need and mechanisms of attaining a mutually supportive work context that facilitates the completion of others' tasks as well as the accomplishment of individual tasks
- Ability to deliver optimal services and products given the constraints of budgeted resources

- Maximization of work unit goals without sacrificing departmental or organizational goals

Numerous other organizationally oriented end results are possible from staff development programs. However, the preceding list captures the essence of the *potential* organizational contribution from staff development. The strategic management problem for health care executives is to orchestrate their staff development efforts to achieve these outcomes (MacStravic, 1984).

The bottom line for staff development as a strategic management issue is its capacity to facilitate attainment of work unit, departmental, or organizational goals. If a staff development program does not demonstrate this ability in today's health care environment, management must seriously assess whether the organization should be making any investment in the program. Most health-care organizations no longer have the luxury of slack resources to waste on programs that do not provide an adequate return on investment. The ultimate validation of this capacity is attainment of prespecified goals.

Although organizational goal attainment is the focal point of staff development, health care executives should realize that conflicts may arise among work unit, departmental, and organizational goals. To a certain extent, staff development programs may help alleviate the contention among work unit, departmental, and organizational efforts. Normally, staff development is designed to help employees perform more effectively in their work unit. Yet, the work unit exists only in a program, departmental, and organizational context. Hence, the thrust of staff development should be to improve employee performance in a specific work unit, yet not at the expense of the department or the organization.

For example, the hospital pharmacy that instructs its staff in inventory control establishes a basis for exceptional cost control as a work unit. However, if the training encourages the pharmacy staff to institute an excessively elaborate system of control, it will be difficult for other personnel—nurses—to obtain patient medications, and the hospital will not want to accept the tradeoff of lower nurse productivity for the sake of a marginal decrease in pharmaceutical losses. In this illustration, the work unit (i.e., the pharmacy) is maximizing its goal attainment (i.e., inventory control) at the expense of other departments and the organization as a whole. Thus, staff development must secure organizational goal attainment within the context of supportive work units and departments.

Individual Goal Accomplishment

If staff development had the sole purpose of helping organizations to achieve their goals, it would be excessively unidimensional. For staff development to be meaningful to both organizations and individuals, it must provide for individual and organizational goal accomplishment (McConnell, 1984). To the extent that staff development can be designed to achieve this mutual goal orientation, employees will be more interested in deriving the most from the experience. This is only natural. When staff development is in tune with what employees want to attain over the course of their careers and employment, workers will be more likely to seriously invest the commitment necessary to improve their skills, behavior, attitude or knowledge.

Does staff development always accomplish both employee and organizational goals in the best programs? Unfortunately, the answer to this question is no. There will always be cases of organizations that need employees to develop in some manner, yet employees are reluctant to make a conscientious effort in growth. For example, nursing staff may resist in-service training on how to handle troublesome or abusive patients because they are inundated by work and believe that the special demands of prima donnas are the responsibility of the psychological or psychiatric service.

Alternatively, the staff may desire training in how to handle the troublesome or abusive supervisor, but the employee training program does not have such a curriculum available, nor can it spare the expense necessary to generate it. In this case, the organization's staff development program conflicts with the individual goals of employees. Realistically, these conflicts between individual and organizational goals are bound to occur. However, it is incumbent on staff development directors to be sensitive to the individual goals of staff members. Organizational intentions should dominate staff development efforts, but they must be properly tempered by attention to individual goals. A harmonious balance will help promote greater commitment to the staff development program in the final analysis.

Integration of Individual and Organizational Goals

A staff development program can be invaluable in integrating individuals and organizations. The most prosperous organizations are those that are able to alleviate the contention between individual self-interest and organizational self-interest (Culbert and McDonough, 1980). Employees who are able to integrate their own goals within the context of an organization and its goals are more likely to perform better on the job. Staff development programs are merely one mechanism for achieving a satisfactory resolution of individual and organizational goals, but there is little question that they are strategic in this respect, given the extent to which staff development can clarify intentions and reinforce the rationale for efforts.

Implementation of Organizational Culture

Finally, staff development is conducive to implementing organizational culture. Such programs transmit the expectations and mores of an organization to new employees, reinforcing these norms throughout the specific programmatic efforts. Through these and other mechanisms, staff development represents a philosophy of management on the part of an organization toward its employees. Such philosophies are beneficial in generating trust and confidence from employees toward the organization and its intended efforts.

To effectively implement organizational culture, staff development programs must be comprised of a full continuum of orientation, training, in-service and continuing education, and career development components. In other words, staff development programs must retain a consistent exposure to employees if they intend to contribute to defining and reinforcing an organization's culture. Incremental efforts will not accomplish results as effective or as durable as a planned and coordinated staff development program.

For example, consider the mental health clinic that has a staff development program consisting of orientation training and continuing education. What opportunities are there to instill or reinforce organizational culture? They are limited because the continuing education is invariably obtained from external sources that are unable to address the specific characteristics or context of the clinic. The orientation training is the only point of departure for explaining the existing culture. In contrast, a mental health clinic whose staff development program incorporates orientation training, training, in-service education, continuing education, and career development is better prepared to continuously underscore its cultural expectations and to thereby facilitate the shaping of a homogeneous culture.

INCENTIVES FOR UPGRADING STAFF DEVELOPMENT PROGRAMS

The incentives for upgrading the effort and resources organizations expend on their staff development programs are numerous, but there is one critical idea: namely, that the emphasis on staff development must undergo a metamorphosis. Staff development particularly should be seen from a strategic perspective. For years health care organizations have approached staff development as a nonessential—an effort that is prevalent but not something to be nurtured, expanded, or otherwise cultivated. These low expectations and investments have yielded mediocre results. However, health care executives are beginning to realize that their organizations are comprised of human resources who make the difference between success and failure. Business organizations have always realized this because the role of the incentive system has always been paramount. Health care organizations are now confronting the matter of incentive systems, and the experience has altered many traditional expectations and perspectives.

The primary result for employee training from a new set of incentives in the health field is the metamorphosis of staff development, as portrayed in Figure 12.2. Most health care executives and their organizations have been unable to rise above a traditional view of staff development according to which there is no strategic emphasis to these programmatic efforts. Staff development has consistently been seen as a support or auxiliary function, subordinate to other critical operations of health care organizations. Essentially, staff development has been designed to address internal goals, that is, training of employees in specific skills to help them perform their work tasks better or to maintain licensure or accreditation. In short, traditionally staff development programs have maintained a limited scope of effort. They have not been directly oriented to performance other than in very superficial ways.

Figure 12.2 indicates that several new incentives are prevalent in the health care field. The rise of competition, tighter operating margins, emphasis on performance, and a new vocabulary (describing new concepts) are forces reshaping the perception of many health care executives of staff development programs. In essence, a metamorphosis has been stimulated by the traditional view of staff development combined with the new incentives. The outcome is a contemporary view of staff development.

The new age of staff development is decidedly a more exciting and more active period. Under the contemporary version, staff development maintains a strategic emphasis because of its performance orientation. In brief, contemporary staff development programs facilitate better performance by personnel and by the organization as a whole. In the contemporary version, staff development is designed to address internal and external goals. It has a much more comprehensive scope. It is directly linked to performance. These unique and meaningful changes in staff development have resulted from the pressures in the health care environment.

Traditional View of Staff Development		*New Incentives*		*Contemporary View of Staff Development*
• Nonstrategic emphasis due to support orientation • Designed to address internal goals • Limited scope • Not directly linked to performance	+	• Greater competition • Tighter operating margins • Emphasis on performance • New vocabulary	Have stimulated a metamorphosis ⟶	• Strategic emphasis due to performance orientation • Designed to address internal and external goals • Comprehensive scope • Directly linked to performance

Figure 12.2. The metamorphosis of staff development.

What are the new incentives in the health care system that demand upgrading staff development programs? Why has staff development gone from insignificance to strategic importance? The answers to these questions are related to several fundamental changes in the health system that threaten every provider.

1. Greater competition is now apparent in the health system because of deregulation, acceptance of marketing, corporate involvement and many other factors (Smith and Reid, 1986). Health care organizations need to form new strategic responses to competition. First, however, staff must be informed of the shift and the plans to become competitive must be formed. In short, a new set of skills (i.e., in marketing) must be communicated to the staff, which must henceforth exercise them.

2. Health care organizations are now facing tighter operating margins, which emphasize efficiency, productivity, and profitability. The fact is that consumers, employers, third parties, and the government are unable to continue funding rising health care costs. This has stimulated a number of policy changes by third-party insurers and the government. Most notable is the use of prospective pricing, which has reduced the possible margin on services (Smith and Fottler, 1985). There is less slack and more control over expenditures. To remain financially solvent, health care organizations must produce goods and services more efficiently. This is possible when staff are better trained because higher productivity and more efficient performance eventually add to profitability. However, fundamental to these achievements is more efficacious staff development, which recognizes the strategic ramifications of employee performance and is programmatically orchestrated to promote such ends.

3. Health care organizations are demanding better performance from staff members because they recognize that investments in staff that are not accompanied by specific improvements are unwarranted. This is evident in a variety of forms. Reduced budgetary slack in organizations implies that employees and organizational programs must produce more output per expenditure. For example, an investment in a non-revenue-generating programs—such as staff development—must show a specific return on investment to justify the original expenditure; otherwise the investment should be reduced or terminated.

4. Staff members need a new vocabulary and new skills to effectively respond to the new environment described above. Consider some of the terms accompanying recent changes in this environment: competition, productivity, cost containment, marginal costs and benefits, prospective payment, efficiency, marketing, multi-institutional arrangements, profitability, and corporate infrastructure. For many health care professionals these are new terms. The related concepts are equally foreign to many health care staff members. Due to training in basic sciences, or lengthy exposure to the minimal demands of the traditional health care environment, staff members are uncertain about these changes or unable to respond to them in the most effective manner. They vitally need a new worldview. They need to understand how to see the changes in terms of possible impacts on their areas of operation. They need to know how to respond to such impacts for the benefit of their work unit, the department, and the organization.

The incentives mentioned above have placed staff development programs at a critical juncture. These programs can either respond in helping their organizations to meet new constraints or they can continue in the traditional mode of complacency and mediocrity. The choice is one of becoming involved in strategic management issues or remaining a supporting program with little or no strategic importance.

The incentive for upgrading staff development programs is that without a specific contribution to organizational goals, such programs should be terminated. The resources should be reallocated to other programs managed by those who are interested in demonstrating a reasonable return on investment. This is a fairly threatening posture, but it is nonetheless a realistic one. Organizations in the health system simply cannot afford to fund programs that do not generate the results that allow goal accomplishment. The only catch to this posture is the licensing requirements of health facilities and professionals. At issue is the accreditation of a facility, which enables it to qualify for reimbursement. Hence, staff development (i.e., in-service education and continuing education) given to providers on the staff is essential for achieving other goals. This may be the salvation of staff development programs that have no intention of contributing to the strategic performance of their organizations. They must be retained to meet other needs.

In any event, the prognosis for health care organizations and staff development programs is certainly clear. They must attain high performance (Medearis and Popiel, 1975). The incentives are numerous and infallible. Without a timely and sufficient response from staff to these constraints, more organizations will face an uncertain future characterized by threatened financial solvency and inability to compete. Properly orchestrated staff development programs can play a significant role in strategically responding to these threats and incentives.

COMPONENTS OF EFFECTIVE STAFF DEVELOPMENT PROGRAMS

Given that health care organizations need to upgrade their staff development programs, what components should be managed in this endeavor? Figure 12.3 provides an overview of the primary components underlying an effective staff development program. Health care executives can use this model from two perspectives: first, to ensure that the content or programmatic infrastructure of their staff development effort includes each of these components, and second, to determine that the components have adequate resources for implementation.

Needs Assessment

Staff development begins and ends with needs assessment (Kirkpatrick, 1977; Moore and Dutton, 1978). Health care executives must adopt the philosophy that programmatic efforts are driven by need. There are at least two primary sources of need as suggested in Figure 12.3. Environmental assessment may identify relevant factors that force an organization to adopt a staff development program or alter the content of an existing program.

For example, a chain of nursing homes faced with new licensing requirements for administrators should prepare to offer its existing and future administrators a method for meeting these requirements. Thus an internal staff development program would be driven by the external regulatory licensing constraints. Or, suppose that a third-party payer proposes to introduce more stringent reviews of patient qualification for reimbursement of home health nursing services. A home health care agency would respond through a staff development program that teaches nurses to identify

Needs Assessment	Decision		Program Preparation	
	Reassess Other Managerial Strategies			
	↑ *No*			
• Environmental assessment	Resolvable by staff development	—Yes→ Set objectives	+ Define curriculum → Implement → Evaluate	+ Prepare materials and sessions
• Organizational assessments				
Work unit				
Departmental				
Organizational				

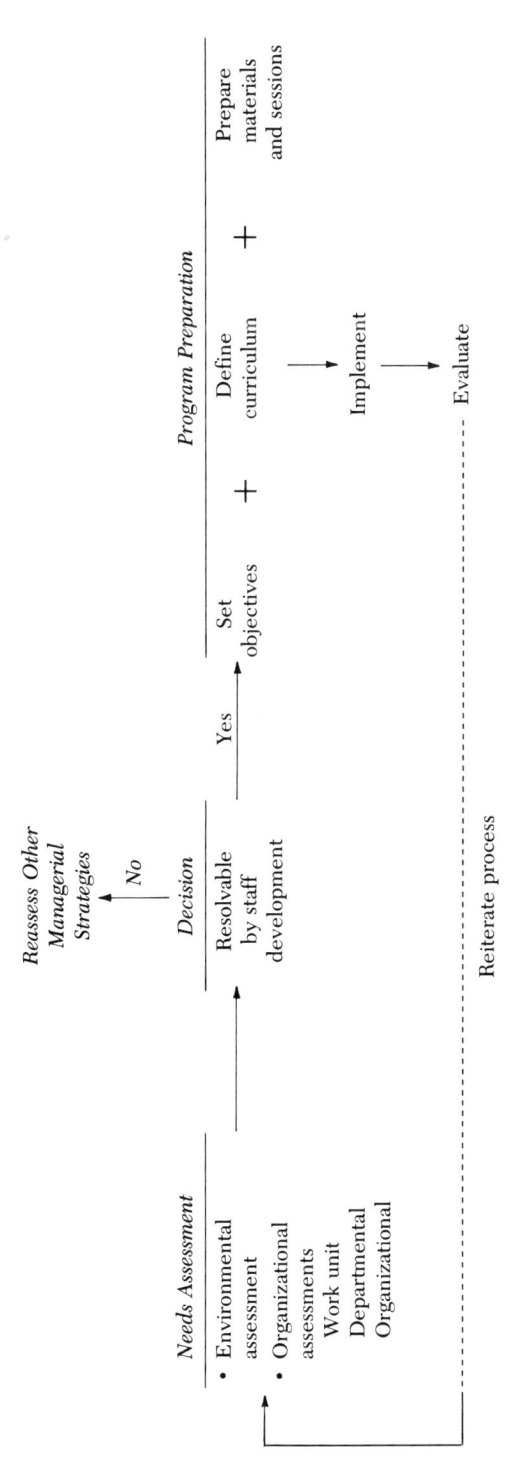

Reiterate process

Figure 12.3. A visual model of staff development. (Adapted from H. L. Smith and N. F. Elbert, *The Health Care Supervisor's Guide to Staff Development*. Reprinted with permission of Aspen Publishers, Inc., copyright © 1986.

reimbursable patients. The point of these two illustrations is that external factors—legislation, reimbursement, technology, the marketplace, labor trends, or any other relevant environmental pressures—may force an organization to alter its staff development efforts. Creation of a staff development program's content should begin by assessing these external factors.

In addition to environmental assessment, organizational assessment is vital to health care organizations that are interested in creating effective staff development programs. This form of assessment is multidimensional and incorporates the work unit, departmental, and organizational levels. The differences in assessment at each of these levels can be summarized as follows:

1. *Work unit level.* Every work unit or team should be guided by a set of specific goals or objectives that determine the extent to which employee efforts are producing the desired return. Each employee in a work unit should also be guided by a set of objectives over a given period of performance. In terms of assessment, the specification of objectives facilitates assessment if the objectives are defined in terms of measurable end results that must be accomplished over a specific period of time. The task for directors of staff development programs or health care executives is to identify objectives that are not being met (e.g., cost control, inventory control, patient satisfaction). Having identified the problems, the supervisor (and in a growing number of cases the employee and entire work unit) can determine why the prespecified objectives are not being met. The reasons that are relevant to staff development thereby set a foundation for the programmatic content.

2. *Departmental level.* Every department should be guided by a set of specific goals or objectives, just as work units are meant to follow specific guidelines. The measures may be different at the departmental level compared to work units but the principle is the same. Having measurable standards facilitates the identification of substandard or deviant performance. Once deviations have been identified, staff development curricula can be formulated to help resolve the causal factors that are amenable to training, education, mentoring, or other learning methodologies. Positive deviations (e.g., remaining under budget) can also be examined to ascertain the causal factor and, where appropriate, generalized through staff development to other department heads.

3. *Organizational level.* Just as every work unit and department should be guided by preset goals and objectives, so too should the entire organization. Obviously, if such standards are not available, an executive will experience great difficulty in defining problems in performance. At the organizational level there is a greater departure in the intention of staff development than at the work unit or departmental levels. The causes of performance deficiencies at the organizational level may be result of the cumulative problems at the other levels. Furthermore, there are fewer managers responsible for performance at the organizational level. This means that staff development will have to focus on a few recipients. Moreover, the type of staff development needed at the organizational level will probably be at the most complex end of the continuum—career development or management development. Only in large multiorganizational firms does staff development at the organizational level approximate that for the work unit and departmental levels. Whatever the case may be, it is important that needs be assessed to guide program context.

In sum, organizational assessment attempts to explain why individuals or components of an organization are unable to perform at desired levels. Where those factors are amenable to staff development, it is possible to conceive a program for resolution.

The Decision to Institute or Modify a Program

The next major component of the staff development model in Figure 12.3 is the decision to institute or modify a staff development program. As the preceding discussion suggests, not every problem in a work unit, department, or organization can be rectified through staff development. The many determinants of effective performance include but are not limited to, the following:

- Organizational structure
- Rewards and compensation
- Leadership style
- Extent of resource support
- Available technology
- Extent of planning (i.e., specification of goals and objectives)
- Control
- Impact of external forces (e.g., market competition)

These are only a few of the categories of variables that might produce an undesired impact on staff attempts to provide products or services. For example, an organization will not want to invest substantial money in altering its staff development program if poor performance is due to inadequate rewards, rather than inability on the part of personnel.

Program Preparation

Assuming that the performance problem of personnel is treatable through staff development, the next step is to prepare the program. Depending on the scope of previous staff development efforts, the degree of fine-tuning required may be limited. In other cases extensive modification in the staff development program may be needed. In any case, the preparation of a staff development program centers on three activities (Kirkpatrick, 1978; Newell, 1976).

1. *Setting objectives.* A staff development program should incorporate a specific set of measurable objectives that it will attempt to achieve over a specified time period. This is difficult because in the past there have been few expectations for employee training programs in health care organizations. Hence, there has been neither goal specification nor measurement of goal accomplishment. Consequently, there may be considerable work to be completed in setting measurable objectives. Remember that staff development is often a rather abstract effort. It is difficult to precisely measure skill or educational (i.e., knowledge) attainment. Furthermore, there is the issue of knowledge decay, which may moderate the measurable results achieved in a training program. The best staff development programs will set objectives that relate not only to individual improvement (i.e., skill or knowledge development) but also to improvement in departmental and organizational performance.

2. *Defining curriculum.* The curriculum of a staff development program should be associated with identified needs and specified objectives. Once these basic guidelines have been clarified, it is possible to determine the feasible alternatives for resolving a

staff development deficit. At this point the issue of using in-house or external trainers is vital to curriculum development. In-house resources are valuable in adapting a program to specific needs. However, the training skills may not exist in the organization, or those who have the knowledge may not be available to conduct the training. For example, assume that a group of physician executives vitally needs training on preparing and using budgets but the chief financial officer or the financial staff are too busy to provide sessions on budgeting. The organization may have to go outside to procure the needed training skills. There are many options and costs associated with using external sources.

3. Preparing materials. The final step in program preparation is to accumulate necessary training materials and to schedule the sessions in an environment that facilitates, rather than hinders, effective training. To a large extent these details depend on the specific instructor or trainer and the methods of instruction chosen to accomplish the objectives.

Program preparation is obviously a significant element in staff development. Entire books have been written on the three phases of program preparation mentioned above (Smith and Elbert, 1986). Interested readers should turn to these resources to clarify any of the programmatic stages.

Implementation

After a program has been defined it must be implemented, and implementation calls for a satisfactory budget to support the program. This is a critical premise that is not automatically fulfilled in today's health care environment. Although health care organizations have traditionally supported staff development, mainly in the form of in-service education, there is no certainty that such support will continue as it has in the past. In fact, the second step in Figure 12.3 assumes that if the employee performance problem is resolvable through staff development, resources will be available to fund the proper programmatic response. This assumption may be unwarranted in health care organizations that are encountering extreme difficulty in meeting budget targets. More often than not the slack has already been cut out of health organization budgets; hence, there are few resources available to expand staff development programs. Therefore, staff development programs must become more efficient and effective in their own stewardship of budgeted allocations.

Evaluation and Control

The final, but certainly not the least important step in the visual model of staff development is evaluation and control of program performance. This is undoubtedly the most abused step in the model. Few executives have been concerned about the ability of staff development programs to produce proper results. These attitudes are rapidly changing, however, as health care organizations confront tighter budgets and as they begin to envision the capacity of staff development programs to offer strategic capabilities.

There are many methods and measures for evaluating the impact of staff development programs. Among the more frequent measures are the following:

- Changes in performance at individual, departmental, or organizational levels
- Changes in goal attainment at individual, departmental, or organizational levels
- Changes in knowledge, skills, or attitudes among participants
- Reactions of participants or recipients to the program
- Reactions of supervisors to participants' changed knowledge, skills, or attitudes

Assuming that an organization has an active planning and control system in which goals are continually restated, performance measured, and actions taken to rectify deviations from preestablished standards, the best measure of staff development centers on changes in goal attainment.

A working knowledge of evaluation methodologies is the fundamental basis for prudent assessment of staff development program efforts. An overview of such methodologies is beyond is the scope of this chapter. Nonetheless, health care managers should recognize that the acid test of a staff development program is the ability to change—to improve—performance toward goals. Without documentation that relies on proper methodology for its conclusions, staff development will remain an expendable activity. Therefore, it is essential that evaluation methodologies be invoked to promote a clear understanding of the relevant accomplishments that justify continued resource allocations to staff development.

TRENDS IN STAFF DEVELOPMENT

What are the future trends in staff development in the health care field? As this chapter has tried to convey, there is a new attitude developing among personnel directors, human resources managers, staff development directors, directors of in-service education, and health care executives in general that a revolution has beset employee training and development. The revolution has already occurred as a result of the forces and incentives mentioned throughout this chapter. The only question remaining is the extent to which each health care organization and its managerial staff will build on these changes to significantly upgrade the content and expectations of staff development programs. A significant challenge is to reshape such programs to make a viable contribution to the strategic management of an organization. In all practicality, this process of reshaping will begin with the program content as suggested in Table 12.2 (Odiorne, 1980; Von Glinow et al., 1980; Wehrenberg, 1983).

There is a major philosophical trend conveyed in Table 12.2 that before the early 1980s would not have been evident. Up until the rise of corporate involvement, competition, prospective pricing, marketing, deregulation, and multi-institutional arrangements in the early 1980s, staff development programs tended to differentiate between the content and orientation of employee training for clinical and for nonclinical personnel. For example, physicians and nurses would receive one orientation while supervisors, operating personnel, and management staff would receive another. In many cases the clinical program content was viewed as rigorous and a significant contribution to professional growth. The nonclinical program content was often seen as less demanding and more oriented to less relevant finishing touches on employee development.

Although the curricula of staff development programs in health care organizations

TABLE 12.2. Trends Reshaping the Content of Staff Development Programs in Health Care

From	To
Past Orientation of Program Content in Staff Development	Future Orientation of Program Content in Staff Development
Clinical staff requires complex training; nonclinicians require less complex development	All staff require complex program context; distinction between clinical and nonclinical staff needs become less meaningful
Clinicians do not cross over into nonclinical areas; nonclinicians do not cross over into clinical areas	Clinicians and nonclinicians frequently share in the delivery of services
Personnel are narrowly focused on one profession or vocation	Personnel are well educated and may be interested in pursuing several careers
Knowledge base is relatively stable except in clinical areas	The half-life of knowledge is rapidly decreasing for many health care vocations

continue to be structured around the clinical-nonclinical difference, it is apparent that the health services sector has experienced changes that dramatically influence the entire approach to staff development. Hence, the programmatic content of staff development programs is also changing. This is conveyed in Table 12.2.

A significant problem for staff development programs in the future is the need to differentiate among categories of personnel. The most prevalent distinction has been between clinical and nonclinical personnel. This differentiating mechanism has been useful because of the professional licensing requirements often existing for medical practitioners and ancillary medical services personnel. However, this method of distinguishing personnel is beginning to break down as nonclinical staff pursue their own credentials and as clinical personnel seek skills formerly not related to clinical training. The end result of this trend, which is likely to continue, is that staff development programs must fill a greater diversity of needs. In short, the staff development program of the future is challenged to incorporate a wide variety of curricula and plans for development or training.

A case in point is clearly seen in many administrative positions, such as nursing home administrators. The professional licensure requirements have been raised gradually, and today's licensed nursing home administrator is well educated. Many of these professionals have a continuing need to expand their knowledge and skills. Fortunately, they are given excellent motivation to pursue additional development from licensure requirements for continuing education. Certainly, many professionals, whether nursing home administrators, hospital administrators, or clinicians, may perceive that they do not require further education or training. For them, the issue is to ascertain the easiest way of fulfilling such requirements. But others are conscientious and concerned about obtaining the highest possible return on their time invested.

The preceding illustration demonstrates that staff development programs will increasingly need to address the career development of professionals—whether clinical or nonclinical. This trend must be integrated with the rising education level of staff in general. The evidence clearly suggests that workers are more highly educated today than ever before. A more highly educated work force is likely to remain interested in developing skills and knowledge over an entire career rather than acquiring just enough information and credentialing to permit their learning to terminate upon employment.

To these trends must be added the decreasing half-life of knowledge in many vocations and professions. It is a blessing but also a problem that knowledge in the health professions is expanding at a phenomenal rate. For staff, the implication is that their knowledge will become obsolete more quickly. Hence, they must continually and conveniently upgrade their knowledge base. It should be clear to all who enter a highly technical field that they are accepting a significant responsibility for remaining current in the future. If there are no licensure requirements, this responsibility is a personal one. But in any environment where knowledge changes rapidly (e.g., computer specialties), to forego continuing development because there are no licensing requirements would seriously undermine potential job effectiveness.

The implications of these trends for staff development planning are numerous. Above all, staff development programs will certainly be forced to grow in their scope of operations. Very simply, staff needs will be much more comprehensive and complex than before. As a result, these programs must begin to develop a more realistic array of options for filling staff development needs from both internal and external means. It is likely that greater emphasis will be given to continuing education and career development. Yet, at the present time these components of staff development are managed on an ad hoc and very rudimentary level by health care organizations. More ofen than not there is no plan other than possibly reimbursing professionals when they obtain external training (e.g., college credits). There is little or no control over where, how, or from whom the education is obtained. The inference is that something must be better than nothing.

Staff development programs will also witness the problem of multiple careers among personnel. Increasingly, professionals and nonprofessionals are trying several careers. This is a distinct challenge to staff development programs that have been built on the traditional concept of one person remaining with one vocation. For example, some physicians and nurses are being attracted to the managerial ranks. How will they perform these administrative, managerial, and executive responsibilities effectively unless they are prepared with appropriate knowledge? There no longer is a guarantee that clinicians will remain clinicians. Nurse and physician executives are increasingly prevalent; to contribute effectively to an organization in a managerial capacity, they must be trained in the basics (of management theory and practice), and their skills must be fine-tuned as they gain experience. Again, this presents a unique and challenging future to staff development.

A STRATEGIC POSTURE

There are many pressures confronting health care organizations. To the extent that these pressures are adequately addressed, it is likely that most organizations will thrive and prosper. Perhaps more than any other variable, it is going to be people—staff— who make or break health care organizations. To this point, however, health care organizations have done little to help prepare their human resources for the present and future challenges. The result is that many organizations cope with their internal and external demands inefficiently and ineffectively.

To many in the health care field, the proposition that staff development programs should become a strategic factor in responding to these challenges would seem ludicrous. After all, staff development programs have occupied the backwaters of most health care organizations. They have been poorly planned and underfunded, and more an ornament than a strategic resource for organizations. If a staff development program maintains this reputation, serious consideration should be given to paring it to the bone. Health care organizations do not have slack resources to waste. On the other hand, it is possible with sufficiently rigorous thinking to transform a staff development program into the sort of strategic tool that supports a vibrant response to present and future challenges presented in the health care environment.

Perhaps more than any other factor, health care managers will have to battle the stereotype that surrounds staff development programs before the strategic relevance of these programs will be understood. Commitment to staff development by top management echelons is a necessary ingredient. Throwing more money into programmatic efforts is only a partial answer. Top managers must reinforce the priority of staff development in the distribution of rewards and incentives (e.g., for participation and subsequent improvement in performance) to personnel as well as by their rhetoric. Through such postures, staff will eventually recognize that training and development are not superfluous extras, but prime ingredients underlying an organization's effort to remain competitive in the challenging environment confronting health care organizations.

SUMMARY

Inservice education, training, staff development and employee orientation are all elements which may comprise an employee training or staff development program. This chapter defines and differentiates each of these programmatic activities to provide managers a better understanding of their role in health care organizations. An effective training or staff development program is predicated on using each of these elements to accomplish specific predetermined objectives.

Training and development programs have seldom lived up to their expectations because health care managers have not envisioned the programs in a strategic sense. For example, inservice education has been viewed as a mechanism to maintain accreditation and professional licensure. Yet, inservice education has seldom been incorporated within the strategic plans of health care organizations. This tendency has developed because managers believe that inservice education, orientation, training and development are activities which contribute in a minor way to accomplishment or organizational goals such as quality of care or cost control. Consequently, training and development programs are seen as expendable programmatic efforts with a low priority.

Prudent use of training and development programs can promote attainment of individual, work unit, departmental, and organizational goals if they incorporate a number of specific components discussed in this chapter, namely:

1. Needs assessment to identify threats, opportunities, strengths, and weaknesses.
2. Consideration of the relevance of staff development in resolving problems identified in the needs assessment.

3. Program formulation including setting objectives, defining curriculum, and preparing sessions.
4. Implementation of the program.
5. Evaluation and control to improve programmatic efforts.

These components represent the basic framework for strengthening the contribution made by training and development programs.

There are many factors which stimulate the need for more effective staff development programs in health care including greater competition, tighter operating margins, an emphasis on performance, and new concepts of health delivery. Health care organizations vitally need their employees to be more productive in delivering quality services at reasonable cost. Staff development programs can facilitate attainment of these goals when they have a comprehensive scope, are designed to address internal and external goals, are directly linked to performance and receive a strategic emphasis because they help employees perform better. This chapter provides several guidelines for improving the structure and use of staff development programs.

REFERENCES

Culbert, S. A., and J. J. McDonough. *The Invisible War.* New York: Wiley, 1980.

Hofer, C. W. "Toward a contingency theory of business strategy." *Academy of Management Journal,* 18:784–810 (1975).

Kirkpatrick, D. L. "Determining training needs: Four simple and effective approaches." *Training and Development Journal,* 31:22–25 (1977).

Kirkpatrick, D. L. "Developing an in-house program." *Training and Development Journal,* 32:40–43 (1978).

Levine, H. Z. "Consensus: Employee training programs." *Personnel,* 58:4–11 (1981).

MacStravic, R. E. S. "Performance auditing for health care supervisors." *Health Care Supervisor,* 2:29–38 (1984).

McConnell, C. R. "Employee training: The shape of things to come." *Health Care Supervisor,* 2:29–38 (1984).

McGehee, W., and P. W. Thayer. *Training in Business and Industry.* New York: Wiley, 1961.

Medearis, N. D., and E. S. Popiel. "Guidelines for organizing inservice education." In *Staff Development.* Wakefield, MA: Contemporary Publishing, 1975, pp. 2–8.

Moore, M. L., and P. Dutton. "Training needs analysis: Review and critique." *Academy of Management Review,* 3:532–545 (1978).

Morrow-Winn, G. "Staff development—More management than training." *Nursing Homes,* 32:6–11 (1983).

Newell, G. E. "How to plan a training program." *Personnel Journal,* 55:220–225 (1976).

Odiorne, G. S. "Training to be ready for the '90s." *Training and Development Journal,* 34:12–20 (1980).

Pascale, R. T., and A. G. Athos. *The Art of Japanese Management: Applications for American Executives.* New York: Warner, 1981.

Peters, T. J., and R. H. Waterman. *In Search of Excellence.* New York: Harper & Row, 1982.

Smith, H. L., and E. F. Elbert. *The Health Care Supervisor's Guide to Staff Development.* Rockville, MD: Aspen, 1986.

Smith, H. L., and M. D. Fottler. *Prospective Payment.* Rockville, MD: Aspen, 1985.

Smith, H. L., and R. A. Reid. *Competitive Hospitals.* Rockville, MD: Aspen, 1986.

Snow, C. C., and J. T. Grant. "Answers to questions about management development programs." *Hospital and Health Services Administration Quarterly,* 25:36–53 (1980).

Swansburg, R. C. *Inservice Education.* New York: G. P. Putnam's Sons, 1968.

Von Glinow, M. A., M. J. Driver, K. Brousseau, and J. B. Prince. "The design of a career-oriented human resource system." *Academy of Management Review,* 5:23–32 (1980).

Wehrenberg, S. B. "Training megatrends." *Personnel Journal,* 62:279–280 (1983).

CHAPTER 13

Performance Appraisal

Charles L. Joiner

Perhaps more so than other organizational executives today, health managers are under extreme pressure to contain cost and improve efficiency of operation. To be competitive in the health industry, these managers must develop and adopt new methods and techniques to improve the performance of their organizations. Before prospective reimbursement systems were implemented, many health executives generally ignored industrial models for management practice. However, much has occurred in recent years to push the management of for-profit and not-for-profit health care organizations into a position of strength regarding the transfer of management knowledge and technology to health care institutions.

One area of management not yet developed to its potential in the health field is performance appraisal (PA). Interestingly, this field holds great opportunity for improving management of health organizations and yielding sizable dividends at both the individual and organizational levels. Through the proper design, implementation, and maintenance of a dynamic performance appraisal system, individual and organizational performance may be monitored and enhanced, resulting in a more efficient and effective organization.

The need to develop a superior and effective performance appraisal system (PAS) in a health care organization can be described in clear and simple terms. Indeed, health care organizations so employee intensive that salaries and wages comprise as much as 60 to 70 percent of the institution's total operating cost are not unusual. Such data reveal the clear linkage between the successful operation of the organization and the effective and efficient performance of its employees. A good performance appraisal system can help the organization attract and retain highly qualified employees. Health administrators should clearly understand the reasons for implementing a performance appraisal system that is effective in promoting organizational goals as well as in developing human resources.

The development of a performance appraisal system should be a key part of management's responsibility for a variety of reasons.

1. Evaluation of employees is an important management task, since managers must be able to make administrative decisions that cause changes in employee status.
2. Management needs to have current information on the performance of individual units and departments. Through the evaluation of individual performance, management can obtain information about the unit of which the individual is a part.
3. Performance appraisal systems provide managers with information about employees' skills and abilities. These data help to validate or change the organization's selection procedures, and they provide the basis for recommendations on employee training programs.
4. The performance appraisal system provides valuable information on the quality of supervision.
5. The performance appraisal system, if successfully implemented, provides useful information for analyzing the role of management in effecting changes in the performance and development of employees.

Performance appraisal is one of the most important processes in any organization. However, from the employee perspective, all too often organizations implement performance appraisal systems as a punishment rather than to assist the employee. If this attitude pervades an organization, it is difficult to realize the positive outcomes possible for employee and organization alike. How, then, can an organization establish a performance appraisal system that will yield desirable results for all? Admittedly, an answer to this question is easy to prescribe but difficult to implement with consistency. In any system, special emphasis should be given to the human side of appraisal since, in the final analysis, the best system may be only as effective as it is *perceived* to be by the employees. If the system is to be perceived by employees as effective, all responsible levels of management must implement it fairly and consistently.

A brief overview of history, methods, evaluation, and problems provides background for discussion concerning strategic performance appraisal, application of management by objectives (MBO) for health care organizations, and linking rewards to performance.

HISTORICAL DEVELOPMENT

The first recorded performance appraisal system in industry was developed by Robert Owens in Scotland around 1800 (Haar and Hicks, 1976). Owens placed a colored block at each workman's place to designate how well the worker had performed the previous day. Different colors indicated various levels of performance.

Formal appraisal systems were first utilized in the United States by the federal government and by certain city administrators in the middle to late 1800s. Frederick Taylor and his work measurement program laid the groundwork for performance appraisal in business, which began shortly before World War I. Soon, in 1916, Walter Dell Scott began the development of the man-to-man rating chart that was widely used to identify and evaluate military leaders during World War I. These early appraisal systems were related to various numerical efficiency factors developed from both work simplification studies and time and motion studies popularized by the work of industrial engineers (Haar and Hicks, 1976).

The graphic rating scale approach was developed in the 1920s. It requires the rater to evaluate an individual on a continuum of "poor" to "excellent" for several characteristics. Human relations was strongly emphasized by management in the 1930s and 1940s, as evidenced by appraisal systems that focused on rating personality and behavior traits of employees.

During the 1950s, the concept of management by objectives began to emerge when firms such as the General Electric Company began to focus on detailed planning processes for their organization. General Electric first identified the elements of management by objectives in its extensive planning for reorganization in the period 1952–1954. Peter Drucker emphasized the need for establishing objectives both for the entire organization and for its individual managers, and for then measuring performance against the objectives. Douglas McGregor added to the impetus for management by objectives when he criticized trait appraisal systems as requiring the rater to "play God" (Haar and Hicks, 1976).

In 1971 the Supreme Court mandated that any form of testing procedure for a specific job must relate directly to the job tasks to be performed (*Griggs v. Duke Power Company* [3 FEP 175 (1971)]. This decision contained major implications for the construction of performance appraisal tools and the use of performance appraisal results. For example, using a test of mental ability for purposes of selection or promotion is illegal if no correlation can be established between the test and the performance of a specific job. When used as tools for selection, transfer, or promotion, performance appraisals are considered tests and the 1971 Supreme Court decision applies (Haar and Hicks, 1976).

Reinforcing the Supreme Court decision in *Griggs v. Duke* are the guidelines issued by the Equal Employment Opportunity Commission (EEOC) in 1970 outlining employee selection procedures. These guidelines require employers using tests to have available data demonstrating that the tests are predictive of, or significantly correlated with, important elements of work behavior (McCormick, 1972).

The topic of performance appraisal is now receiving considerable attention in many organizational settings. This attention has evolved from new demands for performance accountability brought about by cost containment efforts and reduced revenues during times of economic downturn in which expectations of high performance have continued. Additionally, recent EEOC rulings and court decisions have alerted employers to possible discriminatory effects of their performance appraisal systems (Holley and Feild, 1975). As a result, managers are today examining performance appraisal systems to determine the degree to which the objectives of the appraisal systems are met and to ascertain whether any discriminatory effects are present.

Sashkin (1981) points out that performance appraisal is, or should be, a major concern of all middle and upper level management. One reason for his concern is that federal courts increasingly are hearing cases in which plaintiffs argue on behalf of management that EEOC guidelines and those of other agencies actually have contributed to unfair and illegal employment practices. The alleged unfair practices include promotions, demotions, transfers, and terminations based on performance appraisal data.

These developments and others should be considered in any organization's evaluation of its performance appraisal system. It also is important that management have a comprehensive understanding of the most commonly used methods of appraisal. A summary of such methods follows.

COMMON PERFORMANCE APPRAISAL METHODS

Numerous methods exist for evaluating performance; these generally are classified as comparative methods and absolute standards. An organization should choose a method based on two criteria: the factors the organization desires to measure, and the method that applies best to the nature of the organization.

Comparative Methods

The comparative methods compare one employee to another in order to determine performance ranking. *Straight ranking* merely asks the rater to list employees, beginning with the best employee and ending with the weakest employee. In *alternative ranking*, the most common ranking method, the rater repeatedly chooses the best and weakest employees, each time choosing from the names remaining. The process is continued until all employees have been placed on the ranking list, with the last two employees being ranked in the middle (Glueck, 1978).

In *paired comparison* one employee at a time is compared to all other employees. Each time an employee is ranked higher than another employee, a tally is placed by the higher ranking person's name. The employee with the most tallies is considered to be the most valuable, and the others are placed in order according to the number of tallies by their names. *Forced distribution* asks the rater to assign a certain proportion of the employees to one category on each criterion. For example, 10 percent of employees might be in the "superior" category, 20 percent in "good," 20 percent in "average," continuing to "poor" (Glueck, 1978).

Comparison methods are useful in making decisions regarding promotion and selection from within a work unit, and this is their major advantage. However, the use of more than one rater is advisable to obtain more accurate ratings.

Comparison methods are problematic in that they are time-consuming and useful only for relatively small groups of employees. Also, an employee's performance rating is based on other employees' work rather than on desired performance. Comparisons can lead to judgment of personality rather than of performance. Finally, ranking assumes equal distance between employees' ranks, an assumption research shows to be unwarranted.

Absolute Standards

Through the use of standards, each individual is evaluated against written standards, and several factors of performance are measured. In one such method, the *weighted checklists* method (Fig. 13.1), the rater identifies, and assigns a weight to, each of the task to be evaluated. The employee then is scored on each task to determine overall performance. The checklist calls for a simple "yes–no" judgment (Haar and Hicks, 1976).

The major disadvantage to the use of a weighted (or nonweighted) checklist is that it does not reveal the degree with which a specific behavior occurs, requiring only a mere yes-no judgment.

	Delivery of Patient Care	Yes	No	N/A
60%	A. Evaluate and intervene when necessary in nursing care delivered by other non-RN personnel in his or her unit.			
40%	B. Provide skillful and safe care to patients as indicated in patient chart and nursing care plan.			

Figure 13.1. Example of weighted checklists.

In the *forced-choice* method the rater selects statements that best fit the performance characteristics of the individual employee. The rater does not know the value assigned to each characteristic, and consequently forced choice can reduce bias. However, this advantage can become a disadvantage if the rater is offended by the confidential weights assigned to the statements.

Graphic rating is the most commonly used method for performance appraisal. The scale requires the rater to choose a value or statement along a continuum that best fits the employee for each criterion being reviewed. The advantage of a graphic scale is that it shows the degree to which an employee performs a job or task (Douglas et al., 1985).

The *critical incidents* technique was developed in response to the faults of essay and behavioral trait methods. The rater records critical behaviors of employees that are related to both good and poor performance. The critical incidents method, however, does not indicate the frequency with which a particular behavior is performed, nor the degree to which it is performed (Douglas et al., 1985).

As a result of the shortcomings of the critical incidents method, the *behaviorally anchored rating scale* (BARS) was developed. The BARS uses characteristics judged to be critical to job performance and rates the degree to which each characteristic is attained by the employee. The employee's performance is determined by summing the values assigned to each of the critical indicators and/or characteristics (Douglas et al., 1985).

The main disadvantage of the BARS is that its development is time-consuming because separate scales are needed for each job. BARS, however, offers objectivity, which is lacking in such clearly subjective appraisal methods as comparative methods and essays. BARS has been used effectively in hospital work units of nurses and ancillary personnel; such successful usage is due to the dimensionality of the BARS scale, which permits identification of separate components of complex job behaviors. BARS demonstrates a necessary movement toward evaluations that are developmental rather than merely evaluative. Such developmental evaluations, being behaviorally based, then provide the basis for changes in behavior.

A system developed from the BARS model, the behavioral observations scale (BOS), attempts to eliminate some of the disadvantages of BARS. The development of the BOS also begins with the identification of critical incidents, which are then categorized according to behavioral dimensions. The behavioral dimensions usually contain five to eight items each, and these are used to rate employee performance. A frequency format is developed in which the highest number corresponds with "almost always," and the lowest number corresponds with "almost never." A comparison of BOS with BARS reveals that BOS does not require the appraiser to regularly record the occurrence of critical incidents. Instead, BOS asks the rater to evaluate the employee on a variety of behaviors that have been determined to be critical to good or poor performance (Douglas et al., 1985).

Management by objectives (MBO) is a result-based evaluative program in which goals are mutually determined by supervisors and subordinates, and employees are rated on the degree to which these goals are accomplished. MBO stresses the value and importance of employee involvement and encourages discussion of employee strengths and weaknesses. The supervisor acts as a counselor rather than as judge in aiding the employee in reaching his or her full potential (Levitz, 1981).

MBO has become popular for several reasons. First, it promotes better communication and interaction between the superior and subordinate. Additionally, the process of MBO development forces the organization and individual units to recognize and coordinate goals. Also, employees gain understanding of work objectives and come to know what is expected of them. Critical to MBO is the review process in which the supervisor discusses performance with the employee and makes recommendations for future performance (Levitz, 1981).

Table 13.1 and 13.2 may assist managers in reviewing various organizational performance appraisal techniques in terms of cost and applicability (Glueck, 1978).

Although an understanding of the most commonly used methods of performance appraisal is helpful, equally important is a review of significant theoretical and philosophical differences in evaluating methods of performance appraisal to date.

EVALUATION OF PERFORMANCE APPRAISAL

"All organizations, whether in the private, public, or nonprofit sectors, face the problem of engaging the energies of their members in the task of reaching their goals" (Brinkerhoff and Kanter, 1980). To solve this problem, organizations must devise ways to influence and handle the behavior of their members, correcting deviation and rewarding good performance. Performance appraisals constitute one of the major tools used in the organizational control process. At the same time that some form of performance appraisal is a tool in organizational control and decision making, it can be seen as a way of both rationalizing and clarifying the employment relationship, and protecting the individual from arbitrary discipline or the results of non-performance-based favoritism (Brinkerhoff and Kanter, 1980).

Therefore, it is important to assess objectively the process of performance appraisal, both theory and practice. This assessment includes a review of the purpose of

TABLE 13.1. Criteria for Choice of Performance Evaluation Techniques

Evaluative Base	Graphic Rating Scale	Forced Choice	MBO	Essay	Critical Incidents	Weighted Checklist	BARS	Ranking	Paired Comparison	Forced Distribution	Forced Test	Field Review
Developmental cost	Moderate	High	Moderate	Low	Moderate	Moderate	High	Low	Low	Low	High	Moderate
Usage costs	Low	Low	High	High supervisory costs	High	Low	Low	Low	Low	Low	High	High
Ease of use by evaluators	Easy	Moderately difficult	Moderate	Difficult	Difficult	Easy	Easy	Easy	Easy	Easy	Moderately difficult	Easy
Ease of understanding by those evaluated	Easy	Difficult	Moderate	Easy	Easy	Easy	Moderate	Easy	Easy	Easy	Easy	Easy
Useful in promotion decisions	Yes	Yes	Yes	Not easily	Yes	Moderate	Yes	Yes	Yes	Yes	Yes	Yes
Useful in compensation and reward decisions	Yes	Moderate	Yes	Not easily	Yes	Moderate	Yes	Not easily	Not easily	Yes	Yes	Yes
Useful in counseling and development of employees	Moderate	Moderate	Yes	Yes	Yes	Moderate	Yes	No	No	No	Moderate	Yes

SOURCE: J. M. Ivancevich and W. F. Glueck, *Foundations of Personnel/Human Resource Management*, Third Ed. Copyright © 1986 by Business Publications, Inc., Plano, TX.

TABLE 13.2. Recommendations on Evaluation Techniques for Model Organizations

Type of Organization	Graphic Rating Scale	Forced Choice	MBO	Essay	Critical Incident	Weighted Checklist	BARS	Ranking	Paired Comparison	Forced Distribution	Performance Test	Field Review	Assessment Centers
Large size, low complexity, high stability	X	X	X		X	X	X	X	X	X	X	X	X
Medium size, low complexity, high stability	X		X	X	X	X	X	X	X	X			X
Small size, low complexity, high stability	X		X	X	X			X	X	X			
Medium size, moderate complexity, moderate stability	X		X	X	X	X		X	X	X			X
Large size, high complexity, low complexity	X		X	X	X			X	X	X		X	X
Medium size, high complexity, low stability	X		X	X	X			X	X	X			
Small size, high complexity, low stability	X		X	X	X			X	X	X			

SOURCE: J. M. Ivancevich and W. F. Glueck, *Foundations of Personnel/Human Resource Management*, Third Ed. Copyright © 1986 by Business Publications, Inc., Plano, TX.

performance appraisal, a review of the task characteristics important for the PA process, and a summary of performance appraisal system problems.

Purpose of Performance Appraisal

Performance appraisal has served two purposes: evaluation (as the term "appraisal" implies) and development (Kearney, 1977). Brinkerhoff and Kanter (1980) point out that the evaluative function of PA refers to assessment of the extent to which progress toward goals (which we can define as desirable end states) has been furthered by the appraisee's effort. This purpose of PA is historically oriented; Past performance is reviewed in light of results or outcomes. PA for evaluation historically has served as the basis for decision making regarding promotions, transfers, and salary adjustments. It also can be used as a basis for allocating terminations, particularly in an organization that has decided to reduce its work force.

PA's developmental function is forward-looking, aimed at enhancing the future capacity of organization members to be more productive, effective, and/or satisfied (Brinkerhoff and Kanter, 1980). For developmental purposes PA facilitates appraisee improvement in job skills and motivation, as well as career planning and effective coaching between managers and subordinates.

Cummings and Schwab (1973) offer the following useful framework for analysis of the different approaches to performance evaluation: The system is best for the organization and makes its greatest positive contribution in the area of job satisfaction and productivity when the philosophy and operation are that of a developmental program. The time orientation concentrates on future performance rather than on management's assessment of past performance. A developmental system improves performance through personal growth, whereas a judgmental system attempts to improve employee performance through changing personnel and the reward system. A significant aspect of a developmental program is the means by which it achieves its objectives. Specifically, a developmental system includes both a goal-setting process and developmental programs (e.g., management by objectives). Judgmental programs attempt to measure past performance and use a variety of methods such as rating scales and ranking systems.

The methods of operation of these two extremely different systems have strong implications for the way in which the rater is viewed by the employee. In a developmental system, the rater is viewed as a counselor or coach, and the rater views the role as providing encouragement to the employee in the process of personal development. The evaluative system places the rater in the role of judge, one who must assess the value of the employee's performance. Consequently, the individual being rated in an evaluative system frequently is passive or defensive about previous performance. In a developmental system, however, the employee is actively involved in a learning process and in planning for future improvement in job performance.

An assumption of this chapter is that all managers are involved in performance appraisal, regardless of whether it is properly organized and whether it is judgmental or developmental. All managers should be involved actively in organization-wide appraisal systems that are strategically focused and developmental. The role of performance appraisal in health organizations is becoming more and more crucial to institutional effectiveness. Therefore, it seems reasonable that management should spend an appropriate amount of time in analyzing its performance appraisal needs and in developing a system that can be effective in meeting those needs.

Tasks Characteristics

Brinkerhoff and Kanter (1980) stress three characteristics of tasks that have strong impact on the PA process: complexity, clarity, and predictability.

Complexity

As pointed out by Dornbusch and Scott (1975):

> ... the more complex the task, the more complex is the evaluation process required. If the task entails many activities and there are numerous properties of interest in connection with the activities or outcome, then the process of arriving at a valid and reliable performance evaluation is likely to be complicated. (P. 145)

Therefore, the most technically accurate evaluations generally have been limited to jobs that have relatively simple content and unambiguous measures of performance.

Clarity

Clarity concerning tasks and goals is another characteristic that influences the PA process (Brinkerhoff and Kanter, 1980). In other words, knowledge of, and/or agreement on, what is to be done must be clear both to management and to the employee. Tasks with specific objectives on which organizational members can agree lend themselves to implementation and subsequent measurement that forms a good basis for relatively straightforward performance evaluation.

However, tasks characterized as ambiguous pose problems for performance appraisal. Diffused goals, moreover, provide little help in determining what to measure and/or what standards should be employed in the evaluation process. Conflicts and disagreements commonly occur when goals and tasks are vague and ambiguous.

Predictability

Predictability is another factor particularly relevant for evaluations based on outcomes (Brinkerhoff and Kanter, 1980). If the tasks are predictable, the relationships between quality of performance and amount and/or quality of outcome are relatively constant. Typing a letter is a good example of a predictable task in which examination of the outcome provides an accurate picture of performance.

On the other hand, unpredictable tasks do not allow accurate assessments of performance based on outcomes. Such tasks are common in areas where knowledge of cause-and-effect relations is not complete and ambiguous standards for assessing performance are being used (Thompson, 1967). Brinkerhoff and Kanter (1980) have stated that as uncertainty in jobs increases (e.g., difficulties of measuring results, time lag between action and results), so does the tendency to use social characteristics in making decisions about who should occupy those jobs (Kanter, 1977; Scott, 1978). Therefore, it is precisely the difficulty of doing objective appraisals of performance in areas high in uncertainty (such as risky ventures) and higher level management strategic decisions that results in the appearance of bias through reliance on subjective factors such as trust and loyalty.

Brinkerhoff and Kanter (1980) also point out that the PA process is not an independent one but is structurally linked to a variety of other features and processes of the organization. The structural characteristics that have a direct impact on performance appraisal systems include task interdependence, observability of task performance, the structuring of the authority system, power differentials, and the nature of communicated appraisals.

In summary, what develops from a given PAS depends largely on purpose, task features, and characteristics of the organizational structure. Basically, PA data may tend to be unreliable or misleading from the standpoint of the organizational and social-psychological issues. However, most of the appraisal literature, with its technical focus on scientifically developed rating instruments, fails to recognize the realities of practice that contribute to this unreliability. Brinkerhoff and Kanter (1980) conclude that data from formal performance measures tend to be most reliable when the following conditions exist.

1. The purpose of the appraisal is clear.
2. Tasks are simple.
3. Goals for the tasks are clear.
4. Outcomes are predictable.
5. Tasks are relatively independent.
6. Tasks performance is observable.
7. Criteria of performance are set by those later assessing performance.
8. Appraisers feel secure in their own jobs and have no personal stake in hurting the performers.

This conclusion suggests that single measurement systems based on formal checklists and ratings by the supervisor should be used only for the more routine tasks in organizations, which are likely to meet these criteria. On the other hand, as uncertainty grows—or complexity, interdependence, power concerns, and/or multiple appraisal purposes grow—so should the number of additional features and sources of data used for the appraisal system.

Sashkin (1981) developed a brief questionnaire (Table 13.3) to elicit a rough evaluation of an organization's appraisal system. The questionnaire is based on three basic objectives of performance appraisal.

1. Performance appraisal systems should generate information needed for short- and long-range administrative actions, such as salary decisions, promotions, and transfers (all short-range) or human resources planning and managerial succession (long-range).
2. Appraisal systems should let subordinates know where they stand and how well they are doing, as well as any changes in their behavior the superior wants.
3. Appraisal systems should provide a means for coaching and counseling subordinates, to train and develop them to their full potential.

TABLE 13.3. Organizational Performance Appraisal Questionnaire Evaluation (OPAQUE)

Instructions

Respond to the six statements that follow by indicating the extent to which you agree (or disagree) that the statements accurately describe performance appraisal in your organization. Some statements refer to your experiences in appraising your subordinates' performance; others refer to your experiences in being appraised yourself. Try to reflect as accurately as you can the current conditions in your organization based on your experiences.

SA = Strongly agree   A = Agree   ? = Neither agree nor disagree
D = Disagree   SD = Strongly disagree

1. I have found my boss's appraisals to be very helpful in guiding my own career development progress. SA A ? D SD
2. The appraisal system we have here is of no use to me in my efforts toward developing my subordinates to the fullest extent of their capabilities. SA A ? D SD
3. Our performance appraisal system generally leaves me even more uncertain about where I stand after my appraisal than beforehand. SA A ? D SD
4. The appraisal system we use is very useful in helping me to clearly communicate to my subordinates exactly where they stand. SA A ? D SD
5. When higher levels of management around here are making major decisions about management positions and promotions, they have access to and make use of performance appraisal records. SA A ? D SD
6. In making pay, promotion, transfer, and other administrative personnel decisions, I am not able to obtain past performance appraisal records that could help me to make good decisions. SA A ? D SD

SOURCE: Marshall Sashkin, *Assessing Performance Appraisal*. San Diego, CA: University Associates, Inc., 1981.

It is evident that no single system is appropriate or effective for every organization or category of employees. To further assist in the evaluation of an organization's appraisal system, Sashkin (1981) has developed the following 10 questions or rules of thumb.

1. Are managers rewarded for developing their subordinates?
2. Do managers receive skill training and assistance in using the system and, specifically, in being helpers or counselors?
3. Are job descriptions or specific job goal documents based on behavioral characteristics or job-relevant performance?
4. Are employees actively involved in the appraisal process?
5. Does mutual goal setting take place?
6. Do the appraisal sessions have a problem-solving focus?
7. Is the judge role clearly separated from helper/counselor role?
8. Does the paperwork and technical assistance required by the appraisal system place an unreasonable work load on managers?
9. Are peer comparisons a central feature of the appraisal process?

10. Is information that is needed for administrative action accessible and effectively used?

Managers should be equipped with some understanding of the purpose and complexity of valid performance appraisal. They also should be aware of the variety of problems that may be encountered in the development, implementation, and operation of a performance appraisal system.

Performance Appraisal System Problems

The development, implementation, and operation of a performance appraisal system is a major undertaking. Such a system may fail to bring the organization the desired results for a variety of reasons, ranging from problems in the development stage to the problem of reviewing the system.

Developmental Problems

Several problem areas may be encountered when an organization attempts to implement a PA program. One is designing the best possible system to meet the needs of the organization. The design of the "best" system should comprise an objective program that will evaluate employees on the basis of behavior that can be observed and easily measured. Use of the essay format or a comparative method probably will not result in an objective, accurate evaluation of performance.

Second, a PA system ideally should be equally effective in evaluating employees at all levels in the organization. However, use of the same approach to evaluate everyone may not yield the desired result. In a hospital, for example, use of a critical incidents method for housekeeping staff may be appropriate, but for managers and top-level executives this technique would not measure adequately such qualities as goal attainment and leadership ability.

A third factor crucial to the developmental stage is that of ownership of the PA system; those who will use the system—the employees—should own it. Studies show that a system seen as the personnel department's project is less likely to be accepted and to survive in the organization. Therefore, the employees must be involved in the developmental stage.

Finally, the goals and objectives to be measured must be tied into the organization's strategic plan. This linkage ensures coordination of different units, with each working for the long-term advancement of the organization. Unless the organization considers the strategic long-range plan, its work to obtain short-term performance improvements may at the same time be sabotaging future success in reaching long-range goals.

Implementation Problems

Several problems may arise in the implementation stage. For example, a lack of commitment from the organization will cause employees to lose faith in the system, believing that it will be short-lived. Management should show commitment by implementing the system throughout the entire organization, or from top management down, as evidence of top-level and managerial support.

A second possible problem area is training and education. Those who will be using the system should be thoroughly educated in its purpose and process. Such education includes training raters on evaluating employees. Training reduces evaluator resistance to the system while aiding in getting the most accurate information on performance. All raters should get practical, hands-on experience during the training sessions.

Even the most careful construction of an appraisal tool will not eliminate the need for training raters. Training is necessary to minimize rating errors because a biased, distorted, or inaccurate evaluation diminishes employees' motivation and allows management to make inaccurate personnel decisions, which in turn defeat the purposes of the system. To be successful at reducing rater errors, management should recognize the importance of training sessions in which the future evaluators can practice skills taught and receive immediate feedback.

Operational Problems

One of the most common problems in operation of a system is rater error. Identifying rater error can be very difficult if an employee is evaluated solely by one supervisor. Common rater errors include the following.

1. *Central tendency and leniency.* The rater evaluates all employees as average, excellent, or poor. A rater who has this tendency should be made aware that such information is not very valuable and should be asked to rate employees from best to worst for each element of the job. This exercise will aid the evaluator in revealing more accurately the performance of those being evaluated (Lowe, 1986).
2. *Halo effect.* Some raters judge all performance based on one area in which performance is good or poor (Lowe, 1986). That is, the evaluator gives similar ratings for all performance areas even though the performance varies from area to area. This is possibly the most common rater error.
3. *Contrast effects.* The rater evaluates the individual relative to other employees rather than on the requirements of the job. A rater who is tempted to change an individual's rating after rating others is probably making a contrast error. Most people assume that performance in a work unit should follow a normal curve; this is incorrect and can lead to a violation of EEOC regulations.
4. *Similar-to-me.* The tendency to judge more favorably people perceived as similar to the rater should be avoided (Glueck, 1978).

Crucial to the operation of a PA system is prompt feedback and guidance for the evaluated employee. Feedback allows the system to move from merely being evaluative to being also developmental (Glueck, 1978). For example, if feedback is provided 3 months after the evaluation, performance may have changed and the postevaluation may praise or correct behavior that no longer exists. Prompt feedback, when directly associated with the period under evaluation, can help to maintain good performance or to change poor performance. Prompt feedback in the form of a face-to-face meeting between the supervisor and subordinate is crucial to the operation of a performance appraisal system. The rater counsels the employee to help him or her to reach full potential.

Finally, the operation of a system must include periodic review.

A rigid performance appraisal system cannot survive successfully in a dynamic organization. The system will require periodic alterations; a yearly review is advisable.

The review should gather input from participants at all levels to accomplish a dual purpose: to assess the participants' level of satisfaction in the system and to determine changes that can be made to improve system effectiveness.

Critical to the building of a strategic human resources management system (see Chapter 1; Fig. 1.1) is the recognition that the functions of performance appraisal comprise a significant part of the overall strategic human resources management process. These ideas are discussed in the sections that follow, along with the application of management by objectives.

STRATEGIC ROLE OF PERFORMANCE APPRAISAL

Latham (1984) defines strategic planning as the process through which the basic need of an organization is identified, its objectives are set, and the allocations of resources to achieve these objectives are specified. Latham further notes that performance appraisals often are viewed as retrospective because they emphasize what occurred in the past. Yet the success or failure of strategic plans rests, in large part, with management's ability to identify the key actions that must be performed to formulate and use the steps that will lead to the attainment of the organization's long-range goals. Therefore performance appraisal must be the process through which the critical job behaviors of management are identified, the specific objectives of each individual manager are set, and the steps or resources needed to obtain them are agreed on.

A few organizations have formal appraisal systems to evaluate top-level managers on how well they perform against the organization's strategic plan. The prevailing attitude seems to be that good measurement indicators simply do not exist. Frequently, emphasis is placed on managerial style or charisma rather than substance, and/or on whether a given result was achieved, with little or no questioning as to *how* it was achieved. For example, a chief executive officer may appraise a senior vice president primarily on whether a set of objectives was pursued in a manner similar to the way in which the CEO would have pursued them. Creativity and divergence of thinking are consequently stifled. Quite common are appraisal measures that reward managers for attending to the "bottom line" of their respective functions, without taking note of how their actions, or those of their people, affected the operation of other departments. No formal assessment is made to determine whether they and their subordinates behaved in a unifying or integrated way with colleagues so that the organization's overall mission could be achieved. Organizational behaviors of these types emphasize why many strategic plans are not carried out successfully (Latham, 1984).

Fundamental to the understanding of the strategic role of performance appraisal is the recognition of the dual nature or purpose of the process.

Dual Nature of Strategic Performance Appraisal

The purpose of strategic performance appraisal is twofold (Latham, 1984). First, the performance appraisal instrument defines what is meant by implementation and adherence to a strategic plan related to the individual employee. Therefore, when the strategic plan changes, the evaluation instrument should be reviewed for necessary modification and revision. It is through the use of the evaluation instrument that the second objective of performance appraisal is attained: namely, to bring about and sustain effective and/or efficient job performance. This can be done either through self-management or through coaching and counseling other people.

Performance appraisal is the sine qua non of a strategic human resources management system. This is because a performance appraisal system should make explicit what effective and efficient behavior is required of an individual employee for the organization to implement its strategic plan. Just as a nursing services department is concerned with the quality of patient care, the maintenance department is concerned with the operation of equipment, and housekeeping is concerned with maintaining a sanitary environment, the human resources system should be concerned with identification of what the people in nursing services, in maintenance, and in housekeeping must do to be proficient in their respective functions. In a similar way, it should be determined what top management must do to implement the strategic plan once it has been formulated (Latham, 1984). The outcomes of these analyses translate into an appraisal instrument that is valid in that people are being evaluated on areas that are important to the attainment of their departmental and/or organizational objectives. To the extent that valid performance appraisals are done, valid decisions can be made regarding which employees should be rewarded. To the extent that valid performance appraisals are not made, it is impossible to make valid selection and reward decisions.

Valid performance appraisals are also critical to health care training programs because they identify people who lack the ability to perform effectively in their jobs. Use of a valid appraisal instrument makes it possible to identify not only those who need training but also the type of training needed.

The strategic purpose of PA cannot be achieved without the implementation of a PAS designed to assist in blending organizational goals with those of individual employees, to assist in the strategic planning process and in the accomplishment of organizational mission. When a properly designed PAS is implemented with consistency, it becomes the cornerstone of an effective human resources management system. As such, the PAS serves a number of strategic functions, some of which are as follows.

Strategic Functions of the Performance Appraisal System

Douglas et al. (1985) list five strategic functions of the PAS.

1. To provide a major source of human resources planning information.
2. To provide a control mechanism for management.
3. To activate and support the motivation system.
4. To provide a means of employee development.
5. To provide a basis for justifying personnel actions.

Human Resources Planning

If done properly, PAS should form the basis for important records including data on each employee's special abilities. This information can be particularly important in identifying managerial talent. The human resources planning aspect of the PAS is significant in terms of strategic plans as well. Each employee's permanent personnel records should include a complete evaluation of her or his background, ability, and potential. The records also should include a summary of work history, education, and training, as well as assessments of factors such as motivation, leadership skills, and potential for assuming greater organizational responsibilities.

Control and Motivation of Personnel

Performance appraisal systems, which contain the justification for distributing the rewards and punishments that organizations have to offer, should be the heart of any motivation system. In concept or theory, any organization should strive to distribute its available resources so that excellent or good performers are rewarded more than average or poor performers. Accomplishing this objective depends largely on the ability of the institution to develop reasonably objective appraisal measures and methods, and then to relate the rewards directly to these objectives and measurements. If success is achieved in these areas the problems or perceptions of inequity will be reduced, though probably not eliminated.

An extremely important aspect of motivation is goal setting. Much evidence points out that people with explicit performance goals do better than people with vague performance objectives, such as "Do the very best you can." Many studies also indicate that difficult goals produce higher performance than easy goals, although some work suggests that the goals must be limited to what is realistically possible. Unrealistic goals may simply cause people to become discouraged and unproductive.

The idea of control also is related to motivation. Key components of control are goals, measurement, and feedback. Douglas et al. (1985) indicate that good appraisal systems are supported by measurement, progress is noted according to those measures, and feedback occurs to an area in which changes can be made. Control is necessary to enable both management and employees to make associations between behavior and output. The same measures also can be useful in determining rewards.

Personnel Development

One of the primary reasons for PA in any organization should be to assist employees in developing their skills to their maximum potential. The control aspect of PA can help employees make adjustments through better performance. In addition, objective appraisal can help in identifying weaknesses as well as strengths of each employee. Such weaknesses may be dealt with through planning recommendations communicated to the employee at the time of the appraisal interview.

Justification of Personnel Actions

The performance appraisal process is extremely important in providing information. It requires justification for promotion, transfer, and demotion or termination. This is particularly true in view of civil rights legislation and EEOC regulations. However, any good performance appraisal system must have a carefully thought-out plan of objective measurement and documentation process. Although judgment is extremely important in many aspects of management, a performance appraisal system that relies almost exclusively on supervisors' judgment will be very difficult to defend. Documentation of objective measurement is essential. If a record of well-designed and properly conducted appraisals is available and indicates consistency, important personnel actions such as promotions or terminations can be justified on the basis of objective, documented information.

To serve the purpose and functions of strategic performance appraisal, the organization must establish and effect a PAS that has application and acceptability throughout the organization. One system with this potential is management by objectives, presented as a strategic appraisal system for health care organizations.

MANAGEMENT BY OBJECTIVES (MBO): A STRATEGIC APPRAISAL SYSTEM

Although no performance appraisal system that exists will meet all the needs of a given organization, the MBO concept is more broadly applicable and fundamentally sound than most other systems. Of particular relevance to this chapter is the possible linkage of the MBO concept to management's strategic planning functions.

The term "management by objectives" was used first by Peter F. Drucker more than 30 years ago. Drucker (1954) pointed out that in this system, each manager should have clear objectives that are identified with and support those of high levels of management. Individuals thus can develop an understanding of their own objectives, as well as those of their managers, and the organization. Many authors have supported the MBO concept developed by Drucker. Douglas McGregor (1960) brought the concept into the performance appraisal arena by advocating that MBO be used to encourage the discussion of employee strengths and potential, making the superior more a counselor than a judge. Whereas Drucker first viewed MBO as a method of integrating the activities of an organization, McGregor developed the idea of applying MBO as a performance appraisal technique. In recent years, the MBO approach has been recommended as an appraisal technique that should be linked to management's strategic planning process.

Definitions

Although a variety of definitions of MBO programs exist, some common elements are present (Cummings and Schwab, 1973):

1. Goal setting
2. Involvement of managers in participation in the formulation of personal goals and methods to accomplish these goals
3. Periodic reviews of progress toward the accomplishment of these goals
4. Evaluation of performance
5. Self-appraisal
6. Feedback and evaluation
7. Suggestions for development and training

As demonstrated by the common elements listed above, the real process may be viewed as a cycle of events that includes planning, setting objectives and goals, negotiation, performance, review of performance, and evaluation and feedback.

Figure 13.2 shows a model objective setting process and Figure 13.3 presents an overview of the MBO cycles.

McConkie (1979), who further identified common elements from various authors' descriptions of MBO and their explanations of how performance appraisal should be conducted under MBO, found that almost complete agreement existed on three items specific to goal setting in MBO programs. These items indicated that goals and objectives should be specific, should be defined in terms of measurable results, and should reflect both individual and organizational perspectives.

The ideal objective list tells managers and employees:

1. What should be done
2. When it should be completed
3. When it actually is completed
4. What priority each objective has
5. The percentage of time of the total each objective uses
6. What authority level the employee has to implement plans

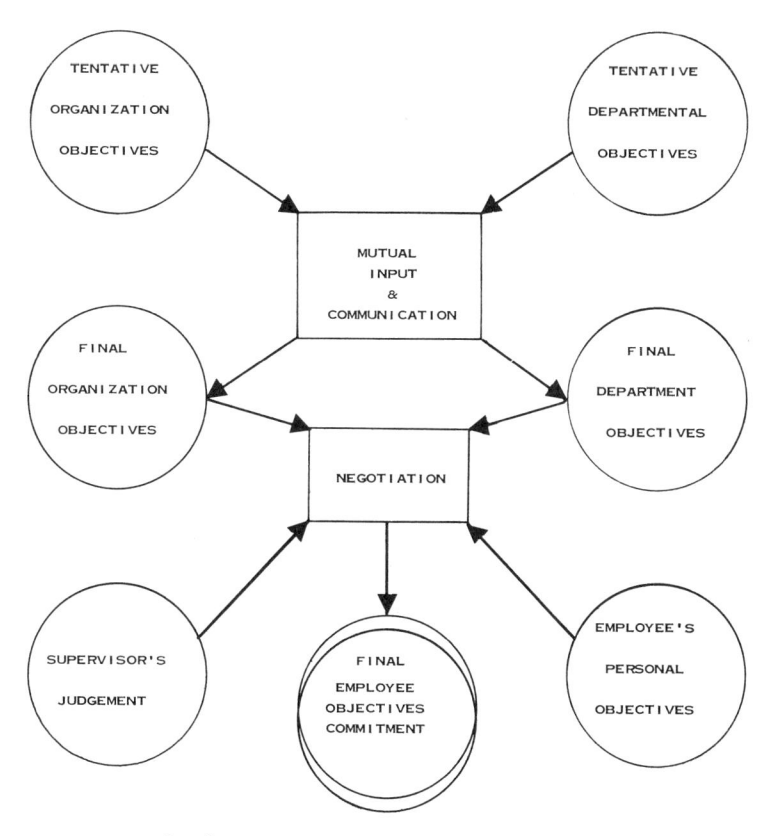

Figure 13.2. The objective setting process.

```
                    ┌──────────────┐
                    │  DETERMINE   │
                    │ ORGANIZATION │
                    │  OBJECTIVES  │
                    └──────────────┘
```

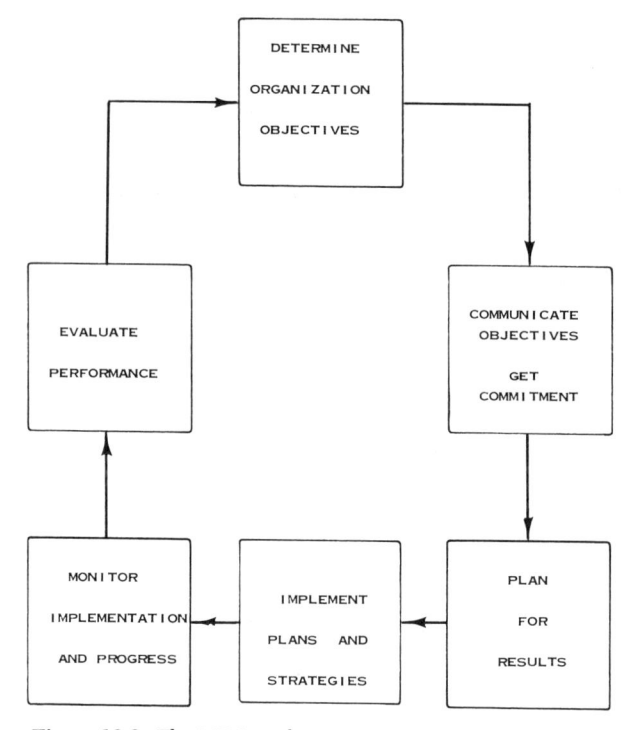

Figure 13.3. The MBO cycle.

7. What is being done to develop each employee
8. How well the employee performs
9. When objectives are to be reviewed
10. The total job expected

It appears that with the linkage of MBO to performance appraisal to establish an effective system that provides a setting for managers and subordinates to set objectives against which performance is measured, management receives better data than are obtained from other traditional methods. Using these concepts, McConkie (1979) concluded that MBO could be defined as follows:

> a managerial process whereby organizational purposes are diagnosed and met by joining superiors and subordinates in the pursuit of mutually agreed upon goals and objectives, which are specific, measurable, time bounded, and joined to an action plan; progress and goal attainment are measured and monitored in appraisal sessions which center on mutually determined objective standards of performance. (P. 37)

It is important to note that the key areas of management involvement that contribute to the success of MBO programs relate to administration's understanding of and commitment to the various features of the system.

Management by Objectives as an Appraisal System

Migliorie (1979) describes a frame of reference for the implementation of the MBO program that is based on identifying purposes, objectives, and desired results, and in evaluating performance in the following nine-step process:

1. Defining the organization's purpose and reason for being
2. Monitoring the environment in which it operates
3. Assessing the organization's strengths and weaknesses realistically
4. Making assumptions about unpredictable future events
5. Prescribing written, specific, and measurable objectives in principal result areas that contribute to the organization's purpose
6. Developing strategies on how to use available resources to meet objectives
7. Making long-range and short-range plans to meet objectives
8. Appraising performance constantly to determine whether it is meeting the desired pace and remaining consistent with the defined purposes
9. Reevaluating purpose, environment, strengths, weaknesses, and assumptions before setting objectives for the next performance year

This system clearly reflects the intent to match the program to organizational goals, and this is particularly important for strategic planning. However, it is sometimes difficult to bring together organizational units in a way that takes advantage of the existence of such interdependencies.

Health care organizations present numerous examples of the interdependencies. Most, for example, bring together persons from different disciplines to perform specific functions that contribute to the overall quality and efficiency of patient treatment or service. However, a successful treatment program, in terms of both health outcome and cost, generally requires a high degree of coordination among employees in different departments. These personnel represent very different backgrounds, training, and understandings of what activities or tasks need to be performed and in what order. Therefore, the treatment program must be linked to the total management perspective so that individual objectives are developed within the frame of reference of the contributions made by other departments and other individuals.

Management by Objectives and Health Organizations

Gerstenfeld (1977) indicates that many MBO programs do not succeed because of failure to recognize that "it is not sufficient to delineate an objective in one portion of the health system while not fully integrating other parts of the system that are concerned with that objective." It is important that systems of performance appraisal, including MBO, reflect an interest in interdepartmental and, where appropriate, interorganizational management and that health managers understand the many benefits that can accrue from their approach to performance appraisal.

Performance appraisal is not yet widely accepted by individuals in health care settings. Health care employees often feel that evaluation should be performed by peers and not be linked necessarily to the operation or management of an organization.

Although such feelings may stem from years of professional training, management of health organizations must have appropriate methods to appraise employees and evaluate the effectiveness of the organization.

In terms of strategic human resources planning and overall management needs, it seems suitable for health care organizations to strive for systems that will contribute to individual growth and development. Some of the traditional methods of evaluation (which are comprised of subjective or trait evaluations, requiring choices on numerical scales or markings on checklists) are not appropriate appraisal tools for many health care employees. Employees simply need an evaluation process in which they participate, one that provides an opportunity to enhance professional skills and will assist workers in making greater contributions to the organization.

An effective evaluation system must be based on observation of skills and performance, not on a laundry list of personal attributes. Job analysis procedures may lead to the development of objective criteria, which then may be used in performance evaluation. These criteria must address both the quality and the quantity of work performed. They also should measure factors over which an employee may have influence, and they should be applied over a realistic period of time (Hand and Hollingsworth, 1975).

MBO can be applied effectively in health organizations where there is an integrated management approach and the motivation to have the system work (Gerstenfeld, 1977; Hand and Hollingsworth, 1975). The primary motivation should be an interest in the development of employees, with rewards to follow for effective and efficient performance. Research also indicates that an MBO program, when combined with other programs such as job enrichment, may enhance both productivity and worker satisfaction (Deegan, 1977; Hand and Hollingsworth, 1975).

Guidelines for Implementing Management by Objectives

When an organization is considering implementation of performance evaluation systems such as MBO, management and employees alike should be aware of the total process from its inception. Through this communication, management and employees can develop an understanding of the commitment needed and the length of time required for a successful program. Listed below are the guidelines Levitz (1981) offers for implementation of an MBO program.

1. Management and employees should be committed to the MBO program and supportive of it. A survey may be necessary to determine how open the organization's climate is.
2. Everyone involved in the process should develop an understanding of the purpose and objectives of the program.
3. Management and other personnel should meet to develop common goals.
4. Departmental objectives should be developed that are consistent with those of the health organization.
5. Job descriptions should be written in result-oriented form, with statements on the measurement of satisfactory performance.
6. Subordinate-superior goal setting should occur at regular meetings.

7. Clear, valid, and measurable objectives for individuals need to be set and agreed upon.
8. Superiors should be trained in evaluation methods, developmental methods, and performance interviewing.
9. Developmental feedback sessions should be scheduled based on individual needs.
10. Employees should view the MBO program as being linked to the reward system.
11. Continual monitoring of the system should occur through a linkage with other management functions.

From the perspective of strategic human resources management, several benefits appear to accrue from the operation of an MBO program, and these make it worthwhile for the management team to invest the resources required for implementation and maintenance of such a system. Levitz (1981) lists several advantages that may result from a successful program:

1. Improved direction and planning of activities toward the accomplishment of organizational goals
2. The linking of institutional management functions to control systems through the development of appropriate work standards
3. A reduction in role conflict
4. A better understanding of how performance is linked to rewards
5. Increased employee satisfaction through the use of objective criteria
6. The career development of personnel

Through an understanding of the total MBO program and its necessary linkage to performance appraisal, compensation, and other management functions, health care managers can exert better and broader control over performance. The end result will be a more effective and efficient health services organization.

Applications of Management by Objectives

Kessler (1981) indicates that the MBO concept has been used in numerous settings by those who found old ways ineffective, those who tended to be experiment minded, and those who were inclined to research the state of the art or current practices in their organization and felt obliged to pursue the conclusions of that research. Two examples of MBO applications are presented: the GE work planning and review process and the case of a community hospital, St. Charles Hospital in Toledo, Ohio.

The General Electric Work Planning and Review Process

An illustration for application of the MBO program comes from the General Electric Company (GE), noted for its long-standing contributions in human resource management. GE addressed the concept of MBO by focusing on a "work planning and review process" based on certain internal research projects indicating significant opportunities for improved performance and working relationships (Aircraft Engine Division, 1986).

The simplicity of the work planning and review process offers a wide range of applications.

Development of the Work Planning and Review Process. To get work done, there must be some form of continuous work planning. Supervisors give tasks and work to their subordinates and have a system to check the work to make sure it is being done properly. However, few supervisors make optimum use of time spent with subordinates because of difficulties inherent in the supervisor/subordinate relationship (e.g., operational problems, pressure of the business, and problems in communication). Concurrently, most subordinates want to do the job and most often feel that they do not understand fully what the boss expects of them. They desire clearer knowledge of how the boss thinks they are doing. Most subordinates want the feelings of accomplishment, interesting work, and additional responsibilities. They want the opportunity to offer suggestions to their superiors and to have those ideas weighed carefully, even though they may not be accepted. In other words, supervisors want to do a better job of managing, and subordinates want to improve their own performance. A question GE faced was, How do we set up conditions whereby people can be helped to do a better job?

GE had hoped that its performance appraisal process would help, but research indicated that it did not. The biggest problem was found to be the conflicting roles of the supervisor as both counselor and judge. Because performance appraisal was tied so closely to salary decisions, the supervisor was forced to play the role of judge with regard to salary action, while at the same time the role of a counselor had to be assumed when the supervisor was advising employees on improving their work performance. These two roles simply did not seem compatible, and in the new approach they were separated. The next question was: How is it possible to provide a climate in which managers can act as helpers for their subordinates in improving work performance? The answer found by GE was "work planning and review."

What Is Work Planning and Review? Kessler (1981) points out that the work planning and review process consists of periodic informal meetings between subordinates and the supervisor. The meetings are oriented toward the daily work and result in mutual planning of the job and solving of problems that arise in the course of getting it done, and a progress review. The process, which does not involve formal ratings, was designed to take advantage of principles that relate to requisite conditions for subordinate motivation and performance enhancement. The three motivational principles are as follows.

1. An employee needs to know what is expected on the job.
2. An employee needs to know what progress he or she is making.
3. An employee needs to be able to obtain assistance when and as needed.

Figure 13.4 shows how the principles apply to the job.

A Community Hospital Example

Large profit-making corporations can use [a work planning and review process], but can it work in small, nonprofit organizations such as a community hospital? Application in such a setting is illustrated by the goal-setting process of St. Charles Hospital in Toledo, Ohio, a

Performance Appraisal 343

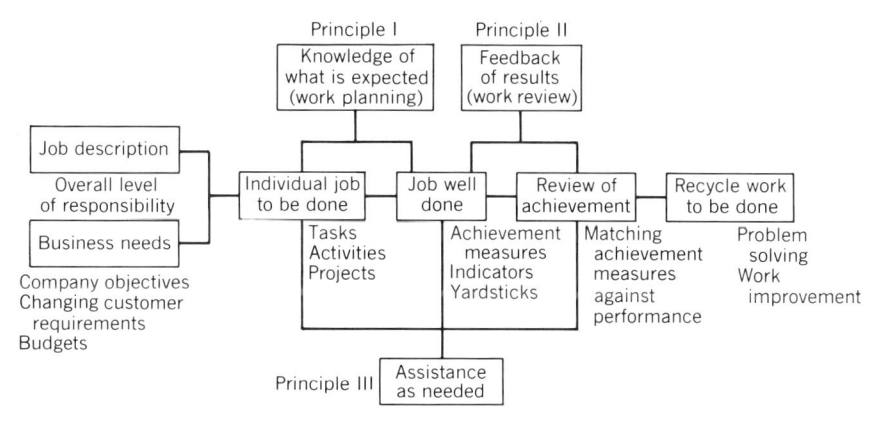

Figure 13.4. Work planning and review diagrammed. From Cincinnati: General Electric Company, Aircraft Engine Division: *Increasing Management Effectiveness Through Work Planning.* New York, General Electric Co., 1986, p. 4.

medium-sized Catholic community hospital involved in an organizational development (OD) program over several years. The goal-setting process is viewed as an integral part of the OD activities and as being congruent with their basic principles and values.

Over the last several years various refinements in the process evolved. Changes were designed to (1) encourage more ownership and commitment to the goals, (2) simplify the paperwork, (3) provide better linkage of individual objectives to key goal areas developed by top management, and (4) focus more attention on results to be achieved rather than having vague work plans or outcomes. More complex goals involving an integrated effort with other functions were developed as the skill and confidence level of managers improved.

The current process at St. Charles Hospital begins with an annual planning meeting in which top hospital management identifies broad major goal categories of prime interest and high priority for the coming year. Subsequently managers complete a more detailed identification of related goals in each broad category, with accountability assigned to a specific top-level manager. The management group reviews budgeted resources to confirm allocations or adjustments. This process recently resulted in identification of six broad goal categories and 46 specific related goals for the hospital.

The six key goal areas are patient services, organizational development, education, physical plant, external affairs, and operational issues. Examples of specific goals in the patient services area are the "development of a program for nursing regarding role and duty to monitor quality of care," and "research feasibility of nurse practitioners and physician assistants to define role, assess feasibility, and determine training needed." Under organizational development, one goal specifies "identification of two *unproductive* norms evident in employee behavior as focus for management skill training." The operational issues include a wide variety of matters such as a goal to "research and evaluate a new hospitalwide telephone system" and to "develop an orientation program for new physicians added to the medical staff."

Department heads in collaboration with their respective division heads select goals applicable to their functional area. Goals requiring interface with other functional goals are planned jointly, with a series of subgoals and work plans developed accordingly. In addition, intradepartmental goals and plans are developed based on needs and priorities as perceived at that management level. For example, a goal to revise certain internal scheduling routines to improve a department's efficiency could be of major importance to the department but not impact others directly. In effect, department heads are responsible for striking a balance between stated hospital goals where they have a participatory role and their own departmental objectives within the framework of allocated resources.

Division heads (top management team) review goals and work plans to assure proper application of criteria for well-stated goals, logical and complete work plans to achieve results (including time frames and accountability), and appropriate measurements of progress, completion, and quality of work.

Communication with all members of management on the final plans is handled through group meetings to share the composite result. Other sharing techniques can be developed so long as adequate information is conveyed to those impacted (where coordinated efforts are required), and a general overview of the coming year's work stimulates support and commitment from the total organization. Further communication usually takes place throughout all levels of the organization consistent with a participatory style of management. Periodic progress reviews and/or some goal modification may be necessary to maintain momentum awareness, and commitment.

(SOURCE: T. W. Kessler, "Management by objectives." In N. Metzger, (Ed.), *Handbook of Health Care Human Resources Management*. Reprinted with the permission of Aspen Publishers, Inc., © 1981.)

Developing and consistently implementing a strategic performance appraisal system to meet the needs of a particular organization can be a very rewarding process. For the PAS to be effective, however, it must be linked properly to a strategically designed reward system. A brief discussion of this relationship between extrinsic rewards and job performance follows.

EXTRINSIC REWARDS AND JOB PERFORMANCE

The findings of management studies vary substantially in rating such factors as pay and other extrinsic rewards in their relationship to job performance. However, many advantages are obvious when rewards are tied to performance, and this section advances strong support for this position. A variety of other rewards such as recognition and praise are equally important and should be used to complement the basic extrinsic rewards.

Lawler (1973) emphasized that the relationship between extrinsic reward level and performance has a crucial influence on organizational effectiveness. When extrinsic rewards are related to performance, the results are higher motivation and a tendency for turnover to be concentrated among the poor performers. Despite the obvious advantages of having rewards tied to performance, many organizations do not adopt this practice. Although there are some situations in which tying rewards to performance is dysfunctional, some organizations do not relate rewards to performance even when doing so would be highly functional.

Lawler's (1973) discussion of extrinsic rewards stresses the importance of who receives the extrinsic rewards given out by an organization. Each organization has a limited quantity of rewards to give. How they are distributed determines who will continue to work for an organization, how hard people will work, and the attitudes of employees toward the organization. Extrinsic rewards represent an investment in people, and such enhancements should be a crucial issue in any organization. Indeed, one effective way to understand an organization is to look at the actual distribution of its extrinsic rewards and how the arrangement is perceived by the employees. Determination of the relationship between extrinsic reward satisfaction and performance in an organization also can provide important information about the eventual impact of an organization's reward distribution system. A strong positive relationship between satisfaction and performance usually indicates a reward system that is functioning well and is rewarding for good performance.

On the other hand, a negative relationship between satisfaction and performance indicates a poorly functioning reward system and should be taken as a warning. Specifically, such relationships mean that motivation is likely to be low because rewards are not clearly tied to performance. They also mean that turnover in the organization is likely to be centered among the high performers, a costly and extremely negative result for an organization attempting to improve its effectiveness or market share. Therefore, a performance appraisal system not linked to extrinsic rewards will not be as successful in motivating employees as one that is appropriately linked to these rewards. Generally, when performance measures are used, productivity will increase, and when pay-for-performance incentives are added, productivity can be expected to increase more significantly. Chapter 14 on compensation management includes more detailed coverage of these relationships.

In summary, a positive relationship between extrinsic rewards and a good PAS should yield big motivational dividends within the organization, laying a solid foundation for other complementary awards available within the organization.

SUMMARY

This chapter has advanced the case of performance appraisal as one of the most important processes in any organization. In some manner, all organizations do it, but the results may be either positive or negative.

Performance appraisal is not new, nor is there a single system appropriate for all organizations. This is particularly true in health organizations, which employ a multiplicity of highly trained individuals with widely diverse backgrounds, skills, and developmental needs.

A review of methods and of theories concerning the evaluation of performance appraisal provided background for discussion focused on the strategic role of performance appraisal within the human resources management system. The MBO method was selected as the PAS with the most comprehensive application for health care organizations, and guidelines for the implementation of such a system were given. Two case examples served to provide insight into the actual application process for two different organizations.

The importance of tying rewards to job performance, although not the primary focus of this chapter, was discussed briefly. If the PAS is to be a positive motivational force, it must be administered consistently by all levels of management and linked directly to the organization's reward system.

Any organization should seek to develop the PAS that best meets its needs. Special emphasis should be given to the human side of appraisal since, in the final analysis, the best system may be only as effective as it is *perceived* to be by the employees.

REFERENCES

Aircraft Engine Division. *Increasing Management Effectiveness Through Work Planning.* New York: General Electric Company, 1986, pp. 4ff.

Brinkerhoff, Derick W., and Rosabeth Moss Kanter. "Appraising the performance of performance appraisal." *Sloan Management Review,* pp. 3–16 (Spring 1980).

Cummings, L. L., and D. P. Schwab. *Performance in Organizations: Determinants and Appraisal.* Glenview, IL: Scott, Foresman, 1973.

Deets, Norman R., and D. Timothy Tyler. "How Xerox improved its performance appraisals." *Personnel Journal,* pp. 50–52 (April 1986).

Dornbusch, S. M., and R. W. Scott. *Evaluation and the Exercise of Authority.* San Francisco: Jossey-Bass, (1975), p. 145.

Douglas, John, Stuart Klein, and David Hunt. *The Strategic Managing of Human Resources.* New York: Wiley, 1985, pp. 391–405.

Drucker, Peter F. *The Practice of Management.* New York: Harper & Row, 1954.

Gerstenfeld, Arthur. "MBO revisited: Focus on health systems." *Health Care Management Review,* 2(4): 51–57 (Fall 1977).

Glueck, William F. "Performance evaluation and development of personnel" and "Performance evaluation and promotion." *Personnel: A Diagnostic Approach.* Plano, TX: Business Publications, 1978, pp. 283–334.

Haar, Linda Pohlman, and Judith Rohan Hicks. "Performance appraisal: Derivation of effective assessment tools." *Journal of Nursing Administration,* pp. 20–29, 37 (September 1976).

Hand, Herbert H., and H. Thomas Hollingsworth. "Tailoring MBO to hospitals." *Business Horizons,* pp. 451–452 (February 1975).

Holley, William H., and Hubert S. Feild. "Peformance appraisal and the law." *Labor Law Journal,* 26(7):423–430 (July 1975).

Kanter, R. M. *Men and Women of the Corporation,* New York: Basic Books, 1977.

Kearney, W. J. "Performance appraisal: Which way to go?" *MSU Business Topics,* pp. 58–64 (Winter 1977).

Kessler, Theodore W. "Management by objectives." In Norman Metzger, Ed., *Handbook of Health Care Human Resources Management.* Rockville, MD: Aspen Systems, (1981), pp. 181–192.

Latham, Gary P. "The appraisal system as a strategic control." In Charles Fombrun, Noel M. Tichy, and Mary Anne Devanna, Eds., *The Strategic Role of the Human Resource Management.* New York: Wiley, 1984.

Lawler, E. E. III. *Motivation in Work Organizations.* Pacific Grove, CA: Brooks/Cole, 1973, pp. 113–147.

Levitz, Gary S. "Performance appraisal in health organizations." In Norman Metzger, Ed., *Handbook of Health Care Human Resources Management.* Rockville, MD: Aspen Systems, 1981, pp. 225–238.

Lowe, Terry R. "Eight ways to run a performance review." *Personnel Journal,* pp. 60–62 (January 1986).

McConkie, Mark L. "A clarification of the goal setting and appraisal processes in MBO." *Academy of Management Review,* 4(1):29–37 (1979).

McCormick, R. R. "Can we use compensation data to measure job performance behavior?" *Personnel Journal*, pp. 918–922 (December 1972).

McGregor, Douglas. *The Human Side of Enterprise.* New York: McGraw-Hill, (1960).

Migliore, R. Henry. "The use of long-range planning/MBO for hospital administrators." *Health Care Management Review*, 4(3):23–28 (Summer 1979).

Sashkin, Marshall. "Appraising appraisal: Ten lessons from research for practice." *Assessing Performance Appraisal.* San Diego, CA: University Associates, 1981, pp. 120–128.

Scott, W. Richard. "Organizational effectiveness: Studying the quality of surgical care in hospitals." In M. Meyer, (Ed.), *Environment and Organization.* San Francisco: Jossey-Bass, (1978).

Thompson, J. D. *Organizations in Action.* New York: McGraw-Hill, 1967.

CHAPTER 14

Compensation Management

Kerma N. Jones
Charles L. Joiner

Compensation management is the structure and process of managing pay. Although effective compensation management is necessary in all businesses, it is of particular importance for health organizations. A changing (and aging) American society, shifting reimbursement and insurance systems, increasing consumerism, and rising costs have combined to exert mounting pressures on health care organizations. More than ever before, administrators of health organizations face a mandate for cost containment. Moreover, compensation often represents a major portion of the total operating budget for a health organization. In such a setting, effective compensation management becomes essential if the organization is to survive.

Compensation management incorporates two primary objectives: distributive justice (equity) and incentives for employees to improve performance and productivity. Equity is feasible only if the management system includes procedures for such considerations as job analysis, job evaluation, and job pricing. Incentive systems involve rewards dependent on performance, but these systems must also be clearly defined components of the total compensation program. A third objective should be consideration of benefits.

A comprehensive program for compensation management must address each of the foregoing areas. Additionally, success for such a program requires attention to internal (employee) pressures and concerns, as well as to external pressures from the general public, governmental agencies, and, at times, stockholders. If the portion of the total operating budget devoted to compensation increases, the validity of the compensation management plan grows in importance.

STRATEGIC PLANNING AND COMPENSATION

The various components of compensation management (e.g., wages, salaries, benefits, incentives, and bonuses) are familiar to human resources and compensation management specialists, as well as to most health care administrators. However, the philosophy that compensation management strategy should be an integral part of the organization's strategic planning process is not so familiar.

Traditionally, personnel specialists have handled the compensation and benefits functions but usually were not involved in the strategic decision-making process. Over the next few years, human resources personnel will need to be more directly involved in both developing and implementing the strategic plans of the organization. Performance appraisal and compensation management are two of the specific functions that need careful evaluation in terms of their mutual relationships and integration into the strategic management formulas of the organization.

Chapter 13 presents the case for the strategic nature of performance appraisal, and this chapter summarizes the significant funtions of compensation management for health care managers. Specifically, compensation and benefit systems must be strategically related to the overall mission of the organization, as well as to the efficiency and quality objectives. It is also important to note the variety of environmental forces that influence the compensation of workers, including the following (Douglas et al., 1985):

1. Tax laws
2. Changes in mandatory retirement laws
3. Social Security retirement plans versus private pension plans
4. Employee attitudes toward work time flexibility
5. Individualized benefit choices
6. The entry of more women into the work force
7. The presence of more dual-earner, dual-career couples who want child care benefits
8. Changing values and norms regarding career mobility

These environmental factors, along with competitive market forces controlling market share and profits, are all considerations that necessitate the continuous designing, implementing, and evaluating of compensation management systems.

It is vitally important that health care executives and human resources managers understand the purpose of compensation management and its role in strategic management of human resources. In addition to such traditional issues as job evaluation, salary surveys, and compensation systems, this chapter addresses the key issues of executive incentive programs, physician compensation, exceptional performance awards, and benefits.

A Strategic Compensation Model

Because so many compensation issues have an effect on productivity goals and the quality of work life, strategy-oriented compensation specialists will likely become an integral part of future strategic management teams. It is helpful to define a strategic model approach to the compensation process, given the relationships of compensation to internal and external environments, as well as to other management systems. Douglas et al. (1985) developed a three-stage model for this process, as shown in Figure 14.1. Various components of this model are developed in this and related chapters.

Compensation of Health Care Personnel

Certain principles of compensation management apply to all organizations; however, unique aspects of the health care industry require special consideration in the development and implementation of a strategic compensation management plan. For example, consumers expect availability of health care—regardless of the extent of the problem, the time of day or night, or geographic location. In very few segments of the service industry does one find such a high degree of consumer expectation, and this places distinct requirements on health care organizations and their personnel. Moreover, the types of personnel involved in health care delivery vary from those involved in housekeeping and maintenance to physicians and top-level managers. When one also considers the federal, state, and local regulations within which such organizations operate, it is easy to see that the complex environment of health care organizations requires a detailed yet flexible plan for compensation management. Additionally, because of these very factors, the necessity for inclusion of human resources management specialists in the development of the total strategic management plan becomes obvious.

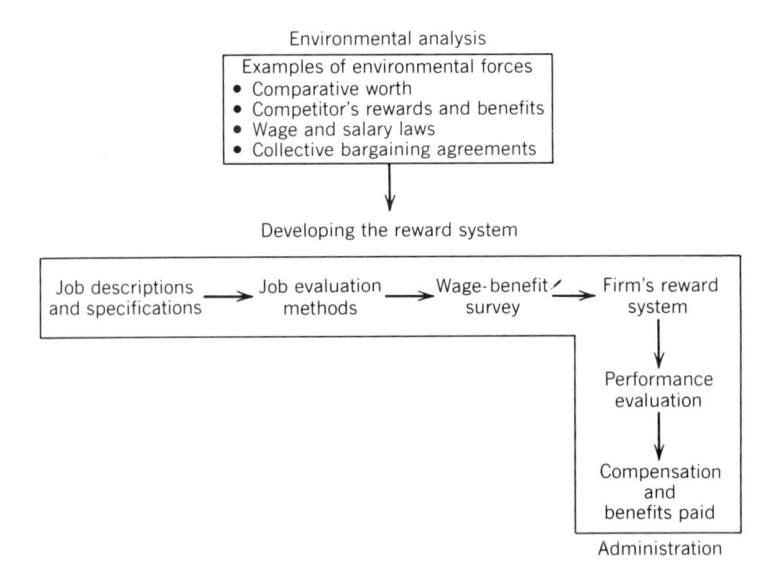

Figure 14.1. A strategic model for the study of compensation systems. From J. Douglas, S. Klein, and D. Hunt, *The Strategic Managing of Human Resources.* New York: John Wiley & Sons, Inc., 1985, p. 422. Copyright © 1985, John Douglas.

Base Components of the Compensation Management System

In developing a comprehensive compensation plan, various factors must be considered. For each position there should be a systematic process for developing a written job design, analysis, and description. Such specifics are essential in evaluating the job and in establishing equitable job pricing. Moreover, the existence of specific job design and analysis is essential to objective and equitable employee performance evaluations and determination of which employees merit consideration for bonuses or special rewards. Job designs and analyses for health care organizations must also incorporate such concerns as comparable worth and physician compensation, topics discussed later in this chapter.

Job Design

Job design encompasses the manner in which a given job is defined and how it will be carried out. This involves such decisions as whether the job will be handled by an individual or by a team of employees, as well as the determination of where the job fits into the overall organization. If properly implemented, job design requires a conscious effort for the organizing of tasks, duties, and responsibilities into a unit of work to achieve a certain objective. From the detailed information incorporated into the job design, a specific job description can be produced, and this will form the basis for recruiting and hiring an employee to fill that position, as well as the foundation for future employee evaluations and compensation schedules. In terms of strategic planning, each separate job design must be formulated to meet the overall objectives of the organization. Where appropriate, each job design must also acknowledge any unique skills possessed by members of a given profession and must incorporate appropriate professional guidelines or limitations for tasks to be performed.

In stressing the need for compensation management to be incorporated into the total strategic management plan, it should be noted that job design is one of the single most important considerations. If each job within a given organization is well defined and well designed, the necessary basis exists for job analysis, job description, and job evaluation, as well as for equitable compensation, performance evaluation, and employee satisfaction. As indicated in Figure 14.1, development of job descriptions and specifications is the first step in developing the reward system. If specific requirements for each job are not clearly defined initially, there can be virtually no basis for assessing the value of a given position or the expertise of the employee who holds it.

Considerations to be included here extend far beyond the organization itself. Obviously, job designs in popular use in comparable businesses should be reviewed, and general satisfaction of both organizations and employees with these definitions should be determined. For a growing number of health care organizations, job design must involve union guidelines—whether contracted or informally outlined. In addition, each job design must allow for the concept of employee status. Employees with high skill levels can be expected to object if assigned the same job titles as employees with little or no skill.

It would not overstate the case to say that the long-term success of a given organization requires that each of these factors be considered. An excellent example of the importance of sometimes subtle distinctions can be seen in the design of nursing positions. Within the health care industry, the most prominent group of employees is nursing personnel, and specializations within nursing have changed dramatically during the past few decades, resulting in an increased need for precise job design and description. As suggested above, the graduate of a baccalaureate nursing program is likely to object to being given the same title and duties as a graduate of a certificate or associate degree program. A nurse with specialty training in emergency and trauma care has acquired skills that should be differentiated from the skills required of the nurse in a long-term care facility or an ambulatory diagnostic center. To disregard these differences in development of job designs (and resulting job descriptions and titles) is to set the stage for employee dissatisfaction, regardless of differentials in compensation.

Precision in job design is indispensable to the organization for even more obviously self-serving reasons. If an organization is to hire and retain only those employees who are essential to the organization's success, there must be a clear understanding of what employees will be needed and what their functions will be. As will be discussed later, employee compensation constitutes an enormous investment by a service industry. To assure that compensation dollars are being spent wisely, top managers must have an accurate view of the role of each employee. It is during the job design stage of strategic planning that duplication of responsibilities must be eliminated, to be sure that all necessary duties will be performed, and to determine the actual skill level that will be required in each position. Care in this endeavor will make an important contribution to future employee job satisfaction and performance, as well as to the financial well-being of the organization.

Job Analysis

Closely linked to job design, job analysis involves the compiling of information about various jobs, their components, and the desired qualifications of persons performing those jobs. It is part of the process to develop concise job descriptions and specifications and to provide a sound basis for job evaluation. Douglas et al. (1985) note that job analysis is essential to the development and recruitment of highly productive and quality-oriented personnel. The necessity for thorough and expert job analysis in meeting the goals of the strategic management plan for any organization is apparent. It is also an intricate part of the development of equitable compensation in that it involves analysis of jobs within the given organization, as well as comparable jobs at other organizations within the field.

To return to the example of nursing personnel, analysis of nursing jobs within the health care industry indicates that different organizations specify varying duties for nurses, with accompanying pay differentials. Although "nurse" seems to be a definite choice for many people thinking about future career plans, job analysis could show it to be an inadequate and inappropriate job title for some organizations. During this phase of strategic planning, assessment of external organizations takes on renewed importance. Planners must review job descriptions and compensation scales at comparable organizations across the country, as well as those for similar positions in organizations within the community, even if the local organizations differ in product or service.

For example, maintenance personnel in a tertiary care facility are likely to confront equipment and standards that are more complex and demanding than those found in primary or general health care settings. Thus, comparing job descriptions for maintenance workers in the tertiary setting to job descriptions for similar workers at a walk-in diagnostic facility or a general hospital would not be as valid as a comparison with similar workers in the computer chip manufacturing industry, which also has exacting standards for sanitation and a complex high-tech environment. As with job design, the emphasis must be on the realistic needs of the organization and the actual work that will be required of a person in this position.

Job Evaluation

Job evaluation involves assigning a *relative* value to each job within the organization. Whether the jobs are within the health care field or in a manufacturing plant, the same requirements apply. As Straus and Sayles (1980) have pointed out, job evaluation is a *system* for exercising judgment, a system that incorporates the value of each job in relation to other jobs. The value assigned to a given job must be equitable in terms of other jobs within the same organization, but it must also be in keeping with comparable jobs in the local job market. For health care organizations with facilities scattered across several states, or even the entire country, an equitable evaluation for the entire system will throw some job values (and, as a result, the pay scale) out of line with local standards if overall pay scales and the cost of living in certain communities are below the national or dominant levels. In such instances, managers must analyze and evaluate specific jobs even more minutely to avoid both internal and external problems.

Objective and consistent assignment of value is essential. [See Straus and Sayles (1980) for a more detailed discussion of specific job evaluation methods.] One of the most common approaches to job evaluation is the development of a point system. This allows compensation managers to assign relative worth to numerous positions without discussing specific pay or benefit costs. It also allows managers to distinguish, when necessary, among positions that may appear to be equal. For example, maintenance positions that require personnel to work in excessively noisy conditions, in extremely hot or cold environments, or at unusual times may legitimately be allocated additional points to compensate for the hardships. Support personnel positions that require job holders to maintain budget figures or personnel records would be assigned more points than comparable positions without such requirements.

Importantly, a point system, when properly developed and managed, eliminates bias from the determination of job evaluation. Job evaluation is of greater significance for new organizations or businesses, but, to some extent, the changing nature of the entire health care industry calls for renewed attention to existing job designs and descriptions. Current employees may hold with tradition ideas of the relative values of given jobs, but the evolving role of professionals within the health care field—and the most recently educated professionals rushing to fill the newly created vacancies—will place new demands on strategic planners in health care organizations in coming years.

Straus and Sayles (1980) note, for example, that sex discrimination has, to some extent, been traditional insofar as jobs more often held by women have been assigned lower relative values. Thus nurses—who must complete training beyond the high school level—may be paid less than truck drivers. Such a system gained acceptance during a time when nurses were most likely to be women and truck drivers were most likely to be men. In keeping with changing social opinion and federal legislation, few such disparities will be acceptable in years to come. In particular, the concept of comparable worth can be expected to be applied in more and more areas of health care. It is reasonable to assume that strategic planners for tomorrow's health care organizations will face even greater burdens to assess comparable worth in the absence of work similarities. Although this will represent quite a challenge, it is an area that offers future planners vast opportunities to develop compensation systems with greater equity than has been feasible in the past.

As will be seen when job pricing is discussed, the value placed on a given position by an individual employee does not always conform to the relative value assigned that position by strategic planners who must evaluate all positions within a given organization. For example, a clerical employee with minimal skills may view his or her position as infinitely "better" than that of an expert plumber. Because the traditional blue-collar jobs have often been held to be inferior to white-collar jobs, such self-assigned status may offset differences in compensation that would belie such a view. In other words, the minimally skilled clerical worker may be entirely satisfied with a wage below that of a highly skilled plumber as long as the perceived status of the clerical position is not challenged.

Another important area of job evaluation concerns the extent to which the evaluation is centralized. Should a job be evaluated in terms of an entire, possibly nationwide, company, or should the evaluation focus on a single facility? If a job is to be evaluated nationally (and the resulting job price tied to the result), should that position incorporate duties not routinely assigned to it in comparable, local companies? As health care organizations expand into regional and national chains, such questions will assume added significance.

Job Pricing

Job pricing is the final planning stage in establishing a staffing and compensation system for a given organization, and it involves exactly what the name implies. Once each job has been designed, analyzed, described, and evaluated, it is assigned a specific financial value, a wage scale, and a range. Because equality across and within the system is vital, job pricing depends on the careful completion of the preliminary stages.

Interestingly, however, in this one area the appearance of fairness is at least as important as the reality, if not more so. For example, employees are most likely to compare their wages with those of persons holding positions they deem comparable, regardless of the organization's formal job evaluation results. Employees who consider themselves to be skilled labor rarely compare their compensation with those they consider to be unskilled; clerical employees are more likely to compare their pay rates with other clerical workers, not with line workers or top managers. To return to the example of the minimally skilled clerical worker, the assigned price for the job must be in keeping with that assigned to other, probably more skillful, clerical workers. Thus, equality within broad categories of workers (or the appearance of equality) is often the determining factor in employee satisfaction with pay. The same is true for fringe benefits and other factors in the total compensation package. Workers in any general category will probably compare their paid holidays to those offered similar workers, even if some workers in other categories receive different holidays.

Job pricing comes under particular scrutiny in the provision of pay differentials for unusual requirements. Just as a point system gives added value for environmental factors, the job pricing system must acknowledge such differences as night or weekend shifts or assignment to a particularly demanding area. Although such differences are addressed during the evaluation stage, care must be taken to implement these distinctions when assigning job prices.

Once relative values have been assigned during the evaluation process, it would seem that the assignment of an actual price to each position would be a relatively simple matter. However, the reality of job pricing is a far more complicated undertaking. First of all, the entire compensation investment cannot exceed the organization's income. This simple concept involves complex income projections, realistic assessment of the external environment, and a comprehensive understanding of regulations and professional standards governing the organization's operation. Before a total compensation plan can be developed and implemented, the organization must account for all factors that will influence its income and its ability to remain in business.

For example, the diagnosis-related groups (DRG) payment system recently implemented by the Medicare program establishes limits for Medicare coverage based on the primary diagnosis at the time of hospital admission. A hospital that habitually expends greater resources for these patients than can be recouped through the DRG system could face financial disaster, regardless of the care with which jobs were originally priced. To take another example, a health care organization serving walk-in clients would be foolhardy to locate in an isolated area, where the likelihood of heavy usage could not be anticipated.

These are but a few indications of the factors to be considered by compensation managers. Another concern is the labor market in the area of operation. A definition of "labor market" usually incorporates the extent to which qualified workers are available for and interested in employment in a given field of endeavor and a given geographic region. Differences in the number and qualifications of workers usually vary over time and from place to place; nevertheless, such an assessment must be made when determining the fair market value of a given service and the jobs involved in providing it. Thus, the entire exercise of job pricing must assume that such homework has already been completed as thoroughly and accurately as possible.

In setting up the job pricing system, compensation managers must define such concepts as a base salary that is hourly, weekly, or monthly and whether to use salary ranges or a flat rate (i.e., a single amount paid to all incumbents regardless of longevity, experience, or special training). The advantages of such a system are the ease of communication of the wage, simplicity of salary planning, ease of cost analysis, and the defensibility of equal treatment by position and job title. The most obvious and problematic aspect of such an approach is that each employee with the same title receives the same base pay, regardless of justifications for differences.

A pay range system involves setting minimum and maximum pay levels for each job title. Such a system has several advantages: No qualified incumbent will fall below a specified level of pay; better-qualified employees can justifiably be paid high wages; a maximum is established beyond which the organization will not go, thus helping to control costs; and the range can accommodate fluctuations in the job market, whether within a specific geographic region or over time due to a changing economy. One major disadvantage of the range system is that it may, in practice, make some positions more difficult to fill. For example, if an organization usually fills positions at the lowest level of a pay range, a shortage of qualified personnel may make those positions more difficult to fill. If long-range planning has been based on filling those positions at the lower end of the pay range, the organization could be forced to choose between meeting staffing needs or meetig budget constraints.

When coming to the actual pricing of jobs, external factors again enter the picture. As with the evaluation process, job rates paid by comparable businesses and by competing businesses in the same area must be reviewed and considered. If one organization consistently sets job prices below comparable rates in the same area, that company will face ongoing employee dissatisfaction and personnel turnover. On the other hand, a company whose overall wage scale is far above the local market rate may be constantly challenged to meet its payroll commitment. As Straus and Sayles (1980) point out, the burden is on the job evaluation system to include sufficient flexibility to accommodate internal variances and external standards.

BASE PAY

The steps leading to job pricing are designed to establish equitable pay scales within a given organization, but base pay is just the beginning of employee compensation. As discussed later in this chapter, reward systems beyond the base pay are often of greater importance to employees. However, the base pay is the employee's initial primary concern. As is the case with job pricing, the perception of the relative ranking of one's base pay may be more important to an individual employee than the pay itself.

For example, a dietician who feels that his or her pay is in keeping with professional standards probably will not be concerned with the base pay of a truck driver or an administrative assistant. On the other hand, the same dietician may be content to work for a base pay below the national average in the field if he or she is aware of regional or organizational differences in pay across the board. It goes without saying that an employee with several years of applicable experience will expect a base pay higher than that given beginning employees with the same job title.

The complex relationships that led to job pricing and the establishment of base pay for a specific employee do not go away once the job has been priced and an employee hired to fill the position. Thus, it is worthwhile to look briefly at the issue of pay over time.

Just as society changes, so do the jobs within any organization. This may be particularly true for certain areas of the health care industry. As increased technology alters the responsibilities of some employees, the very nature of the work day and its requirements may undergo drastic change. As Straus and Sayles (1980) note, employees who feel that their jobs have grown more difficult probably will ask for a new evaluation of their positions. Employers, aware of the changes and the new demands on the employees, are likely to agree. At this point, the whole process of design, analysis, evaluation, and pricing may begin again. But such a reevaluation does not happen in isolation. If changes in the job evaluation and the relative value for one job place its value above others, a new group of employees is likely to request review, beginning the process again in a different area of the organization. It should be apparent that a payment system cannot remain static; at best, it can provide functional guidelines for relative values.

It should also be briefly noted that actual management and implementation of the compensation system are of greater long-term benefit or damage than the system itself. A system with certain inequities may never be challenged (or may never generate employee dissatisfaction) if skilled managers apply it in a way that compensates for the problem areas. Conversely, an excellent system poorly applied or mishandled by inexperienced or inept administrators might just as well be flawed itself. As with each aspect of strategic management, the talents and judgment of top-level administrators must be oriented toward equitable standards fairly applied.

INCENTIVE COMPENSATION PROGRAMS

We now turn to the issue of compensation over and beyond the base. In recent years, numerous approaches have been developed to deal with motivating and rewarding employees who make significant contributions to an organization's success. Such programs are commonly referred to as incentive programs.

Incentive programs exist to serve two primary purposes: to motivate employees and to reward those who exceed certain, defined guidelines. Because of the potential expense involved, as well as the potential for increased productivity, such programs must be developed within the total strategic management plan. Bonus systems have existed for many years, but these have most traditionally applied to hourly paid employees or to those producing directly measurable piece goods. Recently, however, increasing attention has been paid to incentive programs for executives, top-level managers, and (within the health care industry) physicians. Because of the relative newness of these programs, additional care is necessary when incorporating them into the strategic management plan.

Executive Incentive Programs

Executive incentive programs, which have been among the fastest growing, fastest changing forms of compensation in American business, are emerging rapidly in the health care field. More than ever before, health care organizations and their managers are involved in variable compensation plans. Because of the dynamic environment in health care, incentive plans in health care organizations are being developed or are being changed to link them more effectively with the organization's particular business requirements.

Until recently, the health care industry has lagged behind other industries in the area of executive compensation. However, these organizations now find that they must follow the lead of other industries and adopt more sophisticated approaches to executive compensation if they are to be successful. Health care institutions across the country are competing for the highest caliber executives. The health care organization cannot afford mediocre executives, particularly in an environment of changing reimbursement systems, declining census rates, and diversification. Increasingly, the health care industry is competing with other industries for the best executives, and—like the industries with which it must compete—health care is using compensation as a powerful recruitment tool. As the emphasis shifts to performance and leverage costs, the fixed costs of base salaries and benefits are giving way to a variety of incentive and bonus programs.

Executive incentive programs, when designed and administered properly, can reward outstanding performance appropriately and can be an effective management tool in achieving targeted results. When discussing incentive programs, it is important to understand the difference between a bonus and an incentive. An *incentive* is designed to motivate individual workers and to improve the economy and efficiency of organizational operations. A *bonus* is any compensation in addition to regular wages and salary. Bonuses are often related to the performance of a corporation or organization rather than to individual performance. If the corporation exceeds expectations, bonuses may be awarded to all employees, or to all employees within certain classifications. However, incentives are directly tied to executive performance, and rewards are related directly to preestablished individual performance targets. Incentives are developed to motivate executives to actions that accomplish organizational results.

Executive Benefits and Perquisites

Generally speaking, health care organizations provide fewer benefits and perquisites for management than do other industries. Perquisites are employee benefits applicable only to executives. "Perks" may supplement employee benefit coverage, or they may provide coverage that does not exist in the standard employee benefit program. The most common perquisites in health care organizations include company cars, country and lunch club memberships, physical exams, and liability insurance.

Annual Executive Incentive Programs

Annual executive incentive programs in health care organizations often consist of measuring financial targets for operating revenue, operating income, receivable days, and productivity measures. In addition to financial goals, emphasis is placed on measures assuring high-quality standards, as well as on personal, strategic, and innovative goals requiring attention.

Meeting well-established targets and standards of performance provides a return that is greater than the incentive payout. If the executive targets are not met, there is no obligation for the organization to reward the executive with the incentive pay. For example, assume the total operating revenue for a hospital is $65 million. A financial target is established to increase net operating income by 1 percent ($650,000), and in return the hospital's top six executives could receive 20 percent of their base salaries as incentives if the target is achieved. The return on investment would be as shown in Table 14.1.

For best results, incentive payouts should vary with the level of achievement. In the example shown in Table 14.1, the incentive program established could pay 10 percent to executives if they achieve an increase in income of 0.5 or 15 percent if net income increases by 0.75 percent. Financial "windfalls" representing an increase in revenue due to such factors as a technological improvement should be excluded from the incentive agreement.

Personal performance targets should also be established. These targets should not be related to routine, ongoing efforts, but should outline a major new effort that is innovative, strategic, and/or entrepreneurial. A certain portion of the total incentive opportunity should be linked to overall organizational performance rather than to strict departmental activity. This should be weighted based on the executive's ability to contribute to specific organizational goals and effectiveness. This is critical to ensure that health care executives understand the importance of team effort in achieving overall organizational success. The weightings will vary somewhat from one executive to the next, with line managers usually having heavier weightings on financial measures than support staff executives.

Long-Term Executive Incentive Programs

Another trend in incentive programs for health care administrators is the inclusion of top management personnel in both annual and long-term incentive plans. This ensures that management does not maximize short-term objectives at the expense of long-term goals. Long-term incentives for top executives have already become standard practice in other industries, as well as in for-profit health care companies. However, they are just now gaining acceptance among not-for-profit organizations, hence are still limited in number.

TABLE 14.1. Effectiveness of Incentive Program as Return on Investment

Position	Base Salary	20 Percent Incentive
CEO/administrator	$90,000	$18,000
Assistant administrator		
Clinical	55,000	11,000
Nursing	60,000	12,000
Finance	60,000	12,000
Planning/marketing	50,000	10,000
Personnel	50,000	10,000
Total payout		$73,000
Additional operating income	$650,000	
Additional compensation	−73,000	
Return on investment	$577,000 or 890 percent	

The specific purposes of long-term incentive programs are to direct the focus on strategic goals of the health care organization, to build a balance between short- and long-term goals, to create among executives a sense of belonging and ownership, to assist in attracting good executives, to provide retention interest for outstanding executives, and to provide long-term, tax-efficient, capital-building opportunities for executives.

There has also been an increase in the use and formality of special incentive programs. Special incentives targeted for selected groups—exceptional performers, physicians, sales persons, and nonmanagement employees—have also begun to emerge in both for-profit and not-for-profit health care companies (Hay Management Consultants, 1986).

Benefits from Executive Incentive Programs

Well-developed incentive programs can provide opportunities for added value and other significant benefits to any health care organization. These benefits include the following:

- Documentation of improved quality of care
- Early identification of key targets
- Development of a strong sense of managerial direction
- Development of a better balance between high-quality care and financial results
- Establishment of a tracking system for targets on at least a monthly basis so early action can be taken when required
- Promotion of a team spirit and the understanding by executives that each must play a role in achieving group targets
- Focus of managers' attention on organizational, as well as functional or professional, goals
- Identification of outstanding performers for future executive openings

A properly planned and supported executive incentive program can pay dividends for the executives and the health care organization alike.

Group or Team Incentives

Today, more and more emphasis is being placed on programs that motivate and reward groups of employees financially for productivity improvement. Organizations accomplish this by sharing a portion of the savings with the employees directly responsible for the improved productivity. The concept is relatively new in the service industries, including health care. However, several very successful group incentive programs have recently been developed and are drawing attention and focus from other health care organizations.

With the present environment and challenges requiring maximum productivity, it can be expected that group incentive programs will receive much more attention in the health care industry. Such programs are typically very formal, incorporating eligibility and performance measures that are clearly specified in advance. In addition to design issues, the challenge is for management to assess employee attitudes and readiness carefully. This will take on added importance as programs planned and designed to focus activities on improved health care and unit productivity become more prevalent.

For some time, the industrial sector has experimented with variable compensation plans designed to increase productivity. Most of these efforts have tied the reward to the traditional profit-sharing or after-the-fact bonus concept. Others have been more ambitious, approaching productivity improvement through an integrated economic system that links all human resources to a previously defined business objective. Group incentives constitute a similar approach, with incentives available to all employees in a unit or even in the company. The overriding premise of such an incentive program is that the business must produce a measured improvement in performance before any such bonus or profit-sharing takes place. Once this economic reality has been recognized and achieved, team incentives provide a way to increase the personal income of everyone involved.

These programs typically target improvements in defined areas, such as quality, profits, service, or cost reductions. Everybody in the organization is committed to a set of common objectives to increase productivity and, ultimately, profitability. The return received by the company is then shared with all the employees who made it happen. In addition to the shared reward, two other traits are common in successful group incentive programs. First, management philosophy and practices must create a positive climate for excellence and must encourage a high degree of employee commitment and participation. Second, a system and structure must be developed that enable all employees to become more deeply involved in solving problems of productivity, quality, and service. It has been demonstrated that all three factors—management practices, shared reward, and employee participation—have the potential of significantly improving productivity.

Trends in Incentive/Bonus Compensation

All forms of incentive compensation plans are becoming more and more common in all segments of the health care industry, with slower growth found among organizations with religious affiliations. These plans are more prevalent in for-profit health care organizations, but they are continuing to increase at a significant pace in not-for-profit, secular health care organizations. The formality of these plans varies considerably—from being totally discretionary to incorporating strict, formal criteria for eligibility, performance measures, and award size. However, the overall trends are toward increasing formality and toward extending incentive programs to lower echelons within organizations, making such rewards available to greater numbers and classifications of employees (Hay Management Consultants, 1986–1987).

Physician Compensation

Many physicians operate on a fee-for-service basis, contracting with hospitals or other health care organizations for rights to use medical facilities. However, some physicians are recruited and paid directly by health care organizations in a bona fide employer-employee relationship. In the past, hospitals often developed individual employment contracts with physicians based primarily on what the physician wanted and how critical that physician may have been to the hospital's overall success. This system resulted in significant compensation differences among physicians in the same hospital. Because of the disparity of physician salaries in existing contracts, internal equity may be a major problem, not only among physicians, but with other staff positions as well. It has been difficult to survey hospital-based physician compensation because of the different forms of compensation and the reluctance of hospital administrators to share such data. Hospitals with highly valued physicians are now looking for ways to compensate them in a more orderly and moderate manner in order to remain competitive. Many hospitals are now attempting to focus physicians' attention on overall success through incentive programs. However, the difficulty in establishing such programs is in setting up meaningful performance and job content criteria.

Although the approach may undergo additional changes, it can be expected that in the future greater effort will be made to equalize and formalize physician compensation.

Exceptional Performance Awards

Many companies have begun to use incentives to motivate individuals who are particularly crucial to the long-term success of the organization. Special compensation approaches, particularly exceptional performance awards, are gradually increasing in the health care industry because of the usefulness of this method of eliciting high commitment and performance from selected employees. Health care organizations are experimenting with these plans to recognize outstanding contributions made by any employee. Incentives may be appropriate, for example, in projects with major cost-savings implications, or in those that in some way increase the organization's ability to compete in today's rapidly diversifying marketplace. Special incentives are also being used to reward truly entrepreneurial successes. Such plans are designed to acknowledge, motivate, and retain critical employees who are not adequately covered through existing compensation programs. These plans typically provide a limited number of awards for outstanding achievements, and they often include employees who would be very difficult to replace. These employees with ongoing records of significant contributions may have knowledge of a proprietary nature or may be involved in special projects. The award is usually sizable and varies considerably depending on the type of program, the significance of the accomplishment, or the importance of the individual to the organization (Hay Management Consultants, 1986).

Traditional compensation components—base pay and benefits—are designed to meet broad human resources objectives. These, however, often fall short of meeting the demands of a health care business attempting to deal effectively with exceptional performers. Base salaries, even when administered under a merit-based system, typically reflect little difference between average performers and those who are outstanding. Annual merit increases tend to be only slightly greater for the best performers, and these increases are generally much smaller than the value received by the organization from a truly exceptional performer. Annual salary increases are rarely tied to specific outcomes. Furthermore, salaries usually are of limited value in retaining key individual high performers. Substantial salary increases are often offered by competitors to attract or recruit key talents. This situation, coupled with the industry's critical need for innovation and diversification, has caused health care organizations to begin to examine and adopt less conventional compensation techniques aimed at dealing with recognition and rewards based on an equitable remuneration scheme for key performers.

As the health care industry becomes more aware of effective ways to create and implement special compensation plans, it is likely that exceptional performance programs will grow in popularity. Pressure to protect the people and the ideas on which a given organization depends will lead to an increase in both the number and types of formal programs—and the pressure for such action can also be expected to increase. Group incentive plans offer significant leverage in targeting goals and objectives, but plans to recognize and reward exceptional performance are likely to receive greater acceptance.

BENEFITS

One might assume that benefits have been a part of health care employee compensation for many years. In reality, however, benefit programs in all industries are an occurrence of the mid-twentieth century. The first major national development in benefits came in 1935: The Great Depression had made people overtly aware of financial problems associated with illness, old age, disability, and death, and the federal government responded with the Social Security program, to provide a basic level of retirement protection. Until after World War II, there was little voluntary employer action. Because wages were frozen during the war, interest grew in the benefits area. After the war, however, in 1948, the National Labor Relations Board ruled that benefits were subject to collective bargaining, and employee pressure for benefits began to increase. The strong postwar American economy, coupled with substantial employee, union, and employer interest in benefits, resulted in a 30-year boom in employee benefits. Competing favorably with other employers or being the first to develop or implement a new benefit program was a primary objective for many businesses in the 1960s and 1970s. Another major turning point came in 1974 with the enactment of the Employee Retirement Income Security Act (ERISA). This act set guidelines for pension plans and required disclosure of information concerning welfare benefits. ERISA provided increased assurance that employees would actually receive a retirement benefit and set the stage for even more government regulations. Although ERISA added substantially to administrative burdens and to the costs of benefit programs, benefit plans continued to grow and improve throughout the decade of the 1970s. Indeed, there was virtually a benefits race that resulted in most major companies and industries offering a full package of benefits by the middle of the decade (Hay Management Consultants, 1986).

The health care industry was no exception. Although health care organizations tend to spend less on benefits than firms in other industries, the cost is still substantial. For example, among hospitals, the current average cost of so-called fringe benefits is nearly 32 percent of total payroll costs. In 1984 the average hospital investment in fringe benefits was $6146 per employee, compared with an all-firms average of $7842. The investment of hospitals in fringe benefits may lag behind the national average, but it should be remembered that given the figures above, the total expected expenditure for a hospital with 1000 employees would exceed $6 million!

Data released by the U.S. Chamber of Commerce illustrate both the relative and the absolute cost of employee benefits in our economy. Table 14.2 summarizes some of these data. Note that the single largest expenditures category is for time-off benefits such as sick leave, vacation, and holidays. The average hospital spends about $2342 per employee annually for these items. The true cost is much greater, however, since many absent employees in hospitals must be replaced by other employees, often at overtime rates.

TABLE 14.2. Employee Benefits in U.S. Economy Compared with Hospitals

	All Firms Amount	All Firms Percent	Hospitals Amount	Hospitals Percent
Legally required	$2054	26.2	$1598	25.8
Social Security	1408	18.0	1287	20.9
Unemployment Compensation	361	4.6	154	2.5
Worker's Compensation	262	3.3	148	2.4
Railroad Retirement Tax	24	.3	10	.2
Tax exempt	$1721	21.9	$1189	19.3
Life and health insurance	1581	20.2	1025	16.7
Dental insurance	85	1.1	85	1.4
Education	55	.7	79	1.3
Tax deferred	$1460	18.6	$ 879	14.3
Pension	1011	12.9	704	11.5
Profit sharing	219	2.8	67	1.1
Thrift plans	118	1.5	11	.2
Disability	112	1.4	97	1.6
Time not worked	$2381	30.4	$2342	38.1
Breaks, wash-up, etc.	399	5.1	405	6.6
Vacation, holidays, sick leave	1982	25.3	1937	31.5
All other	226	2.9	138	2.2
Total	$7842	100.0	$6146	100.0

SOURCE: U.S. Chamber of Commerce (1985), p. 13.

Legally required benefits, such as Social Security, unemployment compensation, and worker's compensation, have increased dramatically over time and now constitute 26 percent of the cost of benefit programs. Further increases are expected in the future, and health care executives can do little to control expenditures in this area. Therefore it will become even more important for health care organizations to integrate new legally required benefits with existing benefit coverage.

Pension expenditures for the hospital industry average $704 per employee in 1984, compared to $1011 spent per employee for all firms. Hospitals, like all health care organizations, are likely to face growing pressure in the future to increase their pension benefits. Careful restructuring of retirement/annuity benefits and integration with Social Security benefits may be helpful in this regard. Another important strategy is to terminate existing defined benefit plans and replace them with defined contribution plans. This trend, which is already becoming common in other industries, could also be beneficial to the health care industry. Sponsored tax-deferred programs, matching savings programs, and other sound investment strategies can also pay large dividends in the efforts of health care organizations to reduce total pension costs.

Referring again to Table 14.2, it is obvious that expenditures for insurance, especially health insurance, constitute the largest benefit cost that is not required by law. The average hospital spent $1025 per employee for insurance in 1984, compared with the average of $1581 for all firms. However, many hospitals write off as unmeasured costs to their employees or provide free medical services, so the amounts actually expended are probably much greater than they appear.

With expenditures by hospitals as examples, it is clear that benefits have become increasingly expensive for employers in the health care field. However, the benefits race slowed during the economic recession of the early 1980s when cost containment became the watchword and, for many organizations, the key to survival. Today, benefits are clearly no longer "fringe" compensation. Benefit programs, rather than base salary, often keep employees from changing employers. Over the years the proportion of the compensation dollar allocated for benefits has grown so that today's benefits consume more than a third of the total compensation dollar. As a result of this growth, top executives share a great deal of interest in containing benefit costs.

One of the trends already evident in these years of reckoning is the leveling off and slight decrease of benefit costs as a percentage of direct pay. Direct pay and benefits escalated more rapidly than the cost of living in the twentieth century, but benefits grew at a faster rate than direct pay until the late 1970s. A U.S. Chamber of Commerce survey (Fig. 14.2) illustrates this pattern.

Cost containment continues to be the dominant trend in benefit planning and is considered by many employers as one of their top planning priorities. Top management is becoming more involved in benefits cost containment, but there is equal interest in the quality of what is purchased with the benefits dollar.

As health care organizations in all parts of the nation emerge from the recession of the mid-1980s, most continue to hold the line on benefits. Components of the health care industry continue to be challenged with changes in reimbursement systems and guidelines, a growing consumerism, and greater competition. Ambulatory clinics and free-standing diagnostic centers, as well as a broad range of entities of other types, now compete with traditional office-based medical practices or provide functions formerly carried out by hospitals. All these factors exert pressure on the more traditional health care organizations to effect cost-savings measures. With the high price tag on benefits, savings in this area take on added importance.

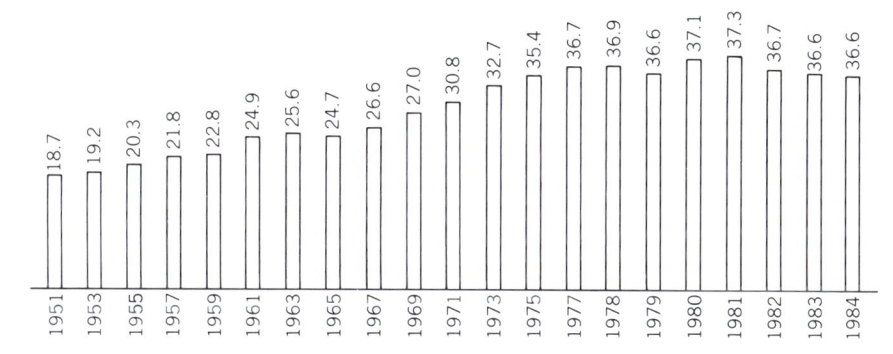

Figure 14.2. Benefits as a percentage of payroll, 1951–1984. Figure is from *Employee Benefits*, a publication of the U.S. Chamber of Commerce.

Cost containment is unquestionably the dominant trend in benefits. Now that the quality of benefits is basically under control, renewed attention will be directed to cost containment. Instead of increasing the level of benefits, health care organizations are already redesigning benefit programs to meet employee needs better while holding the line on costs. It is worth noting that efforts to contain rapidly escalating costs of health care benefits have apparently been successful. Previously skyrocketing costs are beginning to stabilize as premium increases slow, dropping from earlier rises between 18 and 26 percent in the first half of the 1980s to increase rates closer to the rate of inflation. This slowdown has been effected by increased use of cost-sharing measures, encouragement of the use of alternative services, and implementation of peer review programs. Plans designed to share more costs with employees are becoming more common as employers raise deductibles and copayments while increasing the employee's share of the premium.

In conjunction with the efforts of health care organizations to reduce the relative level of benefits, more and more benefit programs will feature choice-making options. More cafeteria-style selections of benefits and perquisites will be offered so that individual employees can choose the options that best fit personal and family needs. Pressure for such flexibility can be expected to increase in the health care industry. The field is already dominated by female employees, and the growing trend of dual-income families can be expected to alter drastically the benefits deemed most desirable by one partner in a dual-income family. As health care organizations continue to emphasize limiting the quantity of benefits and stressing the quality of what can be purchased with each benefits dollar, they will most likely be joined in their efforts by employees with similar goals.

Flexible Benefits

A flexible benefits system is one under which employees have some choice in the form of their total compensation. As a cost containment strategy, flexible compensation helps employers control costs in several ways. First, it redefines the employer's obligation in terms of overall level of expenditure rather than locking an employer into paying for a specific package of benefits regardless of costs. Additionally, flexible plans enable the employer to control the benefit allowance for each employee—that is, the employer is no longer at the mercy of inflation or other factors beyond the organization's control.

A positive side effect of flexible plans is that by giving an employee a choice, an employer can offset the negative effects of benefit cutbacks. This is particularly true with respect to health benefits. Surveys show that employees like flexible plans, even when fewer benefits are provided. Flex plans provide the opportunity for employees to tailor their benefit coverages to meet individual situations. Most benefit programs in the past were designed around the idea of the traditional family (where the father works and the mother stays home), but patterns of family life have changed dramatically during the past few decades. There is now a significant need for benefits designed around the various stages of an employee's career and around the diversity of each employee's individual needs.

Because flex plans provide individual choices, they have been popular with employees, who can decide, for example, whether to contribute toward extensive coverage or to accept smaller benefits, paid for entirely by the employer. The presence of individual choice makes cost containment more acceptable while offering a wider menu of benefits from which to choose. And when cost containment involves added costs to employees, the workers show a greater willingness to help finance the benefits they most desire.

An example of the advantages of a flex plan is found in the area of medical insurance coverage. When employees are given a choice of medical options, the majority of employees select health coverage different from that originally offered. This usually results in cost savings due to decreased plan utilization. Moreover, since employees must choose their benefits under such a plan, they learn more about the various components, leading to increased employee understanding of the true value and cost of the programs. An additional asset is that certain benefits—such as child care, uncovered medical expenses, and legal expenses—can be designed into a flex plan in a way that brings tax savings to both the employer and the employees.

Because of the various positive aspects of flexible benefit programs, the movement toward flex plans is anticipated to continue at a rapid pace in companies of all sizes. By the end of 1986, more than 500 companies were estimated to have implemented a formal flexible compensation program in employee benefits (Fig. 14.3). As illustrated in Figure 14.4, flexible compensation programs have been implemented by companies in all size categories. Much of the recent growth has been fueled by medium-sized organizations—those with 1000 to 10,000 employees. By industry, flexible programs

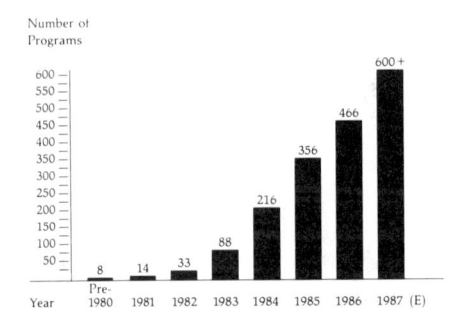

Figure 14.3. Flexible compensation programs. From Hewitt Associates, *On Flexible Compensation*, Lincolnshire, Illinois, January 1987, p. 1.

Compensation Management   367

Figure 14.4. Flexible compensation programs by size of employer. From Hewitt Associates, *On Flexible Compensation*. Lincolnshire, Illinois, January 1987, p. 1.

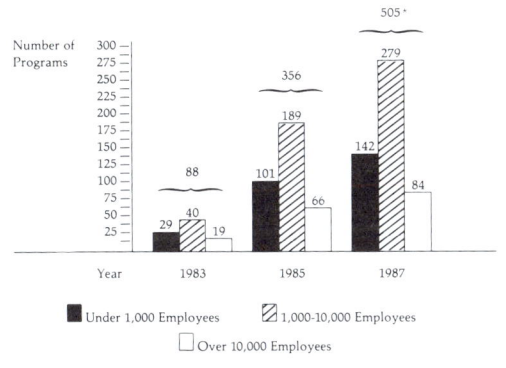

are well represented across industry lines. Figure 14.5, which gives the concentration of flexible programs by industry, indicates that the health care industry has been a leader in the introduction of flexible compensation programs.

Taxing Benefits

When benefit programs started, several tax rulings established the "nontaxability" of most employee benefits. That philosophy was rarely attacked—until recently. Strong movements are now underway to tax employee benefits. Aside from the need to obtain more tax dollars, the pressures for taxation of benefits appear to stem from the perception that it is not fair for employees who work for companies that provide benefits to receive those benefits tax-free while employees of companies that do not provide such benefits must purchase them with aftertax dollars. The opposing philosophy is that by providing families with financial protection in time of illness,

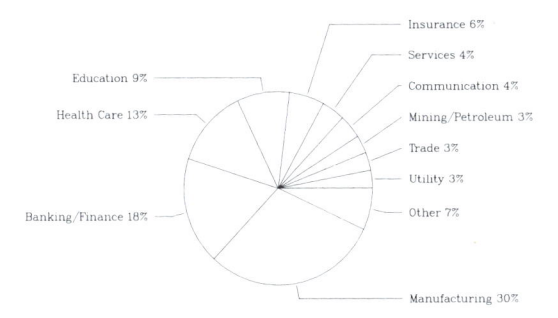

Figure 14.5. Flexible compensation programs by industry, representing a total of 505 flexible compensation programs in effect. From Hewitt Associates, *On Flexible Compensation*. Lincolnshire, Illinois, January 1987, p. 1.

disability, death, and retirement, employee benefits serve an important social goal and, in the long run, will save the government money. As more and more benefits become taxable, it is likely that many health care organizations will be pressured by employees to cut back on benefits and raise base pay. It can be expected, however, that many employees would not use such a salary increase to replace the protection they have lost, and many employees who eventually needed that protection would not have it. It can be argued that in addition to the potentially devastating impact on employees and their dependents, such a scenario would place additional burdens and pressures on social and govenrment programs (Hay Management Consultants, 1986).

Future Benefit Trends

There will continue to be strong pressure from the consumer, the public in general, and the government for health care organizations to hold the line on costs. Health maintenance organizations and preferred provider organizations can be expected to increase in popularity, as can such alternatives to inpatient hospitalization as home health care, walk-in care centers, and surgery centers. Medical and hospital pre-admission authorization programs are likely to become more prevalent, as are employee fitness and wellness programs. Medical plans will continue to be redesigned to shift more cost to employees, and choice making of benefits packages by employees will be more critical as the dollars available for benefits become more limited. Flexible benefit programs will be even more popular as their value and positive acceptance become more apparent. As health care organizations grow more determined to gain more control over the cost of benefit programs, defined benefit pension programs will continue to give way to defined contribution plans, profit sharing, or savings programs, or some combination of the three.

It will also be critical in the future for health care organizations to improve the manner in which they communicate with employees about benefits. Current benefit issues—redesign of the medical plan, 401(k) programs, flexible benefits—are even more complex than traditional programs. In such a setting, employees must better understand both the actual cost of benefits and the significance of their total compensation plan.

SUMMARY

Effective compensation management has come to incorporate a multitude of human and environmental factors. From the establishment of specific jobs and their requirements to the formulation of a company-wide compensation program, decisions made by strategic planners determine the satisfaction of employees and the ultimate success of the organization. Compensation packages that motivate and reward the administrators and other employees responsible for the organization's success can be expected to increase in number and scope as governmental mandates for cost containment and cost effectiveness add to consumer pressure for fiscal restraints.

REFERENCES

Douglas, J., S. Klein, and D. Hunt. *The Strategic Managing of Human Resources.* New York: Wiley, 1985.

Douglas, J., S. Klein, and D. Hunt. "Compensation and benefits." In Hay Management Consultants, Eds., *1986 Health Care Management Company Total Compensation Survey.* San Francisco: Hay Management Consultants, 1986.

Hewitt and Associates. *Flexible Compensation.* Chicago: Hewitt and Associates, Jan. 1, 1986.

Straus, G. and L. R. Sayles. "Wage and salary administration." In *Personnel,* 4th ed. Englewood Cliffs, NJ: Prentice-Hall, 1980.

U.S. Chamber of Commerce. *1984 Employee Benefits.* Washington, DC: Government Printing Office, 1985.

U.S. Chamber of Commerce. *Ideas and Trends,* Issue 111. Washington, DC: Commerce Clearing House, Feb. 21, 1986.

CHAPTER 15

Preventive Labor-Management Relations

Charles L. Joiner

Strategic labor-management relations is concerned with maintaining a positive labor relations climate regardless of whether the organization is unionized. This chapter focuses on the issues necessary for an understanding of how to develop a preventive management program, based on the premise of a management policy to maintain nonunion status. The fundamental principles are applicable, however, even if simply focused on maintaining good relations between management and organized labor.

Chapter 16, which assumes that employees are in a labor organization, is written from the perspective of management's need to understand the total framework and process of negotiating and administering contracts with a union. It was developed in the belief that good acumen of both cognitive and behavioral requirements can lead to a relationship that minimizes the adversarial factors and enhances the probability of mutual trust and respect. Many strategic decisions must be made during the negotiation process either for an initial contract or for a replacement contract. Also, administering the contract fairly and consistently requires a high level attention from management to yield a productive and positive labor-management relationship.

With well over 3 million workers, the health care industry represents one of the largest work force population groups in the United States. The health care industry also represents one of the largest pools of nonunion employees, and, therefore, a prime target for union organizers. This is particularly true in times of economic stress when

---

This chapter is adapted in part from C. L. Joiner and J. O. Morris, "Management's response to the union phenomenon." In the Catholic Hospital Association, Ed., *Hospital Progress.* Copyright © 1978 by the Catholic Hospital Association. Reproduced from *Hospital Progress* by permission.

management decisions must be made concerning the employment status of many employees, both professional and nonprofessional.

DEVELOPING AN EMPLOYEE RELATIONS PHILOSOPHY AND STRATEGY

Management strategy regarding its desired relationship to labor organizations should be formulated as a part of overall policy development. This strategy formulation must, of course, take into consideration the geographic, demographic, and historical factors pertinent to its setting. For example, an organization that is located in an area where unionization is prevalent may find it extremely difficult to prevent unionization of groups of its employees even with the best preventive plan. Nonetheless, it is management's responsibility to develop and communicate to its employees the organization's employee relations philosophy.

The organization's employee relations philosophy should be developed on the basis of its objectives regarding such factors as communication with employees, management rights, and unions. If an organization is not presently unionized, management should consider the array of environmental and organizational issues in the process of determining its policy relating to unions. Specifically, management should consider the available strategic options for developing and maintaining a positive employee relations climate.

One option is to adopt a nonunion policy and begin to implement a preventive management program. This option is explored in detail in this chapter. A second option is for management to implement essentially the same program without communicating a formal nonunion policy, depending on its analysis of circumstances and objectives. Regardless of the strategic option chosen, it is essential for management to do the necessary analysis and adopt an appropriate employee relations program focused on maintaining good communication and positive relations.

MAINTAINING NONUNION STATUS

Maintaining nonunion status depends largely on what managers do to prevent the need for a union. This view is based on the philosophy that unionization is preventable if management is doing enough of the "right things." When management actions do not support a positive employee relations climate, workers may find it necessary to seek external help and, in some situations, they deserve help from a union.

This argument may be supported further by noting that union organizers typically do not attempt to organize an employee group until workers themselves have sought union assistance. Union certification elections seem to suggest that employees really are voting for or against management instead of for or against a particular union. Based on these premises, this chapter seeks to help health care managers by identifying issues important to good personnel relations and the maintenance of nonunion status.

To provide a sound basis for prevention of unnecessary problems, it is essential to understand the historical perspective of the underlying issues including employee perceptions of the need for unionization. The purpose of the chapter, therefore, is accomplished through a review of labor law history and trends, an overview of the fundamental causes of friction between management and labor, a summary of reasons health care employees give for joining unions, an analysis of criteria used by union organizers to evaluate health care institutions and, finally, specific recommendations for establishing a preventive management program and maintaining nonunion status. Since knowledge of the legal framework is essential to any manager who desires to avoid foolish mistakes in the implementation of a well-conceived program, it is appropriate to review labor law history and trends first.

LABOR LAW HISTORY AND TRENDS

The National Labor Relations Act (NLRA) is the foundation for the labor laws of the United States. The NLRA, the so-called Wagner Act, was adopted in 1935 and has been amended by the Taft-Hartley Act of 1947, the Landrum-Griffin Act of 1959, and Public Law 93-360 (the Health Care Amendments) in 1974.

The Wagner Act authorized the formation of the National Labor Relations Board (NLRB) to administer the provisions of the act. The Wagner Act encompassed all institutions that had an impact on interstate commerce. The status of nonprofit health care institutions was left to the interpretation of the courts. Proprietary institutions and nursing homes were considered within the jurisdiction of the act. Since by definition governments are not employers, federal, state, and municipal hospitals were specifically exempted from the jurisdiction of the act (Rakich, 1973).

Under the protection of the Wagner Act, unions flourished in industries of virtually all types, creating a host of problems regarding the regulation of union-management relations. Industries had to contend with many jurisdictional strikes caused by disputes between competing unions. Some labor leaders, because of their new and unbridled power, refused to bargain in good faith (Rakich, 1973). The Wagner Act proved to be inadequate to curb these and other abuses of the bargaining process. Therefore Congress in 1947 passed the Labor Management Relations (Taft-Hartley) Act. It is this legislation that has become the backbone of the nation's labor laws (Rakich, 1973).

The Taft-Hartley Act amended the Wagner Act by listing specific unfair labor practices. In addition, it specifically exempted nonprofit health care institutions from coverage under the act. The status of other types of health care institutions did not change.

In 1959 Taft-Hartley was amended by the Labor-Management Reporting and Disclosure (Landrum-Griffin) Act. Among its many provisions, this act requires employers, including voluntary nonprofit health care facilities, to submit a report to the U.S. secretary of labor detailing the nature of any financial transactions and/or arrangements that are intended to improve or retard the unionization process (Rakich, 1973).

Until 1967 the courts on a case-by-case basis determined which proprietary health care institutions and nursing homes had an impact on interstate commerce and thus came under the NLRA. As a result of several court cases, the NLRB in 1967 determined that proprietary health care institutions with an annual gross revenue of at least $250,000 and nursing homes, regardless of ownership, with an annual gross revenue of at least $100,000 were covered by the act (Rakich, 1973).

With voluntary hospitals comprising the largest sector of the health care industry, it was only a matter of time until they too fell under federal legislation. Their shift in status occurred in 1974, when Congress passed PL 93-360 to amend the labor relations act. These amendments, which extended the coverage of the labor laws to include all health care institutions under nonpublic ownership and control, defined a health care institution as any "hospital, convalescent hospital, health maintenance organization, health clinic, nursing home, extended care facility, or other institution devoted to the care of sick, infirm, or aged persons."

A decade ago, only about 20 percent of the total nonpublic health care labor force was unionized (Pointer and Cannedy, 1974). However, because of their declining rate of growth among blue-collar workers, labor unions are renewing their interest in the organizable group of white-collar employees in the health care industry (Pointer and Metzger, 1975; Reed, 1970). Attempts have been made in recent years to modify the Taft-Hartley legislation to favor union organizational campaign strategies.

The legislative background and prospects certainly point to difficult times for health care managers seeking to stay nonunion. For a realistic perspective on maintaining nonunion status, management should have a good understanding of the fundamental causes of labor problems. Reasons for labor-management friction are summarized next.

CAUSES OF LABOR-MANAGEMENT PROBLEMS

Fundamental differences between the goals and objectives of management and labor create friction that cannot be totally explained in terms of desires for higher wages, shorter working hours, or better working conditions. Two fundamental causes of such friction are the issue of management rights and the issue of efficiency versus human value. Management always will assert its right to prescribe certain modes of action or levels of desired productivity to justify its existence or that of the organization. Yet labor unions question whether management should have complete power over the work force. This is a point of conflict. Organized labor attempts to shift the locus of control by seeking to obtain a voice for employees about working conditions and terms of employment.

The question of management's right to govern is paralleled by the question of human value versus efficiency. If management is to achieve its stated goals and objectives, it must maintain efficiency through increased productivity and cost containment. On the other hand, the union seeks to improve its members' standard of living. Neither side may be totally right or totally wrong in its demands, and unfortunate circumstances often trigger open conflict. For example, a management that wishes to improve the existing fringe benefits package for employees may be prevented from doing so by pressures to contain costs. Evidence of this type of conflict in health services organizations is mounting almost daily, especially as new crises (e.g., malpractice insurance rate hikes) arise and cause even greater cost increases.

With an understanding of the fundamental causes of labor problems, administration can begin developing its philosophy for a preventive management program by reviewing research on employee reasons for joining unions. The analysis that follows summarizes findings from a selected number of such studies.

Why Employees Join Unions

The desire to unionize is thought to be centered on three issues: wages, employee dissatisfaction with work benefits, and employees' perceptions about the organization as a place to work that could reflect perceptions about management or employer. However, other factors have contributed to increased union activity in the health care industry. Over the past three decades social turmoil has precipitated civil rights legislation and stimulated changes in the attitudes and social conscience of many individuals. The idea of being represented by a union is not considered as unprofessional as it once was (Phillips, 1974; Stanton, 1971). The health care industry is just beginning to feel the effects of this turmoil, and the passage of PL 93-360 served only to release the pent-up emotions of the industry's workers and union leaders. Recent labor reform efforts are further evidence of labor's continuing struggle to swing the pendulum in its favor.

In a study pertaining to why employees want unions, Brett (1980) found two main factors.

1. An employee's initial interest in unionization is based on dissatisfaction with working conditions and a perceived lack of influence to change those conditions.
2. The likelihood that a coalition of dissatisfied employees will try to organize a union depends on whether they accept the concept of collective action and whether they believe that unionization will yield positive rather than negative outcomes for them.

According to this study, a significant proportion of employees who were dissatisfied with working conditions—particularly job security and wages—voted for union representation. Initial employee satisfaction with working conditions was measured by interview data collected from employees who were asked the eight questions listed in Table 15.1.

As Table 15.1 shows, the level of satisfaction with wages, job security, fringe benefits, treatment by supervisors, and chances for promotion was significantly correlated with a vote for union representation.

Although study findings vary with respect to factors being measured and the degrees of importance of the factors, a sampling of studies in 11 publications reveals some interesting commonalities in the attitudes of health care workers who seek union representation. Table 15.2 outlines the findings of the search. Although not all-inclusive, the data do show that money and fringe benefits are not always the only issues to employees. Other, less tangible factors such as poor communication, poor supervision, bad working conditions, and inconsistently enforced personnel policies carry considerable weight.

The less tangible issues may be just as important to employees. One reason workers frequently mention for joining unions is failure by management to "treat them fairly, decently, or honestly." Employees view management as fair, decent, and honest if it recognizes the needs of individuals and treats people with dignity. Issues such as wage disputes, respect, and recognition for loyalty and service to the institution are related to the individual need to be recognized and to be treated fairly. A specific example is employees' feeling that management offers no educational opportunities for upgrading their skills as a means to career mobility (Pointer, 1974). Unless these opportunities are available, employees often believe that they are locked into dead-end jobs.

TABLE 15.1. The Correlation Between Job Satisfaction and Voting for Union Representation[a]

	Correlation with Vote[*]
1. Are you satisfied or not satisfied with your wages?	−.40
2. Do supervisors in this company play favorites or do they treat all employees alike?	−.34
3. Are you satisfied or not satisfied with the type of work you are doing?	−.14
4. Do your supervisors show appreciation when you do a good job or do they just take it for granted?	−.30
5. Are you satisfied or not satisfied with your fringe benefits, such as pensions, vacations, holiday pay, insurance, and sick leave?	−.31
6. Do you think there is a good chance or not much chance for you to get promoted in this company?	−.30
7. Are you satisfied or not satisfied with the job security at this company?	−.42
8. Taking everything into consideration, would you say you were satisfied or not satisfied with this company as a place to work?	−.36

[*]$p < .01; r = .08; N = 1004$

SOURCE: "Why Employees Want Unions," *Organizational Dynamics*, American Management Association. Spring 1980, p. 51. Reprinted, by permission of the publisher, from *Organizational Dynamics*, Spring 1980, copyright © 1980 American Management Association, New York. All rights reserved.

[a]The negative correlations indicate that employees who were satisfied tended to vote against union representation.

TABLE 15.2. Reasons Health Care Employees Join Unions: Derived from a Sampling of Studies in 11 Publications

Issue	Publication[a]										
	1	2	3	4	5	6	7	8	9	10	11
Poor communication	×	×	×	×	×	×	×	×	×	×	×
Personnel policies	×		×			×	×	×	×	×	×
Supervision	×	×	×	×		×	×		×		×
Fringe benefits			×	×			×	×	×	×	×
Work conditions		×	×	×	×			×	×		×
Grievances	×	×	×					×			×
Job security	×	×	×		×	×	×				×
Human dignity	×			×	×		×				×
Shift differentials			×	×							×
Wages	×	×	×	×	×	×	×	×	×	×	×

[a]Numbers correspond to the following chapter references.

1. Stanton (1971)
2. Goodfellow (1969)
3. *Hospital Financial Management* (1974)
4. Imberman (1973)
5. Lewis (1974)
6. Phillips (1974)
7. Rakich (1973)
8. Milliken and Milliken (1973)
9. Sibson (1965)
10. Stanton (1974)
11. Metzger and Pointer (1972)

Generally, the reasons health care personnel give for joining unions may be grouped into two broad categories: poor communication and perception of poor treatment. Almost all the reasons are related in some way to communication problems (e.g., little upward communication) or to employees' belief that they are not receiving fair treatment on specific work-related issues. As a generalization, employees may consider the union alternative when they perceive a prevailing attitude that management does not consider meeting employee needs as a primary goal of the institution. Administrators should find an interesting relationship between why employees join unions and what the union organizer looks for. Although this study does not include a correlation analysis, some relationships are evident between the findings of the two.

What the Union Organizer Looks for

Employees usually try to resolve their problems internally before seeking outside help. Typically, union organizers appear on the scene only if they have been invited. In other words, if a union organizer is involved, it is likely that pro-labor activity has progressed to a serious level.

There is no blueprint the health care manager can use to determine how a union organizer will evaluate a given institution. The method of evaluation depends on the organizing team sent into the area and its previous experience or success. Tactics may vary considerably, depending on the contacts from employees and on management's response to the situation. However, the organizer may concentrate in certain areas, including the following.

Employee Loyalty by Work Shift

Normally, the first shift is the most loyal to the organization, the second shift less loyal than the first, and the third the least loyal. This probably is because new employees usually start on the second or third shift. They see top management seldom or never, and the supervisory force usually is smaller. Thus, there often is no one who can provide consistent supervision (e.g., answering employee questions about personnel policies or benefits). These employees tend to feel overlooked and forgotten. They are more susceptible to the pleas of the union organizer, who usually is available on the later shifts (Goodfellow, 1969, 1972; *Hospital Financial Management,* 1974).

Female-Male Employee Ratio

Women historically have been less interested in unions than have men. In the past, many women worked to supplement the family income, but this has changed rapidly. Today, women are prevalent in the work force and frequently earn a primary or major part of the family income.

Nursing personnel, a majority of whom are women, are increasingly recognizing the need to organize to improve their status. The American Nurses' Association is attempting to upgrade and negotiate conditions of employment for its membership. Other professional organizations, such as the American Society of Hospital Pharmacists, the American Society of Medical Technologists, the National Association of Social Workers, and the American Dietetic Association, are also actively seeking more voice in the representation of their members (*Hospital Progress,* 1972; Matlack, 1972; Metzger, 1970; Pointer and Metzger, 1975; Stanton, 1971).

Work Environment

Employees expect management to provide clean working and eating environments. If the health care institution allows the work environment to deteriorate, employees may think that the institution does not care much about them (*Hospital Financial Management,* 1974).

Wage Rates

Traditionally, the health care employee has subsidized health care institutions with low wages. This is an injustice to the employee, who must compete daily in the retail market for goods and services. In addition, the institution must have fair and regular wage differentials. Failure to update these differentials will cause a compression effect between the new employees' base pay and the tenured employees' level (*Hospital Financial Management,* 1974; Goodfellow, 1972; Lewis, 1974; Reed, 1970).

Incentive Pay

In areas in which an incentive pay program has been implemented, employees may complain that some of the rates, or daily quotas, are too high. High quotas obviously breed dissatisfaction if management does not respond by reexamining the quotas periodically (Goodfellow, 1972).

Overtime Practices

Problems arise when overtime is scheduled for employees without their consent. Management assumes that the worker will not object to the extra hours because of the overtime pay, but this often is not a valid assumption. Overtime can be very disruptive to the employee's family life and leisure time. The union organizer will exploit this point of dissatisfaction and force management to hire additional workers. Inequities in the distribution of overtime represent another aspect of the problem (Goodfellow, 1972; Stanton, 1971).

Seniority

Although management may prefer to recognize the skill and health of a worker in assigning a new job, it must not overlook the employees' view of seniority. Seniority to them is job security. If management takes the time-honored seniority concept away completely, it is asking for employee dissatisfaction and unionization, particularly in geographical areas where unionization already is well entrenched (Goodfellow, 1972).

Promotion Policy

When a new job opens up or an employee leaves, present personnel should be given an opportunity to apply for the position. A good job-posting policy can be extremely helpful. Health care institutions also should have education and training programs available to assist employees' vertical or lateral movement (Joiner and Blayney, 1974).

Job Transfers

Most often it is the ambitious employee who would like to be promoted to a more convenient shift or a different job. Research has shown that many times supervisors, who do not want to lose good workers or have to train new ones, sabotage the promotion of ambitious employees. Such practices should be avoided. Frequently, the best employees are those who have worked up the organizational ladder (Goodfellow, 1972).

Fringe Benefits

Research has revealed that most managements underrate the value of fringe benefits to the employee. And, as employers continue to increase the benefits portion of total compensation, the benefits package is likely to increase in relative importance to employees. With the news media and the next-door neighbor discussing the benefits of union representation, it is foolish for health care management to neglect to establish a good benefits program and to adequately explain to employees the benefits offered by the institution.

Discipline and Grievance Procedures

If the institution does not provide employees with written rules covering what is not allowed and what is and to what degree, some supervisors may abuse their authority to reprimand. The grievance procedure serves as a safety valve for employees to release their frustrations about supervisors or other major problems. Management should develop and implement an internal procedure that employees will use instead of resorting to an outside agency to settle disputes. Management also should review the procedures periodically to make sure they are serving workers' needs. Many grievances, for example, are either about or under the direct control of the employees' immediate supervisor. In these cases, the employees probably will not use the procedure unless there is some provision to circumvent that superior (Clelland, 1967; Goodfellow, 1972; Sibson, 1965).

With an understanding of why health care personnel join unions and some of the criteria by which labor organizers evaluate an institution, management should begin to assess its employee relations climate and to plan its strategy for maintaining nonunion status (Keeping Your Employees' Morale, Goodfellow, 1972; *Hospital Financial Management*, 1974).

A PREVENTIVE MANAGEMENT PROGRAM

Assessing an institution's employee relations climate and implementing a program to prevent unionization is a process for which a myriad of management responses are possible. Each institution must carefully design a strategy that is both practical and suited to its own particular situation. Recognizing the significant relationship between the reasons given by employees for joining unions and what an organizer looks for, there is substantial reason to believe that the primary causes of unionization include "communication problems" and the perception by employees of "unfair treatment."

Therefore, a preventive management program should be designed with a primary emphasis on improving communication and dealing with employees and related problems in an honest and fair manner. This emphasis is detailed in several ways in the following recommendations for establishing a preventive management program. These recommendations are an outgrowth of previously described employee-related issues and could serve as the general framework within which each management team builds its own strategy.

Nonunion Policy

If a health care institution intends to be nonunion, it should give careful consideration to the development and publication of such a policy. Good labor counsel should be consulted to assist in the development of an up-front nonunion policy and to advise the best alternatives for communicating the policy to all who wish to work at the institution. All prospective employees should be informed in the screening process and given written evidence of the institutional position regarding unions, along with other significant policies. The prospective employee then has the choice of whether to work for a nonunion institution. This, in itself, should be an indication of fair treatment. Management also should consider publishing the nonunion policy in the employee handbook for reference during orientation and other worker group meetings. This policy should include the following key points (Rutkowski and Rutkowski, 1984):

1. A resolution committing the administration to provide equitable treatment to all employees in their wages, benefits, hours, and conditions of employment
2. A resolution committing adequate funds and time to provide all managers with the information that they need to be effective in employee relations and knowledgeable in ways of avoiding unionization
3. A resolution committing administrators to the philosophy that each employee is important as an individual vital to the optimal functioning of the entire hospital team
4. A resolution expressing a commitment to oppose efforts of outside organizations to unionize employees

Personnel Selection

Management must have effective policies and procedures regarding selection of new employees. Prevention of labor-management problems begins with the proper matching of personnel to specific jobs. A good wage and salary program including job analyses, job descriptions (with performance objectives), and job evaluation is essential. If good procedures are used for selecting on the basis of both the individual's qualifications and the requirements of a specific job, the result is likely to be a better fit for the institution and the employee. Concurrently, the institution is likely to avoid many communication and morale problems. A fair wage and salary system provides at least a basis for establishing an objective employee evaluation system.

Employee Attitude Assessment

Employee attitude surveys, when conducted properly, can provide much valuable data to management at nominal costs. The method chosen should be simple to implement and should elicit concise employee responses. The result should be an accurate assessment of the topics surveyed, clearly differentiating between positive and negative attitudes.

Attitude surveying should be done on a planned, periodic basis so that employees perceive continual concern for their needs and management keeps abreast of fluctuations in worker attitudes. If this procedure is combined with efforts to obtain upward communication through formal or informal channels at all levels of the institution, the result should be a positive change in the attitudes of employees and the development of a management system for dealing with personnel problems before they become sore spots. Once attitudes have been assessed and problems identified, management should be ready to take corrective action, including an appropriate training program.

Probably the single most important part of the attitude measurement analysis process is communication with the employees about:

- Purpose
- How the data will be analyzed and used
- Confidentiality of the individual responses
- Feedback concerning the findings
- What changes, if any, they can expect as a result of their participation in the survey

Management should be careful not to make promises that cannot be fulfilled, but should make a strong effort to do whatever is possible to improve employee relations.

In summary, when management asks employees to take valuable time to participate in a survey, it is extremely important for them to feel that the administration values their input and is doing what it can to meet their needs.

Employee Training

Administration should examine its role and responsibilities in training employees as a function of management rather than as a staff function. If this self-examination indicates that management is assuming little, if any, responsibility for employee training, such abdication is very likely to be related directly to workers' perceptions of poor treatment. For employees to perceive fair, honest, or decent treatment, top management must make the commitment to assume responsibility for training and must transmit it down through all levels to first-line supervisors. This is necessary, for example, before management can develop an adequate performance appraisal and reward system that employees will consider equitable.

Once management has made the commitment to assume its training responsibility, it must determine what type of training program to implement. The following questions may provide evaluative insight into employee needs.

1. Are employee functions and responsibilities agreed upon and clear?
2. Do employees have the ability (technical training and experience) to do what is expected?
3. Do job descriptions contain specific performance objectives?
4. Do employees know what performance standards are being used to evaluate their work?
5. Is there a positive relationship between employee performance and reward?

Management implementation of an appropriate training program should have positive effects on employee attitudes and productivity and should be a major asset in eradicating the dead-end job syndrome.

Employee Value Systems

Management should recognize the different types of value system that exist among various employee groups in both professional and nonprofessional categories. Research has identified as many as seven different employee value systems, varying from tribalistic to existentialist (Hughes, 1976). Some examples of responses to the myriad of value systems and needs include flexible work scheduling, earned time programs, methods of job enrichment, and a cafeteria approach to fringe benefits. Management must develop a variety of imaginative ways to respond to the needs of multiple employee "families."

First-Line Supervisors

Management must recognize the importance of first-line supervisors in preventing serious labor problems. The logic is simple. First-line supervisors represent all of management in the operational contact with nonsupervisory personnel. If these supervisors do not have good management skills, the institution is inviting unionization. Frequently, a problem with first-line supervisors is manifested by the number of grievances filed involving situations that are either about or under the direct control of such persons. Management should evaluate the effectiveness of first-line supervisors' employee relations skills carefully and regularly. When deficiencies are found, management should either assist the supervisor through training or terminate the person, depending on his or her past record and potential.

Performance Appraisal

The institution should establish a performance appraisal policy that reflects management's desire to develop employees to their potential. If management behavior indicates anything else, workers are likely to perceive treatment by supervisors poor or unfair. Performance appraisal must be done honestly and on a regular basis to be effective in improving morale and productivity of all employees.

Management's avoidance of an honest appraisal of the nonproductive employee simply demonstrates to all workers that the reward system is inequitable or that the laggards receive the same rewards as those who are productive. This can be interpreted logically by productive employees as evidence that the nonproductive actually are rewarded more than the productive in relation to their effort. If this attitude prevails, management very likely is "teaching" its employees to move toward mediocrity and union thinking. The implementation of a good performance appraisal system depends largely on the management skills of the first-line supervisors. In other words, the appraisal system used is not nearly as important as the people (managers) who implement it. The best system is as weak as the people who operate it.

Disciplinary Policies and Procedures

Management must take great care in applying disciplinary policies and procedures consistently. Consistent and fair application normally can prevent unnecessary employee relations problems and grievances. One basic principle is that management should have "just cause" for imposing discipline. The definition may vary from case to case, but several basic tests can be applied to determine whether "just cause" exists for disciplining employees.

1. Was the disciplinary rule reasonably related to efficient and safe operations?
2. Were the employees properly warned of potential consequences of violating the rule?
3. Did management conduct a fair investigation before applying the discipline?
4. Did the investigation produce substantial evidence of guilt?
5. Were the policies and procedures implemented consistently and without discrimination?
6. If a penalty resulted, was it related to the seriousness of the event as well as the past record of the employee? (Did the punishment fit the crime?)

Some form of grievance procedure should be viewed as a part of any prevention program, since employees should be able to complain formally about perceived problems without fear of subjective reprisal. Although any grievance procedure is open to problems of interpretation and application, some basic factors can be applied equally in evaluating the system from the employees' perspective.

1. All employees should be able to understand the mechanics of filing a grievance and should know where they can go to ask questions about any step of the system. Thus the procedure should be written.
2. When employees file grievances, they expect prompt action. Promptness is one of the most important aspects of a grievance settlement, and failure to resolve the problem with reasonable speed is likely to lead to adverse employee feelings.
3. The first-line supervisor typically is the first step in a grievance procedure. When

that individual is perceived to be the problem, however, employees need to know that they can access the grievance machinery without going through the first-line supervisor. However, the employees should take every reasonable step to solve the problem with the immediate supervisor before going to someone else with the grievance.

When employees realize that a fair grievance procedure is available and when management is doing what it can to prevent unnecessary problems, the result should be a decreased number of complaints, fair and objective processing of those that are filed, and an employee feeling that management is concerned about employee needs.

Wages

The health care institution should be very careful to stay competitive with regard to wages and should compare its rates at least annually to similar institutions in the same geographical area. Frequently, wage survey data can be found that apply to the local area, but if this is not the case management should conduct its own survey. Even a sample survey of representative jobs will help keep the institution abreast of trend information. Of course, certain shortage points will have to be dealt with on a case-by-case basis and possibly more frequently than every year. Competitive wages are a necessary condition in any preventive management program, but it should not be concluded that being competitive in wages is sufficient for maintaining nonunion status.

As has been indicated, wages is only one of many factors that may enter into employee decisions to seek union help. Health care no longer is as far behind other industries in wages as it was 15 or 20 years ago, and indeed, wages may not be the major motivating factor for a significant portion of employees in a given institution. Although there may not be a great deal that management can definitively conclude from research regarding wages as a motivating factor, the folly of relying totally on competitive wages to prevent unionization can be illustrated best by a review of wage structures in institutions that have had union elections recently.

In summary, the absence of competitive wage levels (particularly in times of double-digit inflation) is a potentially severe problem, but the presence of good wage levels is not sufficient, in itself, to prevent unionization. This is particularly true in multidimensioned institutions that employ a diverse group of employees with a variety of value systems.

MANAGEMENT STRATEGY FOR REACTIONS DURING UNION ORGANIZING CAMPAIGNS

Although many "prevention" steps may have been implemented, managers should not be so naive as to believe that a union organization attempt cannot happen. An extremely important part of a preventive management program is to have a well-planned strategy for reacting if such an attempt does occur.

Brett's (1980) two-point conceptualization of employee reactions during a union organizing campaign holds important implications for both employers and unions.

1. An employer's antiunion campaign that attempts to persuade employees by emphasizing economic control over them and using fear tactics is unlikely to be successful.
2. The employer's most effective antiunion campaign stresses the desire to remain nonunion, provides factual information pertaining to working conditions, benefits, and so on, and indicates that a labor organization cannot guarantee what conditions will exist under union representation.

SUMMARY

Maintaining nonunion status is an attainable goal. Whether it will be achieved is related directly to the behavioral dedication of management in demonstrating its concern for meeting employee needs fairly and equitably. Although the material in this chapter is not all-inclusive and does not offer a formula to guarantee nonunion status, it is suggestive of management practices necessary to prevent communication problems and to avoid employee perceptions of unfair treatment.

The unionization process is highly situational and in some locations may be essentially inevitable. Nevertheless, a positive nonunion philosophy and a preventive management program usually should obviate the need for a labor organization. When employees do not perceive a need for union assistance, the probability is slim that they will elect to begin paying union dues.

REFERENCES

Brett, Jeanne M. "Why employees want unions." *Organizational Dynamics* (American Management Association), pp. 48, 49, 51 (Spring 1980).

Clelland, R. "Grievance procedures: Outlet of employee, insight for management." *Hospitals,* 41(18):60 (Aug. 1, 1967).

Goodfellow, M. "If you aren't listening to your employees, you may be asking for a union." *Modern Hospital* 113(4):88–90 (October 1969).

Goodfellow, M. "Checklist: How the union organizer rates your institution." *Risk Management,* pp. 1–6 (December 1972).

Hospital Financial Management. "Keeping your employees' morale up takes more than money." 28(11):24–29 (November 1974).

Hospital Progress. "Goals and trends in the unionization of health care professionals." 53(2):40–43 (February 1972).

Hughes, Charles L. *Making Unions Unnecessary.* New York: Executive Enterprises Publications, 1976, p. 43.

Imberman, A. A. "Communications: An effective weapon against unionization." *Hospital Progress,* 54(12):54–57 (December 1973).

Joiner, C. L., and K. D. Blayney. "Career mobility and allied health manpower utilization." *Journal of Allied Health,* pp. 157–161 (Fall 1974).

Lewis, H. L. "Wave of union organizing will follow break in the Taft-Hartley dam." *Modern Healthcare*, 1(2):25–32 (May 1974).

Matlack, D. R. "Goals and trends in the unionization of health professionals." *Hospital Progress*, 53(2):40–43 (February 1972).

Metzger, Norman. "Labor relations." *Hospitals*, 44(6):80–84 (March 16, 1970).

Metzger, Norman, and D. D. Pointer. *Labor-Management Relations in Health Services Industry: Theory and Practice.* Washington, DC: Science and Health Publications, Inc., 1972, p. 360.

Milliken, R. A., and G. Milliken. "Unionization—Vulnerable and outbid." *Hospitals*, 47(20): 56–59 (Oct. 16, 1973).

Phillips, D. E. "Taft-Hartley: What to expect." *Hospitals*, 48(13): 18a–18d (July 1, 1974).

Pointer, D. D. "How the 1974 Taft-Hartley amendments will affect health care facilities." *Hospital Progress*, 55(10):68–70 (October 1974).

Pointer, D. D., and L. L. Cannedy. "Organizing of professionals." *Hospitals*, 48(6):70–73 (March 16, 1974).

Pointer, D. D., and Norman Metzger. *The National Labor Relations Act—A Guidebook for Health Care Facility Administrators.* New York: Spectrum, 1975, p. 272.

Rakich, J. S. "Hospital unionization: Causes and effects." *Hospital Administration*, p. 10 (Winter 1973).

Reed, K. A. "Preparing for union organization." *Hospital Topics*, 48(4):30–32 (April 1970).

Rutkowski, Arthur D., and Barbara Lang Rutkowski. *Labor Relations in Hospitals.* Rockville, MD: Aspen, 1984, p. 3.

Sibson, R. E. "Why unions in the hospital?" *Hospital Topics*, 43(8):46, 48, 54 (August 1965).

Stanton, E. S. "The Charleston hospital strikes." *Southern Hospitals*, p. 39 (March 1971).

Stanton, E. S. "Unions and the professional employee." *Hospital Progress*, 55(1):58 (January 1974).

CHAPTER 16

Negotiating and Administering the Labor Relations Contract

Norman Metzger

Negotiators of collective bargaining agreements are skilled practitioners of an art that is little understood. Indeed, collective bargaining is an art, where personalities play a far more important role than any theoretical or academic formats that may be suggested by numerous writers. The art of negotiating has been called a neglected one, and it is far more complex than the mere resolution of the terms of an agreement (Cook, 1972).

Collective bargaining has been described as a poker game that combines deception, bluff, and luck; as an exercise in power politics; as a debating society marked by both rhetoric and name calling; and as a "rational process" with both sides remaining completely flexible (Dunlop and Hely, 1955). Probably all these characteristics at one time or another, in various combinations, are typical of collective bargaining and of negotiating any contract. The hallmarks of successful bargaining are more complex than trite descriptions of the process. In the final analysis, the charade itself is not as critical as the personalities involved in the bargaining, the realistic planning of strategy, and the commitment of top administration and trustees of an organization. These, then, are the hallmarks of successful bargaining.

Successful bargaining is built around and upon the following cornerstones:

1. Advance planning of strategies with pragmatism and minimal subjectivity
2. Selection of a principal spokesperson who is experienced in labor relations and has been delegated full responsibility for presenting management's position in the bargaining
3. Full authorization of management's principal spokesperson to bind management and to make a "deal."

LEGAL DEFINITION OF BARGAINING

Moving from the unilateral determination of policy into the arena of collective bargaining, administrations are faced with the need for a new life style. The National Labor Relations Act (NLRA) of 1935, amended as described in Chapter 15, requires an employer to recognize and bargain in good faith with a certified union, but it does not force the employer to agree with the union. You may, indeed, yield to the union's persuasions, but on the other hand you may also resist, provided you have given the union an opportunity to persuade you. The Taft-Hartley Act definition of collective bargaining is pertinent to further discussion of the mutual obligations:

> To bargain collectively is the performance of the mutual obligation of the employer and the representative of the employees to meet at reasonable times and confer in good faith with respect to wages, hours and terms and conditions of employment or the negotiation of an agreement, or any question arising thereunder, and the execution of a written contract incorporating any agreement reached requested by either party, but such obligation does not compel either party to agree to a proposal or require the making of a concession. [Section 8(d), 1935]

To participate in "good faith" bargaining, the employer must be prepared to receive the proposals of the union and to meet with the union from time to time to discuss such proposals. After an election has been held and a union has been certified as the bargaining agent for a specific bargaining unit, a request to meet is most often presented in a formal letter to the institution. Management of the health care institution is then obligated under federal law to bargain with the union, and to bargain in good faith. This duty, under the NLRA, to meet and negotiate with the representatives of a majority of one's employees, has been interpreted over the years in decisions of the National Labor Relations Board (NLRB) and of the courts. For example, it has been determined that management may not require a union to give rights it possesses as a condition of meeting with management to bargain (*LLR*, 3130, p. 7909).* An express intention not to agree at the onset of negotiations violates the act. This does not preclude "hard bargaining," which is considered to be bargaining in good faith. Except for an outright refusal to negotiate, bad faith is the strongest evidence of a refusal to bargain; indeed, factors indicative of bad faith of themselves frequently constitute refusals to bargain (*LLR*, 3085). It is unlawful to insist that the collective bargaining contract be subordinated to individual contracts or to demand the right to make unilateral changes (*LLR*, 3130.16, p. 7915). Management does not fulfill its obligation to bargain by bargaining individually with employees, or by offering them individual contracts when bargaining has been requested by the majority representative (*LLR*, 3130.6). It is also unlawful to fail to have a representative of management available for conferences with the union at reasonable times and places, or to neglect to appoint representatives with power to reach agreements. Both failures are a violation of the collective bargaining agreements of the act (*LLR*, 3105, p. 7874; 3110, p. 7881). In rejecting union proposals, the employer must submit counterproposals and attempt to reconcile the differences; to act otherwise is considered to be

---

*Throughout this section, references to paragraphs in interpretations found in the *Labor Law Reporter (LLR)* are given in the form shown above.

bad faith. If an understanding is reached, it is an unfair labor practice to refuse to reduce its terms to a written agreement (*LLR*, 3095, p. 7850).

The key then to satisfying the duty to bargain in good faith is approaching the bargaining table with an open mind, negotiating in good faith with the intention of reaching final agreement (*LLR*, 3115, p. 7888). The NLRB determines whether bargaining has been in good faith by the employer's *entire* conduct during the negotiations. A "take it or leave it" approach, a refusal to furnish information requested by the union during the negotiations, and an intensive communications campaign, designed to discredit the union with employees during the negotiations, all are considered unfair labor practices (*Guidebook to Labor Relations*, 1985). When responding to a union demand, the institution must be prepared to back up a rejection by providing relevant information to the union or by agreeing to be audited (*LLR*, 3135.70, p. 7941). The Supreme Court has held that unilateral changes during talks with the union are, in themselves, unlawful without proof of bad faith by the employer (*LLR*, 3143.38, p. 7959).

MANDATORY BARGAINING SUBJECTS

The NLRA states that when a request for negotiation is made by a union representing a majority of employees in an appropriate unit, the employer must bargain collectively with respect to rates of pay, wages, hours, or other conditions of employment, and with respect to questions arising under existing agreements [Sections 8(d) and 9(a)]. (National Labor Relations Act; Sections 8(d) and 9(a), 1935). The National Labor Relations Board has interpreted the term "wages" to include such items of value that may accrue to employees out of their employment relationship as wage rates, hours of employment, overtime, and work requirements. In addition, such mandatory items include procedures and practices relating to discharge, suspension, layoff, recall, seniority, discipline, promotion, demotion, transfer, and assignment within the bargaining unit. It also includes conditions, procedures, and practices governing safety, sanitation, and protection of health in the place of employment. Indeed, vacations, holidays, leaves of absence, and sick leaves are mandatory subjects of bargaining.

The category of mandatory subjects of bargaining has developed from a long line of NLRB and court decisions. Included below are some of these subjects and references to decisions establishing their mandatory nature:

1. Discharge of employees (NLRB v. Baj Chelder 120F2d574, 8LRRM723 (7th Cir. 1941))
2. Seniority grievances and working schedules (NLRB v. U.S. Gypsum Co. 94NLRB112, 28LRMM1015 (1951))
3. Union security and checkoff (NLRB v. Andrew Jergens 175F2d130, 24LRRM2096 (CA9, 1949) cert. denied, 338US827, 24LRRM2596 (1949))
4. Vacations and individual merit raises (NLRB v. Singer Manufacturing Company, 24NLRB444, 6LRMM405 (1940))
5. Retirement and pension and group insurance plans (NLRB v. Inland Steel Co. 77NLRB1, 21LRRM1310, enforced, 170F2d247, 22LRRM2505 (CA7, 1948))

6. Christmas bonuses and profit sharing retirement plans (NLRB v. Niles-Bemont-Pond Co., 199F2d713, 31LRRM2057 (CA2, 1952) and NLRB v. Dicten & Marsch Manufacturing, 29NLRB112, 46LRRM1516 (1960))
7. A nondiscriminatory union hiring hall (NLRB v. Tom Joyce Floors Inc. 353F2d768, 60LRRM2334 (CA9, 1965))
8. Plant rules on rest or lunch period (NLRB v. Miller Brewing Co. 166NLRB90, 65LRRM1649 (1967))
9. Safety rules, even though the employer may be under legal obligation to provide safe and healthful conditions of employment (NLRB v. Gulf Power Co. 384F2d822, 66LRRM2501 (CA5, 1967))
10. Institution-owned houses occupied by employees, as well as the rent paid for the houses (NLRB v. Hart Cotton Mills Inc. 190F2d964, 28LRRM2434 (CA4, 1951))
11. No-strike clauses binding on all employees in the bargaining unit (NLRB v. Shell Oil Co. 77NLRB1306, 22LRRM1158 (1948))
12. Insurance plans. Even though the employer proposed to improve the insurance programs and the expiring agreement contained no provisions concerning the plans (General Motors Corp. v. NLRB, 81NLRB779, 23LRRM1422 (1949))
14. "Most favored nation" clauses (Dolly Madison Industries decision of NLRB 74LRRM1230, Dolly Madison Industries, Inc., Richmond Diary Division, Richmond, Va. and Truck Drivers and Helpers Local 592, International Brotherhood of Teamsters, Chauffeurs, Warehousemen and Helpers of America, Case #5-CA-3475, June 2, 1970, 182NLRB147)
15. A "zipper clause" closing out bargaining during the term of the contract and making the contract the exclusive statement of the parties' rights and obligations (NLRB v. Tomco Communications Inc., U.S. Court of Appeals, 9th Cir., San Francisco, 97LRRM2660; National Labor Relations Board v. Tomco Communications Inc. 76-2178, January 16, 1978)
16. Inplant food services and prices, even where inplant food services are managed by an independent caterer (441US488, 101LRRM2222 (1979))
17. Subcontacting unit bargaining work (Timken Roller Bearing Co. 70NLRB500, 18LRRM1370 (1946) enf. denied on other grounds, 161F2d949, 20LRRM2204 (CA6, 1947))

VOLUNTARY BARGAINING SUBJECTS

The NLRB has distinguished between mandatory and voluntary bargaining subjects. Bargaining subjects of the latter type (i.e., those that may be proposed but not insisted upon as a condition to an agreement) include the following:

1. A clause making the local union the exclusive bargaining agent, even though the international union was the certifying agent (*42NLRB2034 v. Wooster Division of Borg-Warner 356US342* (1985))
2. A clause requiring a secret ballot vote among the employees on the employer's last

offer before a strike could be called (*42NLRB2034 v. Wooster Division of Borg-Warner 356US342* (1985))
3. A clause fixing the size and membership of the employer or union bargaining team (*31LRRM2422 American Newspaper Publishers v. NLRB* 73 Supreme Court 552)
4. A requirement that a contract must be ratified by a secret employee ballot (*38LRRM2574 NLRB v. Darlington Veneer Co.* 236F2d85 Court of Appeals, 4th Cir.), though the method of ratification is an internal union concern (*73LRRM2097 Lear Siegler Inc. v. UAW* Court of Appeals, 6th Cir. 19134).
5. A clause providing that a contract will become void whenever more than 50% of the employees fail to authorize dues checkoff (*38LRRM2574 NLRB v. Darlington Veneer Company*, 236F2d85 Court of Appeals, 4th Cir.)
6. A requirement that the union post a performance bond or an indemnity bond to compensate the employer for losses caused by picketing by other unions (*32LRRM3684, NLRB v. Local 264, Laborers (D&G Construction Co.)* 529F2d778) (49LRRM1831 Arlington Asphalt Company, decision of NLRB 136NLRB67 (1962))
7. A requirement fixing terms and conditions of employment for workers hired to replace strikers (19LRRM1199 Times Publishing Co., decision of NLRB 72NLRB128)
8. Benefits for retirees (78LRRM2974 Chemical Workers vs. Pittsburgh Plateglass Co. 404US157)
9. Interest-arbitration clauses calling for arbitration of disputes over terms of a new contract (*93LRRM3055 NLRB v. Columbus Printing Pressmen*, Court of Appeals, 5th Cir. (1976))

IMPORTANT CONSIDERATIONS THAT AFFECT THE BARGAINING MILIEU

A seasoned negotiator will be familiar with all the many forces, external and internal, that affect the bargaining environment and in the final analysis affect outcomes.

The personalities involved have a critical impact on the style and outcome. Acknowledging this, the selection of the principal negotiator for the institution—the spokesperson—is a threshold decision. Managements often talk about "union irresponsibility" and suggest that union leaders in exercising their responsibilities have a punishing effect on their members. Still another pervasive management criticism of "the other side of the table" is the disconnection of the union representatives from the rank-and-file employees (who are often termed "our" employees). It is essential that the management negotiators be fully aware of their counterparts on the union negotiating team. There should be an attempt at a meeting of the minds between the two principal negotiators. It is not uncommon for the tenor of the negotiations and the outcome as well to be positively or negatively affected by the personalities involved. Friendship of the negotiators is not necessary, but mutual appreciation of integrity is, indeed, critical.

The past history of management-labor relations in the institution will have a decisive effect on the bargain. The climate that is developed during the term of the prior contract through the grievance and arbitration mechanism—the shop steward system, the numerous interfacings between labor relations executives and labor union leaders—can affect the "trust" element so necessary in the bargaining room.

The general economic conditions and the competitive nature of the industry will be a backdrop for the bargaining of a contract. Information about economic pressures on the institution and the new market situation in health care must be exchanged freely in advance of the bargain with the union representatives. The winds of change are blowing in the health care industry. The marketplace is changing; union membership is declining in general in the country, if not in the health care industry. One publication paints the following picture:

> It is no secret that labor unions are losing power, but that doesn't mean companies will ride roughshod over workers. In fact, a remarkable surge of new laws and court rulings are (sic) expanding employees' rights. In all, a revolution is in the making. (*Business Week*, 1985)

Another factor impacting on the bargain is the experience of both parties during the life of the contract. A union that is losing membership, a union that is faced with massive layoffs, a union that is unable to fulfill the pressing needs of employees, will be ineffective in collective bargaining. A management that is uncommunicative, unresponsive to employee grievances, authoritarian in its management style, will be faced with an uncompromising union negotiating committee.

SELECTING A NEGOTIATING TEAM

Management must make certain decisions *before* selecting the members of its negotiating team. One of the first and most important decisions is the character of the approach to the union at the bargaining table. Will the hospital take a "hard-nosed" approach? Will it attempt to contain the union at each turn? Will it attempt to discredit the union during the bargaining sessions? Will it attempt to change major provisions in the collective bargaining agreement?

The decisive factor in determining the eventual settlement is the makeup of the negotiating team. Much depends on the individual skills and judgments brought into the bargaining arena by the negotiators. Logically the major responsibility for negotiating a contract should be with the management executive who has day-to-day responsibility for labor relations. In most institutions this is the director of personnel or labor relations. This person should be familiar with the bargaining unit (its longevity, its wage structure, its grievances and arbitrations over the years) and should understand and appreciate the needs of employees and the needs of the institution. Very often institutions do not employ experienced labor relations executives and, therefore, use labor attorneys to represent the institution in collective bargaining. Such attorneys are well versed in labor law and the realities of collective bargaining. The presence of an experienced labor attorney to direct, guide and, perhaps, plan the strategy for the administrators' position at the bargaining table is truly an asset. Many institutions use both a labor relations or personnel executive and a labor attorney, placing the latter in the role of adviser to the principal spokesperson. In any case, it is essential to have a labor attorney to review the proposed language of the contract before it is signed, whether this specialist is the institution's chief negotiator.

Line administrators normally offer advice and ideas before the negotiations begin. Some are included as members of the negotiating team. The chief executive officer or chief operating officer of the institution does not usually serve on the negotiating committee. It is well to note that the introduction of critical demands and arguments by the union may require the management negotiating team to confer with its principals (i.e., the chief executive officer, the chief operating officer, or the chief financial officer).

It is not unusual to have a key department head on the negotiating team. The smaller the number of members on the negotiating team, the more effective the negotiating.

Most union negotiators are skilled practitioners of their art—at the very least, the union officials. They are well versed in negotiating techniques. The union negotiating team may include the president of the local, vice presidents, and employees of the institution who have been elected by their fellow workers to represent them in the negotiations. This committee is often comprised of the delegates who have been elected to handle the day-to-day problems and, therefore, are well versed in the grievances and arbitrations of preceding years. More often than not they are the institution's most outspoken and militant proponents of the union, and more often than not they were instrumental in bringing the union into the institution. Many union negotiating teams include a labor attorney, who represents the union. In most instances, the principal spokesperson for the union is the local president.

STRATEGIES FOR BARGAINING

At the outset the institution must decide on the issues that cannot be compromised. Strategy must be planned in advance, and experienced labor relations practitioners must supervise strategic planning and actual conduct of the negotiations.

The preparation for bargaining begins long in advance of the actual face-to-face sessions. Management must gather and organize material obtained both from within the institution and from similar institutions and other firms in the community. Such information should include the following:

1. The present wage rates operative in the institution, classification by classification
2. Job descriptions, when available
3. A complete review of the fringe benefits program, showing costs and areas amenable to savings
4. The total number of employees, by classification, in the bargaining unit
5. Hourly schedules for each classification
6. Average amount of overtime by classification
7. Average straight time hourly wage rates
8. Rates of employee turnover by department
9. Seniority lists (i.e., number of employees with length of service in the 5, 10, 15, and 25 year classes)
10. Analysis of experience with grievances and arbitrations

It is also essential to assemble collective bargaining agreements in effect in other institutions in and out of the area, which may be used by either party to the negotiations. Most important are contracts recently negotiated by the same union.

Bade and Stone (1951) provide an excellent list of dos and don't's.

1. Strategy must be planned in advance. Do not play it entirely by ear. Clear-cut decisions must be made as to those issues that (a) cannot be compromised, (b) can be compromised (and to what extent and in exchange for what), and (c) merely represent an antidote to anticipated overreaching in union demands, which one recognizes will be dropped when the union does the same thing with its extreme demands. As part of the preparations, a pragmatic anticipation of union demands should be completed.

2. Do not start with the hard issues and leave the easy ones for the end. It is best to set a mood of compromise. Do not emphasize technicalities and legalisms at the onset. Develop a mood that is conducive to give and take (the hallmark of successful negotiators is the understanding that for every take there may need to be a give).

3. Do not be subtle, pedantic, threatening, or hesitant. Use the right language. Be direct, clear, *calm*, patient, and tolerant. Of course, it may be necessary to play to the bleachers; if so, do not get lost in the feigned emotion.

4. Do not exaggerate or misrepresent the facts. A fact is a fact, and there is no substitute for honesty. It has been said many times that although a union might be able to get away with deception, management cannot. Management's position must always be factually defensible. It may not be the one the union wants management to have, but if it is supported by facts and is rational, it is the only one that should be taken.

5. When responding to the other side's positions, give reasons. A reasoned "no" includes the sharing of how decisions were reached.

6. Do not make commitments at the table that you do not intend to keep. Do not hide behind tricky, vague, or inconclusive language. If making a commitment is not possible, do not gloss over it with murky language.

7. Keep control of the negotiations. It is always best to make proposals or counter-proposals the basis of future negotiations. If an issue gets too difficult, too hot to handle, put it aside for the time being.

8. When there is agreement on a clause or issue, translate the agreement into actual words that both parties can agree on.

9. Do not agree on anything until *everything* has been agreed on. Until the entire contract is negotiated and signed, make sure the union understands that agreement on anything before the entire contract settlement should be considered tentative.

10. Thinking of quid quo pro as the order of the day may well be the backbone of successful bargaining. A granting of a concession by management should be related to the granting of an equal concession by the union. Withdrawals of proposals should be mutual. If one party will not listen to the needs of the other it should be made clear that the party's needs will not be listened to.

11. The myth of "final offers" must be debunked. Never describe a position as the final offer *unless it is*. Never take the position that "this is as far as management will go," then go forward while the pronouncement is still clear in the minds of the union.

SOURCE: Bade, W. J., and Stone, M. *Management Strategy in Collective Bargaining Negotiations*. New London, CT: National Foremen's Institute, 1951. Copyrighted material reprinted with permission of Bureau of Business Practice, 24 Rope Ferry Road, Waterford, CT 06386.

An important part of planning the strategy for negotiations is a clear, objective estimate of strike issues. Are there issues which are likely to be critical ones for the union, thus becoming instigators for a strike? By identifying such issues an institution need not change its position. Reality-based negotiations are productive ones.

It is essential to forecast the impact of a possible strike. In any industry strikes put economic pressure on both parties: The workers lose wages while the employers lose revenue. The key to a successful strike from a union's viewpoint is to inflict inordinate discomfort, expense, and pressure on the employer to effect a compromise or a move toward the union's position. An institution must carefully evaluate the "discomfort," the "expense," and the "pressure" it will be called on to withstand if it takes a strike. The real losers in strikes of health care institutions are the patients and their families, and prospective patients. The patients may well be deprived of services; they may need to be moved from a struck hospital or nursing home; they may be discharged prematurely. Prospective patients will be troubled by the limited beds available; operations will be delayed, and outpatient care discontinued. It is well to state at the onset of discussions of the impact of a strike that such action at a health care facility is the most severe form of labor-management dispute. Many strikes produce mass picketing and still more produce violence.

There are, in any discussion of the impact of a strike, critical factors: the ability of the health care institution to withstand the strike, and the willingness to take a strike. There are at least eight critical indicators that must be evaluated when estimating the impact of a strike in a health care institution.

1. What effect will it have on revenue—will lost revenue be recoverable in the poststrike period?
2. How long will a strike be acceptable—is there a critical point at which pressure on the institution will be unbearable?
3. What support will be available to the institution to make up for the employees who are withholding their services (supervisors, non-bargaining-unit employees, temporaries, bargaining-unit employees who will cross the picket line)?
4. Can strikers be replaced? A policy of whether the economic strikers are to be replaced with permanent new hires must be decided.
5. Will striking employees be able to augment or replace their lost income by finding temporary employment elsewhere?
6. What is the union's policy on strike benefits?
7. What is the hospital's policy on discontinuing benefits coverage for strikers? Health benefits may be covered under a union plan or under the institution's own plan; what is the policy on discontinuing coverage for striking employees?
8. What outside forces may be brought to bear on the institution to avoid or settle a strike?

One other consideration is the use of employees from nonstruck hospitals. The Allied doctrine is a legal doctrine developed from the National Labor Relations Board case law, defines the rights of third parties who provide assistance to the employer involved in a labor dispute. The Allied doctrine affects a secondary employer who, during the course of a labor dispute, performs work that would have been performed by striking employees of the primary employer. In doing such work the secondary employer loses neutral status and, therefore, is subject to the labor organization involved in the dispute, thus extending its economic activity to the secondary employer. As the reports of both houses of Congress (U.S. Congress, 1974) in the deliberation regarding the Health Care Amendments to the NLRA stated:

It is the sense of the Committee that where such secondary institutions *accept the patients of the primary employer,* or otherwise provide life-sustaining services to the primary employer, by *providing the primary employer with an employee or employees who possess critical skills such as EKG technician,* such conflict shall not be sufficient to cause the secondary employer to lose its neutral status . . . (Emphasis added)

In effect Congress intended to affirm that a neutral hospital accepting patients of a primary employer will not lose its status as a neutral. Such a neutral hospital will, however, lose its status if it supplies *noncritical* personnel to a hospital that is experiencing a strike, or if it not only accepts patients from such a hospital but also greatly expands its noncritical staff in the process. Gradually the NLRB's exception to the Allied doctrine for the health care industry has come to depend on the *urgency* of the medical needs of the patients who were transferred from the primary hospital to the neutral hospital (Metzger et al., 1984).

THE 1974 HEALTH CARE AMENDMENTS

The Taft-Hartley Act of 1947 excluded from the definition of "employer" private, not-for-profit hospitals and health care institutions. The NLRB asserted jurisdiction over proprietary hospitals and nursing homes, but it was not until the so-called Health Care Amendments of 1974 that Congress, through PL 93-360, brought the private, not-for-profit health industry within the jurisdiction of federal labor law. The 1974 amendments enacted the following changes.

1. The exemption contained in Section 2(2) of the federal statute that excluded not-for-profit hospitals from the definition of "employer" was removed.

2. A new Section 2(14) was added to define the term "health care institution." It included any "hospital, convalescent hospital, health maintenance organization, health clinic, nursing home, extended care facility or other institutions devoted to the care of sick, infirm or aged persons." This definition is essential in the determination of which employees fall under the special health care provisions and within the scope of the act.

3. A new series of special notices applicable to the health care industry and unions representing employees in that industry was enacted. This Section 8(d) notice period for notices of disputes, which one party is legally obliged to give to the other, was extended from the normal 60 days before contract expiration to 90 days, and notices filed with the Federal Mediation and Conciliation Service (FMCS) from the normal 30 days to 60 days. In the case of initial contract disputes notice must be given 30 days before any strike notice.

4. A new Subsection 8(g) was added. It requires that a union representing employees in a health care institution must give 10 days' written notice to the employer and to the FMCS of its intent to engage in a strike, picketing, or other concerted refusal to work.

5. Broader sanctions under Section 8(d) provide that employees who are represented by labor organizations and do not comply with the requirements of the 90- and 60-day dispute notices or with Section 8(g), Strike Notices, lose their protected status under the act.

6. Mandatory mediation of disputes in the health care industry were provided

under Section 8(d). The FMCS must mediate health care disputes, and the parties involved in such disputes are compelled to participate in that mediation process. It is an unfair labor practice for a party to refuse to participate in such mediation.

7. Section 213, the second special dispute resolution provision, provides that when substantial interruptions of delivery of health care in a community are threatened, the director of FMCS may appoint a special board of inquiry to investigate the issues in the dispute and to issue publicly a written report on the dispute.

8. A new Section 19 provides for an alternate to the payment of union dues by persons with religious convictions against making such payments. It allows contribution to designated 501(C)(3) charities in lieu of dues.

In an interpretation by the Office of the General Counsel of the NLRB, the following guidelines have been established regarding Section 8(g) notices.

1. The notice should be served on someone designated to receive such notice, or through whom the institution will actually be notified.
2. The notice should be personally delivered or sent by mail or telegram.
3. The 10-day period begins upon receipt by the employer and the FMCS of the notice.
4. The notice should specify the dates and times of the strike and picketing, if both are are being considered.
5. The notice should indicate which units will be involved in the planned action.

As with Section 8(d) notices, workers engaged in work stoppage in violation of the 10-day strike notice lose their status as employees. The NLRB will probably interpret violations of the Section 8(g) notice requirement as a separate and distinct unfair labor practice.

In drafting the 1974 amendments to the NLRA, the congressional committee included the 10-day strike and picket notice provision [Section 8(g)] to ensure that health care institutions would have sufficient advance notice of a strike. The committee realized, however, that it would be unreasonable to expect a labor organization to commence job action at the precise time specified in a notice provided to the employer. On the other hand, if a labor organization failed to act within a reasonable period after the time specified in the notice, such action would not be in accordance with the intent of the provision. Therefore, the committee report of the amendments provided that:

> . . . it would be unreasonable, in the committee's judgment, if a strike or picketing commenced more than 72 hours after the time specified in the notice. In addition, since the purpose of the notice was to give a health care institution advance notice of the actual commencement of a strike or picketing, if a labor organization does not strike at the time specified in the notice, at least 12 hours notice would be given of the actual time for commencing of the action.

Thus, absent unusual circumstances, a union would violate Section 8(g) if it struck a facility more than 72 hours after the designated notice time unless the parties agreed to a new time or the union gave a new 10-day notice. Additionally, if the union does not start the job action at the time designated in the initial 10-day notice, it must provide the health care facility at least 12 hours' notice before actual commencement of the action. The 12-hour warning must fall totally within the 72-hour notice period.

The committee report notes that ". . . repeatedly serving 10-day notices upon the employer is to be construed as constituting evidence of a refusal to bargain in good faith by a labor organization," that is, a violation of Section 8(b) of the NLRA. What constitutes ". . . repeatedly serving notice" will have to be defined and interpreted by the NLRB in individual cases. In a memorandum, the board's general counsel provided the following guidelines to regional offices regarding the handling of intermittent strikes or picketing situations:

1. Where the facts and circumstances of the labor organization strike or picketing hiatus support the reasonable conclusion that the activity has not indefinitely ceased and that it is reasonable to assume that it will commence again, no new notice will be required if the activity recommences within 72 hours of the start of the hiatus; but 12 hours' notice to the institution will be required if the activity is to recommence more than 72 hours from the start of the hiatus.
2. Where the facts and circumstances of the hiatus support the reasonable conclusion that the activity has ceased indefinitely and that it will not be resumed in the near future, 12 hours' notice to the institution will be required if the activity is to resume within 72 hours of the start of the hiatus, but a new 10-day notice meeting all the requirements of Section 8(g) will be required if the activity is to resume more than 72 hours from the start of the hiatus.

Exceptions to the requirements that labor organizations provide Section 8(g) notices are indicated in two situations. First, if the employer has committed serious or flagrantly unfair labor practices, notice would not be required before the initiation of the job action. Second, the employer is not allowed to use the 10-day notice period to "undermine the bargaining relationship that would otherwise exist." The facility would be free to receive supplies, but it would not be "free to stock up on the ordinary supplies for an unduly extended period" or to "bring in large numbers of supervisory help, nurses, staff and other personnel from other facilities for replacement purposes." The committee reports held that employer violation of the foregoing principles would release the union from its obligation not to engage in a job action during the Section 8(g) notice period (Metzger et al., 1984).

KEY CONTRACT CLAUSES

Among the many clauses that will be included in a collective bargaining agreement with the union, four have specific interest to any practitioner (for a full list of clauses see Appendix A). These clauses are:

1. Union security
2. Management rights
3. Seniority
4. No-strike

Union Security

Once a union has been certified as the collective bargaining agent for the employees in a hospital, one of its primary aims is to get a contractual provision that gives it (the union) maximum protection in its continued existence in that hospital. Therefore, the union attempts to secure by bargaining some form of compulsory union membership. Six forms of union security will be defined: an open shop, a maintenance of membership shop, a union shop, a modified union shop, a closed shop, and an agency shop.

The *open shop* offers a maximum choice to employees in the bargaining unit. They can join or not join the union, and they can remain in the union or drop out of the union without losing their jobs.

In the *maintenance of membership shop*, members of the bargaining unit are not required to become members of the union to keep their jobs, but employees who do decide to join the union must maintain membership for a specific length of time: either one year or for the duration of the contract. Such a union security arrangement often includes an escape clause that clarifies the times at which union members may drop out of the union without losing their jobs. The following is a typical maintenance of membership clause:

> Any employee who is a member in good standing of the union at the end of thirty (30) days from the date the provision becomes effective or who thereafter joins the union during the term of this agreement shall remain a member of the union in good standing as a condition of employment with the hospital.

Another form of union security is the *union shop*. It is found in the majority of collective bargaining agreements and provides that all employees in the bargaining unit must join the union within a specified period of time after hire, usually at the end of a probationary period. In addition, these employees must remain members in good standing during their employment in the hospital. A sample union shop provision follows.

> All employees on the active payroll at the time of the signing of the contract who are members of the union shall maintain their membership in the union in good standing as a condition of continued employment. All employees on the active payroll as of the time of the signing of the contract who are not members of the union shall become members of the union within thirty (30) days after the effective date of the contract. All employees hired after the effective date of the contract shall become members of the union no later than the thirtieth day following the beginning of such employment and shall thereafter maintain their membership in the union in good standing as a condition of continued employment.

A *modified union shop* provides an option to employees presently on the payroll. Those who are members of the union must maintain their membership, but those who have not joined the union need not join. New employees, however, must join the union within a specified period after hire, again usually after the probationary period. This form of union security is a compromise that protects present employees who do not wish to join the union. The movement from a modified union shop is sometimes affected by a clause stating that after a certain percentage of employees have joined the union, all those remaining out must join the union.

The *closed shop* (which was largely outlawed by the Labor-Management Relations Act of 1947 but is permissible under certain state laws for hospitals that are so covered) requires applicants for employment to be members of the union before they can be hired. This form of union security is usually accompanied by a union hiring hall provision. The union is notified of employment needs and sends applicants to fill the jobs. A sample closed-shop clause follows.

The hospital shall hire only applicants for employment who are members of the union. The union shall furnish such applicants for employment provided that, however, if the union is unable to fill such request, the hospital may hire applicants who are not members of the union, but such applicants must become members of the union immediately upon being hired.

The *agency shop* was a comparatively rare form of union security that had a rebirth as a counterposition of unions to right-to-work laws (operative in several states), which outlaw mandatory or compulsory union membership. In an agency shop the employee can join or not join the union and remain a member or drop out, but all employees in the bargaining unit that is serviced and represented by the union must pay a service fee to support the union as a condition of employment.

In all forms of union security, whether they are partial or total in compelling employees of the bargaining unit to join the union, such employees are covered by all conditions of the contract if they are indeed part of the bargaining unit. If the employees are given the option to remain out of the union, they are nonetheless covered by all provisions of the collective bargaining agreement including the grievance procedure; however, they may not attend meetings and they may not vote on union issues.

Labor leaders argue in favor of compulsory union membership, and their principal argument revolves about the "free rider" inequity. A "free rider" is a worker who refuses to join the union and is permitted to maintain this position based on a union security clause. This worker, the labor leaders argue, reaps all the advantages hard won by the union in its negotiations with management. The union leaders maintain that the philosophy of compulsory union membership is rooted in a basic democratic principle, the rule of the majority. It is their position that given free choice, the employees will join the union, but usually remain out because of fear of reprisal or promise of reward from the employer. Management defends its position for optional membership on the basis of the inherent right of the individual to make choices and to withhold membership from any organization. Actually agreement to a compulsory union security provision offers some advantages to management. Often when employees are given the option of joining or not joining a union, only the most outspoken critics of management join immediately. In a modified union security arrangement, however, the union continues its drive to enroll employees, often undertaking this activity during the normal workday and throughout the year. A compulsory agreement, of course, not only finesses such militance, but may also mollify the extremists, who are unlikely to be influenced by more conservative employees in any event.

Management Rights

Management and labor have usually agreed that any rights not restricted by the collective bargaining agreement reside in management. The question of whether management rights clauses should be expressly included in a contract has been hotly debated over the years. Most managements feel that the right to manage the business or administer the hospital is solely theirs. *But it should be clear that when the administration enters into a collective bargaining agreement with a union, it no longer has the sole authority to administer the hospital.* There are limitations clearly outlined in each of the contract clauses.

The Supreme Court has ruled that management prerogatives can be exercised only in the cases "over which the contract gives management complete control and unfettered discretion." The court's decision points out that management retains only those rights specifically outlined in the labor agreement. Almost three-quarters of all collective bargaining agreements contain a management rights clause. For years many practitioners presumed that the employer could exercise the right to manage the enterprise and that such a clause was unnecessary. They further warned that in an effort to construct a management rights clause that would outline those rights reserved for management, it was impossible to anticipate all the contingencies that should be included. In supporting their position against the inclusion of the management rights clause, some practitioners felt that in negotiating such a clause, which was indeed the prerogative of managements in any case, they would have to grant certain concessions to the union. Yet, the majority of labor negotiators were convinced by the arguments for inclusion. They felt that the institution needs as much protection as it can get and that the clause outlining management prerogatives is most helpful in adjudicating day-to-day grievances. The battle over the pros and cons of including a management rights clause in a collective bargaining agreement has become an academic one in light of the decisions of the Supreme Court.

The collective bargaining agreement does indeed limit management's action in the cases outlined in each of its clauses. Management does well to define those areas in which union rights are waived, since most arbitration awards and court decisions have indicated that any management decision or action that affects the employment relationship and is not clearly reserved to management should be discussed with the union representing the employees. If management wishes to retain its freedom to act in specific areas without recourse to the grievance and arbitration procedure, the management rights clause is the proper place to define those areas.

There are two major categories of management rights clauses. One is a brief clause dealing not with specific rights but with the principle of management rights in general. The other is a detailed clause that clearly lists areas of authority that are reserved to the management. The key areas of authority that would be included in such a clause involve the rights to:

1. Manage and administer the hospital and direct the work force
2. Hire, discipline, and transfer
3. Introduce new or improved methods or facilities
4. Promulgate rules of conduct
5. Set quality standards
6. Discontinue jobs
7. Decide employee qualifications
8. Subcontract work

An example of a brief encompassing clause follows.

The management of the hospital and the direction of the work forces are vested exclusively with the administration, subject only to the restrictions and regulations governing the exercise of such rights as are expressly provided in this contract.

Seniority

One of the most important clauses in the collective bargaining agreement and one that often cuts deepest into management prerogatives is the seniority provision. The principle of seniority should no longer be at issue between management and the union. Unilateral interpretation and application of seniority, on the part of management, is no longer possible. Under the seniority principle, the employee with the longest service is given preference in such areas as promotions, transfers, layoffs, rehirings, choice of shifts, and choices of vacation periods. Unions tend to seek straight seniority, recognizing no factor other than period of service in gaining preference. To them, strict seniority means job security and the basic protection against the possible biases or acts of favoritism on the part of the administration. They maintain that with the application of strict seniority measures, a more stable, experienced, efficient, and loyal work force will develop. Of course the administration argues that operating efficiency requires compromise on the position of straight seniority. Management would prefer more emphasis on merit and ability. In its earliest application the National Railroad Adjustment Board held that seniority is a personal right and "the keystone upon which many rights of individuals under collective bargaining agreements are based."

The right to preference on the basis of seniority exists only by virtue of contractual clauses contained in the collective bargaining agreement. Hospital negotiators to focus on two main areas when bargaining a seniority provision, namely: the role of ability as it modifies the strict application of seniority, and the definition of the seniority unit. The first area has dramatic implications in instances of promotions, layoffs, and rehires. Many agreements provide for a qualifying clause, either the "seniority plus ability test" or the "relative ability test." The *seniority plus ability test* provides preference in promotions or other job rights on the basis of seniority, but the most senior employee must be able to do the work involved: Just being able to do the job—not necessarily better than other employees—will suffice under this provision. The *relative ability test* provides for a comparison of the abilities and skills of the employees claiming rights to a job on the basis of seniority. Here, in instances of layoffs, recalls, promotions, or demotions, ability and seniority will be considered, and where ability is relatively equal, seniority will be the governing factor. If ability is not relatively equal, however, the more able employee will be granted preference.

The definition of the seniority unit is key to the application of the provisions of the seniority system. The broader the unit, the more difficult for the administration to maintain its control over the efficient operation of the hospital. The narrower the unit, the more strict is the limitation on the exercise of an employee's seniority rights. Where there is a wide variation in skill requirements for jobs (such as is evident in a hospital), it is more practical to negotiate a departmental or occupational unit. Departmental seniority, for example, sets up separate seniority units for each department and is applied within an employee's specific department. Since the unit is smaller and contains many interchangeable jobs, the advantages are obvious.

The most efficient seniority unit for a hospital is occupational seniority, that is, separate units for employees performing the same type of job. This is far more practical for departments that contain a broad spectrum of noninterchangeable occupations. Hospital unions will usually agree to a compromise between the limited occupational seniority unit and the broader departmental system. This approach is sometimes called "noninterchangeable occupation groups." When the occupational groups to which seniority may be applied have been defined, the groups are broadened to include occupations that are clearly interchangeable and, therefore, should form one occupational family.

No-Strike and No-Lockout Clauses

A no-strike clause is probably one of the most important provisions of the agreement for the hospital. It is the administration's assurance of peaceful and uninterrupted operation during the life of the contract. No-strike clauses fall into two categories: unconditional clauses and conditional or contingent clauses. An *unconditional no-strike clause* elicits a firm pledge from the union and its rank-and-file members that they will in no circumstances be involved in a work stoppage or slowdown during the life of the agreement. A *conditional or contingent no-strike clause* prohibits stoppages or slowdowns unless certain clearly defined conditions have been met, such as the exhausting of the grievance procedure including arbitration and the refusal of either party to abide by the decision of an arbitrator. These clauses often provide that strikes or slowdowns are prohibited in matters subject to arbitration and that the union must take certain positive action against any unauthorized (wildcat) strikes if it is to avoid any liability. Such actions may include informing its membership in writing that they must return to work and posting notices on union bulletin boards or taking ads in local newspapers stating that the strike is unauthorized and workers should return to work. A union that calls workers out in violation of a no-strike pledge in a contract may be sued for damages the employer suffers as a result. Nearly all no-strike clauses in collective bargaining agreements contain a quid pro quo for the union, a comparable ban against lockouts by the company.

ADMINISTERING THE CONTRACT

Once a collective bargaining agreement has been completed and negotiations concluded, breathing life into the agreement is a key responsibility of the parties. This includes the disciplinary, grievance, and arbitration procedures.

Discipline

The collective bargaining agreement imposes limitations on the disciplinary powers the administration may exercise. However, any limitations of a management right may be counterproductive to the primary goal of establishing an efficient work force. To minimize such limitations, an effective and constructive approach to discipline must be designed and implemented. The limitations imposed on management's right to discipline do not, in themselves, remove from management the right to impose discipline on its employees. The key point about such limitations is that management *may* discipline up through discharge only for sufficient and appropriate reasons. This requires the development of a sound procedure based on due process in the application of discipline toward unionized employees. The right to discharge, suspend, or discipline is clearly enunciated in contractual clauses and in the adoption of rules and procedures, which may or may not be incorporated in the collective bargaining agreement.

The union's role in the disciplinary process is to vigorously defend the employee in the face of what could be considered unfair management actions. Employees expect the union to come to their assistance. It is the heart of the disciplinary process to ascertain the propriety of the management action. The best way to judge the propriety of a management action in a disciplinary case is to "program" the case to its ultimate conclusion in arbitration. Arbitrators will normally support a management action if they find progressive discipline, which includes verbal reprimands, full explanation of what is necessary to remedy the situation, followed by a written reprimand for a second infraction with a clear warning of future penalties that may be imiposed. Suspensions may precede the final disciplinary action of termination. In essence, the review of a disciplinary action, which terminates in arbitration, should be directed to ascertaining whether the employee was fully aware of the standards against which his or her behavior was to be measured. These standards include basic rules and regulations outlining offenses that will subject employees to disciplinary action, and the extent of such disciplinary action. The critical question in all disciplinary actions is: Is there sufficient reason for such action? Frequently supervisors find themselves on the defensive after having administered discipline, and often such actions are overturned because of insufficient evidence or improper remedies, which do not fit into the concept of progressive disciplining. To the arbitrator, discharge is economic death; therefore, the burden of proof lies with the management that such action was the only proper avenue of recourse.

Discipline to be effective—to be fair—must be corrective in nature. Justin (1969) lists some basic rules of proper disciplinary procedures.

1. Discipline to be meaningful must be corrective, not punitive. Corrective discipline encourages the wrongdoer to correct himself and leads to self-discipline.
2. When you discipline one you discipline all.
3. Corrective discipline satisfies the rule of equality of treatment by enforcing equally among all employees established plant rules, safety practices, and the responsibility of the job.
4. It is the supervisor's job to make the workers toe the line, or to increase efficiency, not the shop steward.
5. Just cause or any other comparable standard for justifying disciplinary action under the labor contract consists of three parts:
   a. Did the employee breech the rule or commit the offense charged against him?
   b. Did the employee's act or misconduct warrant corrective action or punishment?
   c. Is the penalty just and appropriate to the act or offense as corrective punishment?
6. The burden of proof rests on the supervisor to justify each of the three parts that make up the standard of just cause under the labor contract.
7. Prove the misconduct or offense charge by sensory facts. Avoid opinions, feelings, or conclusions.
8. In drafting rules or writing disciplinary letters, avoid stating opinions, inferences, or conclusions. Base your rule upon neglect of job; support your disciplinary action by giving the facts of how, when and where the employee neglected his job.
9. To prove cause for disciplinary action, the circumstantial evidence must permit two reasonable inferences or inescapable conclusions to be drawn:

a. Does the circumstantial evidence point to or compel one reasonable finding or conclusion?

b. Does the circumstantial evidence exclude all other reasonable findings or conclusions? (Pp. 294–295)

It is imperative that management keep a record of each reprimand, warning, layoff, and discharge. Annotations of verbal warnings should be maintained by the supervisors in the dpeartment. The procedure for written warnings should include the writing of warnings as soon as possible after the incident. It is important to note on the warning notice the rule violated, or the specific provision of the union contract in question. Explanations should be specific and comprehensive. One helpful device on a warning notice is a special notation, where indicated, that immediate satisfactory improvement must be shown and maintained or further disciplinary action will be taken. The specific action to be taken by management in the event of further violations should be clearly noted on the form.

The right to discipline or discharge for proper cause and the right to make rules and regulations governing the conduct of the employees are primary management prerogatives. These rights notwithstanding, the union will attempt to protect the disciplined employee. In the main, collective bargaining agreements permit the employer to discipline or discharge for cause, but they also permit the union to protest such actions through the grievance and arbitration procedure. Therefore, it is essential that "just cause" be established, that the reasonableness of the action be clear, and that the documentation be complete.

Administering the Grievance Procedure

The grievance procedure is the real heart of the collective bargaining agreement. It is a useful and productive management tool when it makes use of the finding of facts, objective evaluations, and equity. Consistent and fair adjudication of grievances is the hallmark of sound employee-employer relations.

Most grievance procedures contain four steps. The first step always involves the presentation of the grievance by the employee and/or his or her representative to the immediate, first-line supervisor. This is an essential cornerstone of a grievance procedure. Since most grievances refer to actions by first-line supervisors, a second step must be taken outside the department or at a higher level in the department. The second step involves the employee, his or her representative where indicated, and a department head or an administrator from outside the department. The underlying rationale for the third step is the final in-house review of the management decision by an individual outside the department with the twofold responsibility of insuring objectivity and evaluating the need for consistency. The third step involves the employee and/or his or her representative and a union official and, in most instances, a labor relations representative of the personnel department. To ensure such objectivity and the evaluation of the need for consistency, it is important to understand the following caveats.

1. Objectivity and consistency cannot be achieved if the last step remains at the same operating department from which the grievance arose.

2. Objectivity and consistency cannot be achieved if the management representative at the third step has neither the power nor the inclination to overrule an improper decision.
3. Objectivity and consistency cannot be achieved if anyone representing management is perceived as other than neutral.

Six basic areas make up the manager's responsibility for the grievance procedure:

1. Hearing the complaint
2. Getting the facts
3. Making the decision
4. Communicating the decision
5. Preparing a written record
6. Minimizing grievances

The management hearing officer should be guided by the following general principles for effective grievance handling:

1. A strong desire to resolve dissatisfaction and conflicts
2. An empathy toward employees, an understanding of their problems, and an ability and willingness to listen and probe for hidden agendas
3. A complete knowledge and understanding of personnel policies, procedures, and practices, and the union contract
4. A personal commitment to the interest of the institution, side by side with a sense of fair play on behalf of the employees

Very few grievances can be adjudicated as soon as they arise. Indeed, the critical nature of sound administration of the grievance procedure is a thorough investigation rather than a quick evaluation. It is essential in dealing with employees' grievances that management gain employee commitment and reduce employee dissatisfaction. Therefore, the facts are the key. By uncovering the facts it is possible to discover the underlying causes of grievances, rather than their surface appearances.

Rules for Grievance Administration

The following checklist is provided as a guide for managers involved in administering the grievance procedure (Metzger, 1979):

1. *Listen.* Permit the presentation of the full story by the employee and/or the delegate.
2. *Try to understand.* Uncover the how, who, what, when, where, and why of the grievance.
3. *Separate fact from emotion.* This requires painstaking investigation.
4. *Refer to policy and contract provisions.* These are the rules of the road. Do not attempt to rewrite policy and the bargaining agreement at this stage of the game. The role of the supervisor is to interpret such policy and contract provisions as they pertain to the grievance at hand.

5. *Remember that your decision may set a precedent.* A decision in one case has a direct bearing on subsequent cases.
6. *Consult with others.* It is not a sign of weakness to check with other supervisors who may have had similar grievances, and to check with the personnel department.
7. *Explain your decision fully.* It is essential to get employee commitment and understanding. Therefore, be explicit and honest in communicating your decision. (P. 265)

Along with the foregoing checklist are the following key points to effective handling of grievances (Metzger, 1978).

1. Employees deserve a complete and an empathetic hearing of grievances they present.
2. The most important job in the handling of grievances is getting at the facts.
3. Look for the hidden agenda.
4. Hasty decisions often backfire.
5. While you are investigating a grievance, try to separate fact from opinion or impression.
6. After you have come to your decision, promptly communicate that decision to the employee.
7. Remember that you have to sell your decision.
8. There is no substitute for common sense in arriving at a decision.
9. Written records are most important.
10. Follow-up is essential. (Pp. 91–92)

Arbitration

Provisions for the use of arbitration to resolve contract interpretation disputes during the life of the contract appear in more than 90 percent of all collective bargaining agreements. Arbitration is the final step in the grievance procedure.

A voluntary arbitration is judicial in nature. When two parties are unable to resolve a dispute by mutual agreement, they submit the particular issue to an impartial person for solution.

There are several types of voluntary arbitration clause dealing with contract administration disputes. The most common provides for both parties to select an arbitrator each time a dispute arises. Many contracts state that the arbitrator shall be selected through the American Arbitration Association (AAA). When an issue is in dispute, the aggrieved party petitions the AAA, which, in turn, submits a list of arbitrators to both parties. If the two parties cannot agree on which arbitrator to select, the AAA designates the arbitrator. The selection of a separate arbitrator for each issue has the marked advantage of securing a qualified individual to rule on a particular type of dispute (Metzger, 1970).

A less widely used voluntary arbitration clause provides for a permanent arbitrator system. That is, the parties agree on the identity of an impartial arbitrator, who will handle *all* arbitrations during the life of a specific agreement. This has the distinct advantage of providing the parties with a carefully selected individual, who has earned the respect of both labor and management and who, because of repeated experiences with the parties, can develop a thorough understanding of the problems and unique difficulties of a particular institution.

An arbitrator's award is binding on both parties. The arbitrator's decision is final. The first principle of effective arbitration in contract administration is joint agreement between management and the union on the type of arbitration desired: Will it be the ad hoc arrangement (i.e., selecting a different arbitrator for each case), or will a permanent arbitrator be selected for the life of the contract? Even more critical is the question of the limits or nonlimits on what matters shall be submitted to the arbitrator. In deciding which matters should be submitted to the arbitrator and the arbitrator's authority, contractual definitions must be considered. The arbitrator's authority is limited by the agreement. Typically he or she is responsible for interpreting or applying a specific clause in the contract. The arbitrator is precluded from adding to or deleting from any clause contained in that agreement. Some contracts limit the arbitration process to specific issues and not to all clauses contained in the collective bargaining agreement.

In submitting contract interpretation disputes to an arbitrator, both parties risk losing the decision on matters they deem to be important, but this is a small price to pay for uninterrupted service. The advantages of continuing operations while deciding disputed claims arising from interpretation of the contract are obvious. Arbitration neither diminishes nor detracts from the collective bargaining process. Rather, it is an extension of the bargain arrived at during the negotiations. It is an administrative tool for living with the contract. Having adopted the arbitration process voluntarily, both parties are more likely to accept the decision which emanates from the arbitrator.

The AAA (1981) provides us with a good 50-word arbitration clause.

> Any dispute, claim, or grievance arising out of or relating to the interpretation or the application of this agreement shall be submitted to arbitration under the Voluntary Labor Arbitration Rules of the American Arbitration Association. The parties further agree to accept the arbitrator's award as final and binding upon them. (P. 11)

A grievance procedure without voluntary arbitration at its terminal step is ineffective. All grievances should be heard regardless of whether they are covered by the definition in the contract. When grievances cannot be settled internally, the disputed matter should be submitted to a disinterested party. In this way, both the union and the aggrieved employee understand that management's decision can be question or reviewed by a disinterested and unbiased tribunal.

SUMMARY

Successful bargaining requires advance planning of strategies. Such planning can only be successful with a pragmatic and minimally subjective view. Good faith bargaining is a requirement indelibly inscribed in the National Labor Relations Act. The key to satisfying the duty to bargain in good faith is entering the bargaining arena with an open mind and with the full intention of reaching a final agreement.

This chapter reviews the mandatory and voluntary subjects of collective bargaining, it addresses the critical decision of selecting a negotiating team, and thoroughly outlines strategies for bargaining. A most useful section includes Key Contract Clauses with sample language.

Once the contract is negotiated its administration, which is breathing life into the words so negotiated, is critical. A review of the disciplinary mechanism, the grievance procedures, and the arbitration process follows. Appendix A provides a checklist of items to be included in the first collective bargaining contract.

REFERENCES

American Arbitration Association. *Labor Arbitration, Procedures and Techniques.* New York: American Arbitration Association, 1981.

Bade, W. J., and M. Stone. *Management Strategy in Collective Bargaining Negotiations.* New London, CT: National Foremen's Institute, 1951.

"Beyond Unions." *Business Week,* July 1985.

Cook, S. A. "The neglected art of negotiation." *The Daily Record,* Baltimore, January 19, 1972, p. 1.

Dunlop, J. T., and J. J. Hely. *Collective Bargaining,* rev. ed. Homewood, IL: Irwin, 1955, p. 53.

Guidebook to Labor Relations. Chicago: Commerce Clearing House, 1985, p. 315.

Justin, J. J. *How to Manage with a Union.* New York: Industrial Relations Workshop Seminars, Inc., 1969, pp. 294–295.

Metzger, N. "Voluntary arbitration in contract administration disputes." *Hospital Progress,* 51(9) (1970).

Metzger, N. *The Health Care Supervisor's Handbook.* Rockville, MD: Aspen Systems, 1978.

Metzger, N. *Personnel Administration in the Health Services Industry,* 2nd ed. New York: Spectrum, 1979.

Metzger, N., J. Ferentino, and K. Kruger. *When Health Care Employees Strike.* Rockville, MD: Aspen Systems, 1984, pp. 17–18.

U.S. Congress: S. Rep. #93-766, 93d Cong., 2d Sess. 5 (1974); H. R. Rep. #93-1051, 93d Cong., 2d Sess. 7 (1974).

CHAPTER 17

New Developments in Human Resources Management: A Summary for the Future

Myron D. Fottler
S. Robert Hernandez

This chapter integrates material found throughout the book while examining the future of human resources management in health care organizations. To predict likely changes in human resources management, we need to first consider likely changes in the external and internal environment of health care organizations.

Then we discuss the major human resources challenges that will confront health care executives in the future, including the integration of strategic planning and human resources planning, enhancement of employee performance and productivity, and the utilization of multiskilled allied health practitioners. The reader should remember that *all* health care managers are human resources managers, since all are responsible for recruiting, training, evaluating, and rewarding their own subordinates.

ENVIRONMENTAL CHANGE

Recent

The external environment of the health care industry has shifted dramatically during the present decade. Among the more important changes have been the active encouragement of competition by the federal government; the shifting balance of power between managers and physicians; increasing cost-consciousness on the part of government, employers, and third-party payers; the implementation of a prospective pricing system for reimbursing hospitals under Medicare; and the growth and increasing dominance of multihospital systems (Arthur Andersen and Company and the American College of Hospital Administrators, 1984; Fottler and Vaughan, 1987).

The increased competition stems from the efforts by all third-party payers to encourage greater utilization of less expensive alternatives to acute inpatient hospital care. This change in attitude toward competition is manifest by a greater willingness to provide reimbursement for services received in such alternative settings and fewer restrictions on their entry into the marketplace. Among the major alternatives that have grown rapidly during this decade are health maintenance organizations (HMOs), emergi-centers, surgi-centers, wellness centers, outpatient clinics, employee assistance programs, and preferred provider organizations (PPOs) (Ermann and Gabel, 1985).

The environmental changes listed above as well as the growing surplus of physicians have reduced the power of physicians vis-à-vis health care managers (Fuchs, 1982; Morone and Dunham, 1984). The pressures for institutional survival and cost containment have caused managers to attempt to gain firmer control over what doctors do. This has created inevitable manager/physician conflict in many situations. Among the reasons for the increasing relative power and control of managers have been the need for capital to finance growth, the growth of government regulation, and the ability of managers to integrate technology and human resources (Fuchs, 1982).

In addition to the federal government's attempts to contain health costs, major employers have been active in aggressively promoting health care cost containment (Fox, et al., 1984). These activities have included regulatory approaches to cost containment (Egdahl, 1984), as well as promotion of incentive approaches such as cost sharing by employees, wellness programs, employee rewards healthful behavior, and health maintenance organizations (Fottler and Lanning, 1986).

Since hospital reimbursement is now based heavily on payment *per case* (rather than *per day*), there is no longer any incentive to lengthen the patient stay to maximize revenues. The previous reimbursement system emphasized retrospective reimbursement (payment after service delivery based on costs); consequently, there was no incentive to provide cost-efficient or cost-effective service. Now there are incentives to minimize patient stay as well as the resources devoted to patient care. Prospective payment has impacted all areas of health care management in ways that are just beginning to be understood (Boerma, 1983; Crawford and Fottler, 1985; Smith and Fottler, 1985).

Finally, the growth of multihospital systems has been both a response to environmental changes and a major environmental change in and of itself. Such systems represent the ultimate outcome of the "corporatization" of health care. The reasons for the development of systems are similar to reasons for development of large, multiunit national and international corporations in other industries: greater access to capital, easier response to government regulation, economies of scale in the provision of some services, and easier implementation of sophisticated management practices (Ermann and Gabel, 1984).

Future Environmental Changes

The expert panels convened by Arthur Andersen and the American College of Hospital Administrators (1984) have predicted a number of future trends, some based on present trends and some not. Although the amount of money spent nationally on health services will continue to grow, it may grow at a slower rate, and emphasis will shift to ambulatory services, less expensive alternatives to inpatient hospital care, and alternative delivery systems. Both multihospital systems and investor-owned hospitals will continue to grow. Hospitals will have to create new corporate structures and business ventures to compete in the capital market.

In addition, the anticipated oversupply of physicians and the continued trend toward practicing in hospital-based positions and alternative delivery systems will mean for these professionals a continued decline in influence. Indicators of such a decline will include prescribed patient care protocols and increased fiscal restraints on physician expenditures. There will be increased potential for both conflict and collaboration (Arthur Andersen and the American College of Hospital Administrators, 1984).

In sum, all these environmental changes indicate a need for health care organizations to be more entrepreneurial in order to simultaneously achieve both the higher quality and the lower costs necessary to compete in increasingly competitive markets. The achievement of these partially incompatible objectives requires the utmost care in the recruitment, selection, development, appraisal, and compensation of human resources. This is the subject of the next section.

HUMAN RESOURCES IMPLICATIONS

General

As a result of all these environmental changes, a number of observers have described what management changes are necessary for success in the new environment. For example, the Delphi study of health care managers by Arthur Andersen and Company and the American College of Hospital Administrators (1984) concludes that future success will depend on strategic and financial planning, refined management skills, risk identification and analysis, integrated clinical and financial cost accounting, prudent application of new technology, predictive market analysis, and computerized decision support systems.

In the area of human resources, the same study predicts that the 1990s will be an era requiring considerable managerial ability to harness the talent and skill of human resources in health care. The major challenges will be to recruit and develop competent governing board members, to nurture cooperative arrangements with medical staff, to develop compensation programs to attract and retain competent managerial and ancillary staffs, and to achieve greater employee productivity. In addition, CEOs will need to improve their own skills as they move from managing single-entity organizations to managing more complex enterprises. Among the specific recommendations were trustee training, trustee compensation, incentive compensation, opportunities for continuing education of physicians, and recruitment of managerial talent from other business sectors.

Several other commentators have also examined recent environmental trends and their managerial implications. For example, Pointer (1985) predicts an increasing emphasis on cost containment, on doing less more effectively, and on product management, creativity, innovation, and product marketing. He recommends more experimentation, pilot testing, internal venture capital programs, creativity (innovation workshops), and an internal incentive system that encourages creativity and innovation. Finally, Robbins and Rakich (1986) predict a modified human resources perspective including intervention in organizational processes (for organizational survival), creative problem solving, a participative and competitive internal environment, and development of innovative strategies.

Two Models for the Future

One perspective on this evolution is shown in Table 17.1. Prior to about 1975 the "Simple Personnel Model" or the "Labor Relations Model" were dominant. Under these models, the human resources function was low-level, reactive, and not influential in health-care organizations. According to the authors, the human resources model (column 3) describes the present for most hospitals. For a select few—those in transition to the matrix model—it represents the immediate past. However, for most small health organizations such as nursing homes, ambulatory care facilities, and physician's offices, *both* the human resources model and the matrix model represent the future.

The human resources perspective requires an interventionist role for the human resources function (Robbins and Rakich, 1986), which is responsible for initiating and facilitating organization changes. Despite continuance of human resources acquisition and retention personnel activities from earlier models, there is now an emphasis on synergistic, organization-wide outcomes and a direct linkage to organization goals. Previously, these personnel functions were performed in a mechanical, reactive way.

The organization's philosophy is based on a collaborative and cooperative arrangement between management and employees, in contrast to the conflict and adversarial relationships of earlier models. In addition, the human resources function experiences enhanced influence on organizational policy. The specific activities of the human resources function include employee needs assessment, attitude surveys, job satisfaction surveys, interventions to change structure and task design, performance appraisal systems linked to positive behavior modification, training in both technical skills and social skills, career counseling, long-range human resources planning, and employee assistance programs.

The key differences between matrix model for Table 17.1 and the human resources model stem from the need to deal with risk-survival issues (Robbins and Rakich, 1986). In multiunit systems this means the creation of a corporate-level human resources office with responsibilities to coordinate and direct the overall program as a matrix responsibility to the individual institution. As policies, benefits, and wage systems became more centralized and standardized, their impact will be greater. A new attention to quality will result in an overall upgrading of the entire human resources function. Cost savings should also be achieved through such obvious outcomes as the ability to be more sophisticated and efficient in benefits administration.

The human resources department will also assume a greater role in productivity enhancement. The matrix department will be called on for advice and support in changing work methods, standards, and staffing practices. The organizational philosophy will be both participative and competitive. Employees will have the opportunity to participate in organization and work systems, and they will also have an opportunity to participate in the fruits of such efforts, perhaps through incentive programs. However,

TABLE 17.1. Hospital Personnel Administration Models

Attribute	Pre-1965: Simple Personnel Model	About 1965–1975: Labor Relations Model	About 1975–1985 (the present): Human Resources Model	Post-1985 (the future): Matrix Model
Approach to personnel administration	Benign neglect; indifference	Containment of external forces	Human resources perspective (emphasis on people as the most important resource)	Modified human resources perspective
Role	Recordkeeping	Cope	Intervention (for change)	Intervention (for organizational survival)
Process	Simple functional (not integrated)	Fully functional	Integrative functional	Creative functional
Organizational philosophy about employees	Neutral (employees are just another resource)	Conflict and confrontation	Collaborative-cooperative (organization and employees have mutual interests)	Competitive; participative
Predominant strategy	Inactive compliance	Reactive	Proactive	Innovative
Influence on organizational policy	Minimal	Increasing	Enhanced	Greater
Acceptance by managers and staff	Minimal	Necessary	Enhanced	Greater

SOURCE: Stephen A. Robbins and Jonathan S. Rakich, "Hospital Personnel Management in the Late 1980's: A Direction for the Future," *Hospital and Health Services Administration*, Vol. 31, July/August 1986, p. 27. Ann Arbor: Health Administration Press. Copyright © 1986, Foundation of the American College of Healthcare Executives.

they will be at risk in terms of employment security as organizations alter their character to cope with competition.

The matrix model will function in a single-institution setting that is either expanding or contracting as well as in a multi-institutional arrangement in which activities are centralized beyond the single institutional level. In all cases, the human resources role will be interventionist, with the purpose of designing and implementing programs to ensure organizational survival. The predominant strategy should be not only proactive, but also innovative. Influence and acceptance of the human resources function are also expected to be greater under the matrix model.

It should be noted that most health care organizations have a long way to go at the time of this writing (1987) to reach the human resources model, much less the matrix. Most are operating on the basis of some combination of the "simple personnel model" or the "labor relations model" as noted in Table 17.1. Their basic approach is benign neglect of human resources or containment of external forces. Performing record-keeping or simple functional tasks is viewed as the human resources role. The philosophy toward employees is either neutral or based on conflict. The predominant strategy is either inactive or reactive. The influence and acceptance of human resources is either minimal or not fully integrated into the strategic processes of the organization. This description is, of course, less descriptive of larger health institutions and those associated with multiunit systems.

INTEGRATION OF STRATEGIC AND HUMAN RESOURCES PLANNING

Strategic planning is described earlier in this book as a management philosophy designed to provide the organization with a long-term collective purpose and direction (Zallocco et al., 1984). The purposes of strategic planning are to assess and respond to the opportunities and threats in the external environment, to provide for appropriate allocations of resources, and to develop methods for evaluating organizational performance. In a recent survey conducted by American Hospital Publishing and Kurt-Salmon Associates, hospital chief executive officers ranked strategic planning as their critical concern, today and in the future (Moore, 1984). This survey suggests that the planning emphasis has changed from "bricks and mortar" to a new qualitative and financial orientation that encompasses all areas of hospital operations. In addition to the area of strategic planning, 80 percent of the responding CEOs listed experience in finance and accounting, marketing, and human resources as the top areas of importance for today's and future CEOs.

Zallocco et al. (1984) conducted a study to determine the extent to which strategic management and specific planning techniques are employed in hospitals today. From a sampling of midwestern hospitals, these authors reported that 60 percent use a formal strategic planning process. Additionally, larger hospitals (more than 200 beds) are more likely to have a formalized strategic planning process; 55 percent employ an individual or a department with planning responsibilities; 38 percent make budget allocations directed to a strategic planning function; and 66 percent have planning cycles of three years or less. It could be inferred from these findings that prospective payment has made a definite impact on the type of planning used by today's health care facilities and that the contemporary economic climate has substantially decreased the length of long-term planning from the traditional 5- and 10-year plan.

Although this study suggests an increased use of a strategic planning process, it also reported that implementation of the plan was problematic. The process often ends with goal and objective development, but without strategies or methods of implementation or performance monitoring design. Perhaps this is a result of other findings, which suggest that strategic planning is a top-level administrative function with exclusion of middle- or upper-level management personnel or staff representation (including the human resources function). Without participation by those it would benefit most, this planning is not likely to develop into an ongoing process. For example, if the manager responsible for developing human resources is not actively involved, forecasting, recruitment, selection, and development of required personnel are not likely to occur effectively (Zallocco et al., 1984).

One method of improving implementation is to integrate strategic planning and human resources management through human resources planning (HRP), a process that looks at the organization's current and future human resources needs based on external as well as internal environments. HRP involves a plan or strategy to address the gap between what the current human resources picture is and what is indicated it will be. It requires some level of integration with the organization's strategic planning and should serve as a "design and implementation link" for all other human resources programs and systems.

As one writer has commented:

> Strategic planning is generally geared to the gaining of systematic advantage over competitors. If human resource strategy is to be maximally useful to the corporation, it must share the corporation's plan. Too often, the linkage is missing. But without it, the human resource plan cannot be truly strategic. (Goodmeasure, Inc., 1982, p. 1)

The benefits of linking human resources planning to strategic planning have been demonstrated in other industries. During the 1981–1982 recession, employers who had the most highly developed human resources planning systems linked to strategic planning were able to minimize their employee layoffs through hiring freezes, attrition, and other types of advanced action. Almost three-quarters of the companies engaging in such planning were certain that it improves profitability (Mills, 1985). It has also been found to foster organizational survival in the health care industry (Eisenberg, 1986).

Although there are significant benefits to be derived from an integration of strategic planning and human resources planning, most health care organizations are not yet ready to implement such integration. Table 17.2 indicates a format by which a particular health care organization might assess its own readiness to adopt such an approach. Obviously, the greater the number of "high" values that can be assigned along each of the dimensions, the more ready the organization is to integrate strategic and human resources planning.

Assuming the organization is convinced that such an approach would be beneficial and is ready to adopt it, how can integration be accomplished? Miles and Snow (1984) suggest that the human resources department needs a comprehensive understanding of the language and practice of strategic planning. Moreover, appropriate human resources representatives must continually participate in the planning process to assess the probable demand for their unit's services and to help line executives trace the human resources implications of their strategic decisions. The department should also pursue appropriate strategies on its own to match the organization's business strategies.

Different organizations will emphasize building ("defender"), acquiring ("prospector"), and allocating ("analyzer") human resources. Table 17.3 indicates how these

TABLE 17.2. Assessing Readiness for Human Resources Planning

Design Factors	Readiness		
	Low	Moderate	High
	Organizational Context		
Environment	Stable	Moderate change	Significant change
	• Predictable	• Manageable	• Uncertain
	• Limited threat to firm	• Moderate threat to firm	• Major threat to firm
Strategy	Opportunistic	Directed	Disciplined
	• Unclear direction	• Generally clear direction	• Clear mission and measures
	• Driven by market product	• Pattern of resource allocation decisions	• Consistent resource allocation decisions
	• Inconsistent resource decisions		
Structure	Simple	Moderate	Complex
	• Few clear business units	• Centralized major functions	• Many diverse independent units
	• Direct functional reporting	• Decentralized business units	• Indirect reporting relationships
Values regarding people	Incidental	Important	Critical
	• Will always be available	• Need attention but can usually fulfill requirements	• Need effective management and development
	• Skill requirements are usually transferable	• Some specialization needed	• Specialized skills required

Management Requirements

	Financial	Marketing	Strategy
Planning process	• 1–2 year horizon • Budgeting focus	• 1–3 year horizon • Program focus	• 1–3–5 year horizon • Portfolio focus
Commitment to HRP	Low • Minimal to limited support • Resistance to involvement	Medium • Visible support • Usually responsive to HR initiatives	High • Demonstrates commitment • Managers initiate on own

Human Resources Function

	Limited	Traditional	Strategic
Role and Capabilities	• Basic functions • Little influence	• All key functions • Informed and involved	• Broad and extensive • Strong influence
Information systems	Limited • Mixed quality • Inaccessible	Moderate • Usually reliable • Generally accessible	Sophisticated • Broad and reliable • Immediately accessible

SOURCE: Thomas B. Wilson, *A Guide to Strategic Human Resources Planning for the Healthcare Industry*, published by the American Hospital Association, copyright © 1986.

specific strategies might be implemented in terms of specific functions such as recruitment and performance appraisal. It should be noted that most successful health care organizations will be of type B ("prospector") and will need to develop human resources management systems consistent with that approach. This means that such organizations will "buy" the skills they need (rather than develop them) and provide "incentive" compensation (rather than compensation based upon position).

Finally, the human resources department should act as a professional consultant to line units (Miles and Snow, 1984). In addition to their expertise in strictly personnel matters, human resources specialists should be knowledgeable about organization structure, management communication and control processes, and organizational change and development.

What generic categories of strategies are available to a health-care organization? According to Porter (1985), there are three competitive strategies that organizations can use to gain competitive advantage: innovation, quality enhancement, and cost reduction. The innovation strategy is used to develop services different from those of competitors. Enhancing service quality is the focus of the quality enhancement strategy. In the cost reduction strategy, organizations typically attempt to gain competition advantage by being the lowest cost producer.

Health-care organizations could simultaneously pursue different strategies, particularly in different subunits. Moreover, as noted in Chapter 3, certain strategies are more or less appropriate at different points in an organization's life cycle. Finally, it is crucial that the human resource strategies be compatible with the business strategy of the organizations.

Human resource strategies should be different with different business strategies because the expected role behaviors of employees are different. For example, an innovation strategy requires that employees exhibit a high degree of creative behavior, a longer-term focus, a relatively high level of cooperative behavior, a greater degree of risk-taking behavior, and a higher tolerance for ambiguity and unpredictability as compared to quality-enhancement or cost-reduction strategies (DePree, 1986). A quality-enhancement strategy requires that employees exhibit relatively repetitive and predictable behaviors, a high concern for quality, and a high concern for how services are delivered. Finally, the cost-reduction strategy requires a short-term focus on results, low risk-taking activity, and a high concern with quantity of output (Schuler and Jackson, 1987).

What are the implications of these different employee role requirements for human resources management? The innovation strategy requires jobs that allow employees to develop skills useful in other positions, performance appraisals that reflect longer-term and group-based achievements, pay rates that tend to be low but allow employees to choose the mix of pay components (including incentive pay), and broad career paths to reinforce the broad range of skills required (Schuler and Jackson, 1987).

The quality enhancement strategy requires high levels of employee participation in decisions, performance appraisals based on short-term results, and extensive and continuous training and development of employees. The cost-reduction strategy requires narrowly designed jobs and narrowly defined career paths that encourage specialization and efficiency, performance appraisals based on short-term results, market wage rates, and little training and development (Schuler and Jackson, 1987).

All components of a human resource management system need to work together to stimulate and reinforce particular employee behaviors. They need to be implemented or changed simultaneously. Failure to a health-care organization to invoke a particular practice, such as broad career paths, implies invoking another (i.e., narrow career paths and high specialization) that may reinforce the opposite behavior of that desired. Under such conditions employees may become frustrated.

ENHANCEMENT OF EMPLOYEE PRODUCTIVITY

The Delphi study of Arthur Andersen and Company and the American College of Hospital Administrators (1984) identified a number of CEO skills critical to success in the increasingly competitive health care environment. In addition to strategic planning skills, leaders of vision are needed to maximize institutional productivity.

Several of the guidelines identified by Shortell (1985) in his analysis of high-performing health care organizations relate to various methods of enhancing employee productivity. For example, such organizations tend to stretch themselves, to maximize learning, to manage ambiguity and uncertainty, and to exhibit a well-defined culture. Organizations can stretch themselves only if they allow employer participation in jointly setting ambitious goals for individual and/or work group performance. To maximize learning, the organization has to have a very well-developed training and development function. The management of ambiguity and development of a well-defined culture requires implementation through appropriate recruitment, selection, and compensation policies.

Health care executives have become increasingly concerned with productivity as cost containment, prospective payment systems, and increasing competition have become environmental realities. The public has become convinced it is spending too much for health care and receiving too little. The era of "open-ended" finding is past. One method of dealing with this new situation is to generate more output from existing resources, that is, to increase productivity.

The principal difficulty encountered by health care organizations in addressing the question of productivity is that productivity analysis stands as a relatively under-developed management tool (Fottler and Maloney, 1979; Margulies and Duval, 1984). Most health care institutions have not established productivity tracking systems or methods for measuring and analyzing productivity improvement (Mannisto, 1980). However, several recent publications have provided guidelines and systems for such measurement (Eastaugh, 1985; Margulies and Duval, 1984; Williams, 1973).

Specific management approaches to raising employee productivity fall into two general categories: technical and human resources. The technical approaches include changes with respect to capital investment, new technology, and subcontracting. Human resources approaches include changes in work organization, work scheduling, work rules, and participation processes.

A recent review article (Fottler, 1987) has examined a number of structure and process variables related to productivity. Some are related to human resources and some are not. Among the structural factors related to human resources that seem to be associated with higher employee productivity were the use of contract management in the provision of certain services, little or no unionization of the labor force, structural mechanisms to involve physicians in decision making, use of physician extenders, and a close fit between structure and technology.

Process variables associated with higher employee productivity included significant physician, nurse, and department head participation in strategy development and

TABLE 17.3. Business Strategies and Human Resources Management Systems

Organizational/Managerial Characteristics	Type A (Defender)	Type B (Prospector)	Type AB (Analyzer)
Product-market strategy	Limited, stable product line Predictable markets Growth through market penetration Emphasis: "deep"	Broad, changing product line Changing markets Growth through product development and market development Emphasis: "broad"	Stable and changing product line Predictable and changing markets Growth mostly through market development Emphasis: "deep" and "focused"
Research and development	Limited mostly to product improvement	Extensive, emphasis on "first-to-market"	Focused, emphasis on "second-to-market"
Production	High volume–low cost Emphasis on efficiency and process engineering	Customized and prototypical Emphasis on effectiveness and product design	High volume–low cost; some prototypical Emphasis on process engineering and product or brand management
Marketing	Limited mostly to sales	Focused heavily on market research	Utilizes extensive marketing campaigns
Organization structure	Functional	Divisional	Functional and matrix
Control process	Centralized	Decentralized	Mostly centralized, but decentralized in marketing and brand management
Dominant coalition	CEO Production Finance/accounting	CEO Product research and development Market research	CEO Marketing Process engineering

Business planning sequence	Plan → Act → Evaluate	Act → Evaluate → Plan	Evaluate → Plan → Act
Basic strategy	Building human resources	Acquiring human resources	Allocating human resources
Recruitment, selection, and placement	Emphasis: "make"	Emphasis: "buy"	Emphasis: "make" and "buy"
	Little recruiting above entry level	Sophisticated recruiting at all levels	Mixed recruiting and selection approaches
	Selection based on weeding out undesirable employees	Selection may involve preemployment psychological testing	
Staff planning	Formal, extensive	Informal, limited	Formal, extensive
Training and development	Skill building	Skill identification and acquisition	Skill building and acquisition
	Extensive training programs	Limited training programs	Extensive training programs
			Limited outside recruitment
Performance appraisal	Process-oriented procedure (e.g., based on critical incidents or production targets)	Results-oriented procedure (e.g., management by objectives or profit targets)	Mostly process-oriented procedure
	Identification of training needs	Identification of staffing needs	Identification of training and staffing needs
	Individual/group performance evaluations	Division-corporate performance evaluations	Individual-group division performance evaluations
	Time-series comparisons (e.g., previous years' performance)	Cross-sectional comparisons (e.g., other companies during same period)	Mostly time-series, some cross-sectional comparisons
Compensation	Oriented toward position in organization hierarchy	Oriented toward performance	Mostly oriented toward hierarchy, some performance considerations
	Internal consistency	External competitiveness	Internal consistency and external competitiveness
	Total compensation heavily oriented toward cash and driven by superior-subordinate differentials	Total compensation heavily oriented toward incentives and driven by recruitment needs	Cash and incentive compensation

SOURCE: Reprinted, by permission of the publisher, from "Designing Strategic Human Resources Systems," Raymond E. Miles and Charles C. Snow, *Organizational Dynamics*, Summer 1984, pp. 48–49, copyright © 1984 American Management Association, New York. All rights reserved.

implementation; an emphasis on inter- and intradepartmental communication and coordination; use of incentive compensation; use of many indicators of employee performance; efforts to reward creativity and innovation; use of quality circles and other employee participation programs; development of data to measure and enhance employee satisfaction; active efforts to modify physician behavior through a combination of strategies; use of flex time and flexible staffing; provision of many opportunities for employees to voice dissatisfaction; organization-wide development efforts; development of physician protocols where process/outcome relationships are fairly certain; and the recruitment and selection of managers who emphasize outcomes and the constant necessity to adapt to environmental change (Fottler, 1987).

Not all these structural and process human resources variables will necessarily be appropriate or productivity enhancing in all situations. Clearly there are contingencies that will make some more appropriate than others in a given situation. From the viewpoint of the health care executive, the key is to keep up with research results that provide guidelines to productivity improvement in both the human resources and the technical areas. Then the approaches that seem to be most appropriate should be adapted to the particular circumstances of the particular organization.

MULTISKILLED HEALTH PRACTITIONERS

The rapid advancement of technology in the health services has been accompanied by a rapid proliferation of allied health specialties, each of which has identified its own unique role. Specialty training programs have tended to be modeled on whatever the members of the specialty desired. Consequently, an elongation of curriculum with associated increases in expected salaries has occurred.

Financial limitations and a limited pool of trained specialists in rural areas have made it impossible for every physician's office, clinic, or hospital to recruit and hire specialists in major areas of need. Consequently, many individuals perform as generalist health technicians in small and rural facilities. However, one result is a lower quality of health services as a result of these employees' lack of formal training for such roles. Even in facilities that can afford the specialist technicians, the inevitable result is higher than necessary costs due to idle time and high wages.

The training of a new health worker variously labeled "multiskilled worker," "multiple-competency clinical technician," or "expanded-skill health practitioner" represents one approach to solving the cost/quality problem identified above. Interest in this concept is growing rapidly, not only in the United States, but in foreign countries as well (W. K. Kellogg Foundation, 1986). In fact, progress in implementing the concept is much greater abroad.

There is a growing interest in the multiskilled movement in the United States, due mainly to cost-containment pressures, competition, and the emergence of outpatient alternative delivery settings. However, relatively few formal educational programs exist here to provide training for the multiskilled health worker. In 1987 the two major programs were at the University of Alabama at Birmingham and Southern Illinois University.

For the health organization, the multiskilled worker can provide cost benefits by reducing the idle or "down" time of employees and the number of employees needed. Productivity can also be enhanced as a result of the greater versatility of the multiskilled worker. In addition, the multiskilled worker should benefit as a result of job enrichment, higher job satisfaction, a potential for increased pay, increased job security, and increased marketability.

If the idea of multiple competency is so timely, why is it not more widespread? Among the obstacles are human nature's resistance to change; estblished disciplines that oppose encroachment; allied health and nursing schools, which have been encouraged by federal grants and accreditation bodies to train only specialized technicians; and accreditation. No established multiple-competency associations exist to accredit programs (Blayney, 1986). Legal restrictions and liability issues associated with the use of multiskilled personnel also limit utilization.

Health care executives have a vested interest in working for multiple-skilled technician programs and other approaches that make the manpower credentialing process more flexible. Otherwise they will continue to have little control over their staffing, particularly in terms of matching credentials to institutional needs. Calls for reevaluation are becoming increasingly common in the health care literature (Hofmann, 1984).

SUMMARY

The major environmental changes impacting health care have significantly increased competition in the industry. Since health care executives have not fully utilized human resources management as a competitive tool, this remains the challenge for the future. In particular, human resources planning needs to be integrated into the strategic planning process. Then a continual search for productivity improvement should utilize a variety of human resources approaches such as quality circles.

Finally, the development and utilization of human resources needs to be modified to allow greater flexibility. In particular, the trend toward extreme specialization has become dysfunctional in terms of efficiency, and the credentialing process needs to become more flexible. The use of multiskilled health technicians offers a promising option for the future.

REFERENCES

Arthur Andersen and Company and the American College of Hospital Administrators. *Health Care in the 1990s: Trends and Strategies.* Chicago: Arthur Andersen and Company and the American College of Hospital Administrators, 1984.

Blayney, K. D. "Restructuring the health care labor force: The use of the multiskilled allied health practitioner." *Alabama Journal of Medical Sciences,* 23(3):277–278 (1986).

Boerma, H. *The Organizational Impact of DRGs: DRG Evaluation.* Princeton, NJ: Center for Health Affairs, Health Research and Educational Trust of New Jersey, 1983.

Crawford, M., and M. D. Fottler. "The impact of diagnosis-related groups and prospective payment systems on health care management." *Health Care Management Review,* 10:73–84 (1985).

DePree, H. *Business as Unusual.* Zeeland, MI: Herman Miller, 1986.

Eastaugh, S. R. "Improving hospital productivity under PPS." *Hospital and Health Services Administration,* 30(4):97–111 (1985).

Egdahl, R. H. "Should we shrink the health care system?" *Harvard Business Review,* 62:125–132 (1984).

Eisenberg, B. "Strategic human resource plans help providers survive changing conditions." *Modern Healthcare,* 16(6):154–155 (March 28, 1986).

Ermann, D., and J. Gabel. "Multihospital systems: Issues and empirical findings." *Health Affairs,* 3(1):50–64 (1984).

Ermann, D., and J. Gabel. "The changing face of American health care: Multihospital systems, emergicenters, and surgery centers." *Medical Care,* 23(5):401–420 (1985).

Fottler, M. D. "Health care organizational performance: Present and future research." *Journal of Management,* 13(2):179–203 (1987).

Fottler, M. D., and J. A. Lanning. "A comprehensive incentive approach to employee health care cost containment." *California Management Review,* 29(1):75–94 (1986).

Fottler, M. D., and W. F. Maloney. "Guidelines to productivity bargaining in the health care industry." *Health Care Management Review,* 3:59–70 (1979).

Fottler, M. D., and D. G. Vaughan. "Multihospital systems." In L. F. Wolper and Jesus J. Pena, Eds., *Health Care Administration: Principles and Practice.* Rockville, MD: Aspen Systems, 1987.

Fox, P. D., W. B. Goldbeck, and J. J. Spies. *Health Care Cost Management: Private Sector Initiatives.* Ann Arbor, MI: Health Administration Press, 1984.

Fuchs, V. R. "The battle for control of health care." *Health Affairs,* 1:5–13 (1982).

Goodmeasure, Inc. *Strategic Planning for Human Resources.* Cambridge, MA: Goodmeasure, 1982.

Hofmann, P. B. "Healthcare credentialing issues demand increased attention." *Hospital and Health Services Administration,* 29(3):86–93 (1984).

W. G. Kellogg Foundation. "Restructuring the health care labor force: The rise of the multiskilled health practitioner." International Conference, Birmingham, AL, Feb. 24–25, 1986.

Mannisto, M. "An assessment of productivity in health care." *Hospitals,* 54(18):71–76 (1980).

Margulies, N., and J. Duval. "Productivity management: A model for participative management in health care organizations." *Health Care Management Review,* 9(1):61–70 (1984).

Miles, R. E., and C. C. Snow. "Designing strategic human resource systems." *Organizational Dynamics,* pp. 36–52 (Summer 1984).

Mills, D. Q. "Planning with people in mind." *Harvard Business Review,* 63(4):97–105 (1985).

Moore, B. W. Survey shows CEOs' priorities are changing. *Hospitals,* 58(24):71–77 (1984).

Morone, J. A., and A. B. Dunham. "The waning of professional dominance: DRGs and the hospitals." *Health Affairs,* 3:73–84 (1984).

Pointer, D. D. "Responding to the challenges of the new healthcare marketplace: Organizing for creativity and innovation." *Hospital and Health Services Administration,* 30:10–25 (1985).

Porter, M. E. *Competitive Advantage.* New York: Free Press, 1985.

Robbins, S. A., and J. S. Rakich. "Hospital personnel management in the late 1980s: A direction for the future." *Hospital and Health Services Administration,* 31:18–33 (1986).

Shortell, S. M. "A more total approach to productivity improvement." *Hospital and Health Services Administration,* 30(4):7–35 (1985).

Schuler, R. S., and S. E. Jackson. "Linking competitive strategies with human resources management practices." *Academy of Management Executive,* 1(3):207–219 (August 1987).

Smith, H. L., and M. D. Fottler. *Prospective Payment: Managing for Operational Effectiveness.* Rockville, MD: Aspen Systems, 1985.

Williams, N. *The Management of Hospital Employee Productivity.* Chicago: American Hospital Association, 1973.

Zallocco, R., B. Joseph, and N. Furey. "Do hospitals practice strategic planning? An empirical study." *Health Care Strategic Management,* 2(2):16–20 (1984).

APPENDIX

Checklist for a Collective Bargaining Contract

I. RECOGNITION. This clause defines the employees who will be covered by the collective bargaining agreement. It usually contains those jobs to be represented by the union and those to be excluded from the bargaining unit.

SAMPLE

Bargaining Unit Recognition

Section A. The Hospital recognizes the Union as the sole and exclusive representative of the employees of the Hospital as hereinafter defined for the purposes of collective bargaining with respect to rates of pay, hours of employment, and other conditions of employment within said bargaining unit.

Section B. Except as hereinafter limited, the term "employee" as used herein shall apply to and include the following classifications:

Diet Aides
Food Preparer, C, D
Mechanic A, B, C, D, E
Maintenance Worker A, B, C
Laundry Worker A, B, C
Nursing Auxiliary A, B, C
Surgical Technician A, B
Orderly

Pharmacy Technician
X-Ray Technician A, B, C
Telephone Operator C, D

Section C. Except as hereinafter limited, the term "employee" when used in this Agreement shall exclude all other classifications not so included in Section B and shall without limitation, however, exclude the following:

All professional personnel
Administrative personnel
Supervisors
Physicians
Psychologists
Physicists and similar professionals
Registered and graduate nurses
Practical nurses
Student nurses
Social Service workers
Dieticians

II. UNION SECURITY. This clause states the extent to which employees are required to join, maintain their membership, and pay union dues.

SAMPLE

Union Security

1. All Employees on the active payroll as of July 1, 1988, who are members of the Union shall maintain their membership in the Union in good standing as a condition of continued employment.
2. All Employees on the active payroll as of July 1, 1985, who are not members of the Union shall become members of the Union within thirty (30) days after the effective date of this Agreement, except those who were required to become members sooner under the expired Agreement who shall become members on the earlier applicable date, and shall thereafter maintain their membership in the Union in good standing as a condition of continued employment.
3. All Employees hired after July 1, 1985, shall become members of the Union no later than the thirtieth (30th) day following the beginning of such employment and shall thereafter maintain their membership in the Union in good standing as a condition of continued employment.
4. For the purposes of this Article, an Employee shall be considered a member of the Union in good standing if he or she tenders his or her periodic dues and initiation fee uniformly required as a condition of membership.
5. Subject to Article XXVII [not given in this appendix], an Employee who has

failed to maintain membership in good standing as required by this Article, shall, within twenty (20) calendar days following receipt of a written demand from the Union requesting his or her discharge, be discharged if, during such period, the required dues and initiation fee have not been tendered.

III. CHECK-OFF. This clause will outline management's obligation to deduct dues and initiation fees from union members and remit them to the union.

SAMPLE

Check-Off

1. Upon receipt of a written authorization from an Employee, the Hospital shall, pursuant to such authorization, deduct from the wages due said Employee each month, starting not earlier than the first pay period following the completion of the Employee's first thirty (30) days of employment, and remit to the Union regular monthly dues and initiation fee, as fixed by the Union. The initiation fee shall be paid in two (2) consecutive monthly installments beginning the month following the completion of the probationary period.

2. Employees who do not sign written authorization for deductions must adhere to the same payment procedure by making payments directly to the Union.

3. Upon receipt of a written authorization from an Employee, the Hospital shall, pursuant to such authorization, deduct from the wages due said Employee each pay period, starting not earlier than the first period following the completion of the Employee's first thirty (30) days of employment, the sum specified in said authorization and remit same to the [Local] Credit Union to the credit or account of said Employee. It is understood that such check-off and remittance shall be made by the Hospital wherever feasible.

4. Upon receipt of a written authorization, the Hospital shall, pursuant to such authorization, deduct from the wages due said Employee once a year the sum specified in said authorization and remit same to [the Local] Fund as the Employee's voluntary contribution to said Fund. It is understood that such check-off and remittance shall be made by the Hospital wherever feasible.

5. The Hospital shall be relieved frm making such "check-off" deductions upon (a) termination of employment, or (b) transfer to a job other than one covered by the bargaining unit, or (c) layoff from work, or (d) an agreed leave of absence, or (e) revocation of the check-off authorization in accordance with its terms or with applicable law. Notwithstanding the foregoing, upon the return of an Employee to work from any of the foregoing enumerated absences, the Hospital will immediately resume the obligation of making said deductions, except that deductions for terminated Employees shall be governed by paragraph 1 hereof. This provision, however, shall not relieve any Employee of the obligation to make the required dues and initiation payment pursuant to the Union constitution in order to remain in good standing.

6. The Hospital shall not be obliged to make dues deductions of any kind from any Employee who, during the dues month involved, shall have failed to receive sufficient wages to equal the dues deduction.

7. Each month, the Hospital shall remit to the Union all deductions for dues and initiation fees made from the wages of Employees for the preceding month, together with a list of all Employees from whom dues and/or initiation fees have been deducted.

8. The Hospital agrees to furnish the Union each month with the names of newly hired Employees, their addresses, Social Security numbers, classifications of work, their dates of hire, and names of terminated Employees, together with their dates of termination and names of Employees on leave of absence.

9. It is specifically agreed that the Hospital assumes no obligation, financial or otherwise, arising out of the provisions of this Article, and the Union hereby agrees that it will indemnify and hold the Hospital harmless from any claims, actions, or proceedings by an Employee arising from deductions made by the Hospital hereunder. Once the funds are remitted to the Union, their disposition thereafter shall be the sole and exclusive obligation and responsibility of the Union.

IV. PROBATIONARY PERIOD. This clause defines the period of time a newly hired employee shall be considered to be on probation.

SAMPLE

Probationary Employees

1. Newly hired Employees shall be considered probationary for a period of two (2) months from the date of employment, excluding time lost for sickness and other leaves of absence.

2. Where a new Employee being trained for a job spends less than twenty-five percent (25%) of his or her time on the job, only such time on the job shall be counted as employment for purposes of computing the probationary period.

3. The probationary period for part-time Employees whose regularly scheduled hours are fifteen (15) or less shall be twice the length of the probationary period of full-time Employees.

4. Notwithstanding the foregoing, the probationary period for social workers including part-time social workers, according to custom, shall be six (6) months.

5. During or at the end of the probationary period, the Hospital may discharge any such Employee at will and such discharge shall not be subject to the grievance and arbitration provisions of this Agreement.

V. SENIORITY. This clause outlines the application of seniority to eligibility for holidays, vacations, promotions, transfers, overtime, layoffs, shift and shift preferences.

SAMPLE

Section A. Bargaining Unit Seniority is defined as the length of time an Employee has been continuously employed in the Hospital. An Employee shall have no seniority for the first three (3) months of employment, or for the probationary period whichever is longer; but upon successful completion of this probationary period, seniority shall be retroactive to the date of hire. Bargaining Unit Seniority shall apply in the computation of vacation eligibility, holiday eligibility, free day eligibility, sick leave eligibility, and pension eligibility.

Section B. An Employee shall have seniority to be known as Classification Seniority, in each classification in which he or she has completed a probationary period of 3 months retroactive to the date of his or her employment in that classification.

1. An Employee's Classification Seniority shall be used for the purposes of transfers, promotions, demotions, shift preferences, vacation scheduling, lay offs, recalls, and rehires.

Section C. In case of an indefinite layoff, Employees shall be laid off and recalled by classification within their Department.

1. In the event of a layoff, recall, or a promotion within the Bargaining Unit, the following factors shall govern:
    a. Ability to do the work
    b. Classification Seniority
2. Where factor (a) above is relatively equal, factor (b) shall be the governing factor.
3. The Hospital shall be the sole judge of the ability of an Employee to do the work.

Section D. The Hospital may make any transfer deemed by it to be expedient, either within the Department, to another department, or to another shift. When an Employee is transferred to another classification or to another shift, he or she shall maintain seniority in his or her original classification until that Employee has completed the three-(3)-month probationary period in his or her new job. Thereafter his or her Classification Seniority shall be retroactive to his or her date of transfer.

Section E. Seniority shall be broken when an Employee:

1. Terminates voluntarily.
2. Is discharged for cause.
3. Exceeds an official leave of absence.
4. Is absent for three (3) consecutive working days without properly notifying the Hospital, unless proper excuse is shown.
5. Fails to report for work within three (3) working days after being notified by telegram or mail to do so, unless proper excuse is shown.
6. Is laid off for six (6) consecutive months.

Section F. Bargaining Unit and Classification Seniority shall not accumulate during the period of an official leave of absence exceeding one month.

Section G. Employees whose pay is charged to a special or nonbudgetary Fund, and who are informed at the time of their hire or at the time of transfer that their employment is for a special nonbudgetary or research project, shall be excluded from the provisions of this Article. Such Employees may be laid off or transferred without regard to seniority.

Section H. Proper notification of absence for purposes of this Article shall be by telephone call to the Employee's supervisor immediately at the start of the work shift.

Section I. Notification of recall from layoff for purposes of this Section shall be by a telegram or a registered letter to the Employee's last known address, as shown by the Hospital's records.

VI. HOURS. This clause states the hours of work, time for lunch, rest periods, and number of days in a regular work week.

SAMPLE

Hours

1. The regular work week for all full-time Employees shall consist of the number of hours per week regularly worked by such employees as at June 30, 1988. The regular work week for part-time Employees shall not exceed five (5) days. Such hours, not to exceed forty (40) per week, shall be specified in a Stipulation (Stipulation H) between the Union and each Hospital, to be annexed hereto. Employees shall receive two (2) days off in each full calendar week except in the event of overtime.

2. The regular work day for all full-time Employees covered by this Agreement shall consist of the number of hours in the regular work week as above defined, divided by five (5), exclusive of an unpaid lunch period, except for those Employees who received a paid lunch period as of June 30, 1988.

3. The scheduling of weekends off shall be negotiated on a Hospital by Hospital basis.

VII. OVERTIME. This clause states when overtime premium pay shall be paid and in what amount. It also covers how overtime hours are to be distributed and the requirements for working such overtime.

SAMPLE

Overtime

1. Employees shall be paid one and one-half times their regular pay for authorized time worked in excess of the regular full-time work week for their classification as set forth in Article I, Section B.

2. The following paid absences shall be considered as time worked for the purposes of computing overtime: holidays, vacations, jury duty days, condolence days, paternity day, marriage days, and sick leave days. Unpaid absences shall not be considered as time worked.

3. The Hospital will assign, on an equitable basis, "on call" duty and required prescheduled overtime among qualified Employees. Employees shall be required to work overtime when necessary for the proper administration of the Hospital.

4. There shall be no pyramiding of overtime.

VIII. SHIFT AND SHIFT DIFFERENTIALS. This clause includes any provisions for special premium pay for work performed outside the regular day shift.

SAMPLE

Shifts and Shift Differentials

1. Employees working on shifts whose straight time hours end after seven (7:00) P.M. or begin prior to six (6:00) A.M. shall receive the following differentials:

a. Licensed Practical Nurses. An amount equal to three-fourths (¾ths) of the dollar amount of shift differential paid to Registered Nurses working at the same institution on the same shifts.
b. All other Employees. A shift differential of ten percent (10%) of salary, including specialty differential.

2. Employees shall work on the shift, shifts, or shift arrangements for which they were hired. The Hospital may change an Employee's shift only for good and sufficient reason, and any such change shall apply to the Employee with the least classification seniority qualified to do the work.

Whenever the Employee requests a change of shift, approval of such request shall not be unreasonably withheld if a vacancy exists in the classification in which he or she is then working and if more than one Employee applies, such change shall apply to the Employee with the most classification seniority qualified to do the work. Notwithstanding the foregoing, Employees shall have preference in filling vacancies on another shift in the classification in which he or she is then working over new Employees.

IX. DISCIPLINE. This clause may include work rules, the mechanism for discharging or suspending Employees.

SAMPLE

Discharge and Penalties

1. The Hospital shall have the right to discharge, suspend, or discipline any Employee for cause.
2. The Hospital will notify the Union in writing of any discharge or suspension within twenty-four (24) hours from the time of discharge or suspension. If the Union desires to contest the discharge or suspension, it shall give written notice thereof to the Hospital within five (5) working days, but no later than ten (10) working days from the date of receipt of notice of discharge or suspension. In such event, the dispute shall be

submitted and determined under the grievance and arbitration procedure [set forth in the full agreement].

3. If the discharge of an Employee results from conduct relating to a patient and the patient does not appear at the arbitration, the arbitrator shall not consider the failure of the patient to appear as prejudicial.

4. The term "patient" for the purposes of this Agreement shall include those seeking admission and those seeking care of treatment in clinics or emergency rooms, as well as those already admitted.

5. All time limits herein specified shall be deemed exclusive of Saturdays, Sundays, and Holidays.

X. MANAGEMENT RIGHTS. This clause outlines those activities where management is free to act subject only to the limitations of the contract.

SAMPLE

Management Rights

Section A. The management of the Hospital and the direction of the working forces are vested exclusively with the Hospital. The Hospital retains the sole right to hire, discipline, discharge, lay off, assign, and promote, and to determine or change the starting and quitting time and the number of hours to be worked; to promulgate rules and regulations; to assign duties to the work force; to reorganize, discontinue, or enlarge any department or division; to transfer Employees within departments, to other departments, to other classifications, and to other shifts; to introduce new or improved methods or facilities; to reclassify positions and carry out the ordinary and customary functions of management whether or not possessed or exercised by the Hospital prior to the execution of this Agreement, subject only to the restrictions and regulations governing the exercise of these rights as are expressly provided in this Agreement.

Section B. The Union recognizes that the Hospital has introduced a revision in the methods of feeding patients, which has and will provide a revision in job duties and a reduction in personnel in the Food Service Department. The Union agrees that nothing in this Agreement contained shall prevent the implementation of this program and of the specific reductions or of any other similar program to be hereafter undertaken by the Hospital.

Section C. The Union, on behalf of the Employees, agrees to cooperate with the Hospital to attain and maintain full efficiency and maximum patient care and the Hospital agrees to receive and consider constructive suggestions submitted by the Union toward these objectives.

XI. SEPARABILITY CLAUSE. This clause, sometimes referred to as a savings clause, protects the agreement from the possibility that any part of it is found to be contrary to the law.

SAMPLE

Effect of Legislation—Separability

It is understood and agreed that all agreements herein are subject to all applicable laws now or hereafter in effect; and to the lawful regulations, rulings, and orders of regulatory commissions or agencies having jurisdiction. If any provision of this Agreement is in contravention of the laws or regulations of the United States or of [the state of this agreement], such provision shall be superseded by the appropriate provision of such law or regulation, so long as same is in force and effect; but all other provisions of the Agreement shall continue in full force and effect.

XII. ENTIRE AGREEMENT CLAUSE. This clause, sometimes referred to as a zipper, provides that once negotiations have been completed, no further negotiations are necessary. Both parties mutually waive the right to negotiate on any further subject during the term of the Agreement.

SAMPLE

Duration

Section E. The Union, in consideration of the benefits, privileges, and advantages provided in this Agreement and as a condition of the execution of this Agreement, suspends meetings in collective bargaining negotiations with the Hospital during the term of this Agreement with respect to any further demands except as may be dealt with as a grievance . . . or except as may be dealt with under [a section not given here].

XIII. SICK LEAVE. This clause states the eligibility requirements for and amount of sick leave negotiated by the parties.

SAMPLE

Sick Leave

1. Employees, after thirty (30) days employment, shall be entitled to paid sick leave earned at the rate of one (1) day for each month of employment, retroactive to date of hire, up to a maximum of twelve (12) days per year. Employees, after one (1) or more years of employment with the Hospital, shall be entitled to a total of twelve (12) additional days of sick leave as of the beginning of his or her (sic) second and each subsequent year of employment, provided that at no time will an Employee be entitled to accumulate more than thirty-six (36) working days of sick leave during any one year, including the days earned or to be earned in the current sick leave year.

2. Pay for any day of sick leave shall be at the Employee's regular pay.

3. To be eligible for benefits under the Article, an Employee who is absent due to illness or injury must notify his or her supervisor at least one (1) hour before the start of his or her regularly scheduled work day, unless proper excuse is presented for the Employee's inability to call. The hospital may require proof of illness hereunder.

4. Employees who have been on sick leave may be required to be examined by the Hospital's Health Service physician before being permitted to return to duty.

5. If an Employee resigns or is dismissed or laid off and has exceeded his or her allowable sick leave, the excess sick leave paid shall be deducted from any moneys due him or her from the Employer at the time of resignation, layoff, or dismissal.

XIV. HOLIDAYS. This clause will include the requirements for eligibility for holiday pay, the number of holidays granted, and method of payment for such holidays.

SAMPLE

Holidays

Section A. Employees shall be entitled to eight (8) paid Holidays each year as follows:

1. New Year's Day
2. Martin Luther King's Birthday
3. Washington's Birthday
4. Memorial Day
5. Independence Day
6. Labor Day
7. Thanksgiving Day
8. Christmas Day

Section B. To be paid for a Holiday, an Employee must have worked that last complete scheduled shift prior to and the next complete shift after such Holiday unless the absence is authorized or excused.

Section C. Recognizing that the Hospital works every day of the year, and that it is not possible for all Employes to be off duty on the same day, the Hospital shall have the right, at its sole discretion, to require any Employee to work on any of the Holidays, provided however that such Employee shall be given a day off in lieu of such Holiday at the convenience of the Department. Employees who work on a Holiday and cannot be scheduled for a compensatory day off, at the discretion and option of the Hospital shall in lieu thereof be paid an additional day's pay at straight time, in addition to time and one-half their regular straight time rate of pay for all hours worked on the Holiday.

Section D. Employees shall be entitled to four (4) "free days" with pay in the course of a calendar year in addition to the seven (7) paid Holidays listed above. Free days shall be scheduled at the convenience of the Hospital and shall not be taken immediately preceding or immediately following vacation time or a Holiday. Request for scheduling a "free day" must be made by the Employee at least two (2) weeks prior to the date requested.

Section E. For the purposes of computation of pay for Holidays and "free days" under this Article, an Employee will be paid for his or her regular scheduled work day at his or her regular straight time hourly rate.

Section F. Employees will be entitled to time off with pay to vote at regularly scheduled city, state, or federal elections, in accordance with applicable state laws. Such time off will be granted only if the employee does not have at least four (4) hours time between the time the polling places open and the start of his or her work schedule or between the close of his or her work schedule and the time the polling places close. Such time off shall not exceed two (2) hours and shall be granted only if the Employee notifies his or her supervisor not more than ten (10) nor less than two (2) days before the day of the election.

A new series of clauses will be included in the new contract covering wages and minimums, vacation provisions, paid leave provisions, no-discrimination requirements, severance pay, uniform allowance, and any agreement on past practices.

Index

Absenteeism rates and organizational commitment, 193
Absolute standards of performance appraisal, 322–324
Acceptance, need for, 211–212
Acquisitions:
 maturity phase of product life cycle and, 54
 multihospital systems and, 57–58
Acuity patient classification systems, 158
Adaptive cultures, 208–209
 culture gaps and, 215
Adhocracy, 116
Administrative dominance organizational form, 187–189
Adverse impact and selection process, 275–276
Advertising and recruiting, 253, 255
Affirmative action, 246
Age discrimination in Employment Act of 1967, 275
Age and organizational commitment, 191
Agency shop, 399
AIDS, 286
Allied doctrine, 394–395
Alternative ranking performance appraisal, 322
American Arbitration Association (AAA), 406, 407
Analyzer organizations and human resources management systems, 418–419
Anticipated results, analysis, 25
Applicant file, 155, 156
 employment systems and, 159–160
Application blanks, 282–283
 example, 284–285
Aptitude tests, 277

Arbitration, 406–407
 disciplinary procedures and, 403
Assessment centers, 283, 286
Attitude assessment, 380
Autonomous structure, 186–187

Bargaining unit recognition contract clause, 426–427
Base pay, 356, 362
Base rate criteria for a selection device, 288
Behavioral approaches to performance appraisal, 102–104
 defined, 102
Behaviorally anchored rating scales (BARS), 102, 323–324
Behavioral strategy, development based on, 234
Behavioral systems, overview, 10–11. *See also* Commitment of professionals; Corporate culture; Motivation of professionals; Organizational change; Professionals in health service organizations
Belief in people, 224–225
Benefits, 362–368
 cost containment and, 364–365
 flexible, 365–367
 history, 362–363
 HRIS and, 163
 legally required, 364
 medical insurance, 366
 pension, 364
 as percentage of payroll, 364, 365
 table, 363
 taxing, 367–368
 trends in, 368
 union organization and, 378

437

Bona fide occupational qualification (BFOQ), 255, 260, 261, 276
Bonuses, 357
Boston Consulting Group (BCG) business grid, 35-36
  drawback, 44
  strategy identification and selection and, 40
Budgeting, 21-23
  defined, 21
  described, 21
  flow of steps in, 22
  planning and, 21-22
Build strategy, 26
Bureaucracies, 115-116
  machine, 116
  professional, 116
  professionals working within, 184-186
    conditional loyalty of, 184-185
    increasing conflict of, 185-186
    resolution of conflict of, 189-190

Career development, 299. *See also* Training and development
Cash cows, 35-36
Centralization of structure, 114
  influence on human resources management, 116-137
Central tendency and leniency error, 332
Change, *see specific types of change*
  leadership for, *see* Transformational leadership
Change agents, 223
Charisma, 223
  characteristics, 223-224
Check-off contract clause, 428-429
Civil Rights Act of 1964, 255, 260, 275
Clarity of task and performance appraisal, 328
Classical theory of motivation, 200
Clinical information systems, 141
Closed shop, 398-399
Cognitive strategy, development based on, 234
Collapse of niche, 61
  strategic and tactical responses to, 69-70
Collective bargaining, *see* Labor relations contracts, negotiating and administering
Commitment of professionals, 190-198. *See also* Motivation of professionals
  characteristics, 190
  control systems and, *see* Control systems
  defined, 190
  determinants, 191-192

  during the early employment period, 192, 194
  exchange concept and, 190, 194
  managerial applications, 195-198
  outcomes of, 192-195
    absenteeism, 193
    job performance, 194-195
    physician commitment, 195
    tardiness, 193
    tenure, 193-194
    turnover rates, 192-193
  overview, 10-11, 181-182, 190
  references, 204-205
  summary, 204
Committees, review, 196
Comparable worth, 353
Comparative methods of performance appraisal, 322
Compensation, 348-369
  base pay, 356, 362
  benefits, *see* Benefits
  defined, 348
  HRIS and, 162-163
  incentives, *see* Incentive compensation programs
  maturity stage of product life cycle and, 46-47
  objectives, 348
  organizational strategy and, 44
  organizational structure, human resources management and:
    in the group medical practice, 125-126
    in the hospital, 133-134
    in the solo medical practice, 120-121
    summary, 137, 138
  overview, 14, 348
  pay range system, 355
  preventive labor-management program and, 383
  references, 369
  rewards, *see* Rewards
  strategic planning and, 349-355
    components of management system, 351-355
    environmental forces and, 349
    health care personnel and, 350
    job analysis, 352-353
    job design, 351-352
    job evaluation, 353-354
    job pricing, 354-355
    model for, 349-350
  summary, 368
  time factor and, 356

union organization and, 377
Competition, 409, 410
  maturity stage of product life cycle and, 46
Competitive advantage, 27–28
  cost leadership and, 27
  differentiation and, 27–28
  focus and, 28
Competitive analysis, 25
Competitor analysis, 32
Complexity:
  ability to deal with, 224
  evaluation process and, 328
Complexity of structure, 114
  influence on human resources management, 116–137
Computers, 86
  HRIS and:
    evolution and, 143–147
    software, see Software, computer
    trends in, 147–148
Concurrent validity model, 174–175
Conditional loyalty, 184–185
Connecticut v. *Teal*, 275
Consolidation strategies, domain, 65
  contraction of niche and, 68
  erosion of niche and, 67–68
  examples, 66
Consumer groups, analysis, 32
Content theories of motivation, 200–201
  classical theory, 200
  hierarchy of needs, 200–201
  learned-need theory, 201
  two-factor theory, 201
Contingency theory, 90
Contingent punishment, 223
Contingent reward behavior, 222–223
Continuing education, 299. See also Training and development
Contract, union, see Labor relations contracts, negotiating and administering
Contract clauses, 397–402
  agency shop, 399
  checklist for, 426–436
    check-off, 428–429
    discipline, 432–433
    entire agreement, 434
    holidays, 435–436
    hours, 431
    management rights, 433
    overtime, 431
    probationary period, 429
    recognition, 426–427
    seniority, 429–431
    separability, 433–434
    shift and shift differentials, 432
    sick leave, 434–435
    union security, 427–428
  closed shop, 398–399
  "free rider", 399
  maintenance of membership shop, 398
  management rights, 399–400, 433
  modified union shop, 398
  no-lockout, 402
  no-strike, 402
  open shop, 398
  seniority, 401–402, 429–431
  union security, 397–399, 427–428
  union shop, 398
Contraction of niche, 60, 61
  strategic and tactical responses to, 68
Contrast effects error, 332
Control, 53–54
Control systems, 195–198
  case study on, 197–198
  committees, 196, 198
  expectations and, 196–197
  interaction with peers and, 198
  performance appraisal and, 335
  pricing systems, 196
  reaction to, 196
Corporate culture, 206–218
  adaptive cultures, 208–209
  challenging, 210–211
  culture gap and, see Culture gap
  culture rut and, 207, 209, 215–217
  culture shock and, 207–208
  definitions, 206–207
  described, 207
  dysfunctional cultures, 209, 213–214, 217
  fit of HRIS system objectives and design with, 174–175
  folklore and, 210
  formation, 209–211
  maintenance, 211–212
  need for acceptance and, 211–212
  norms and, see Norms
  overview, 11, 206–208
  references, 218
  resistance to change and, 212–217, 229
  as a separate entity, 210–211
  staff development and, 304–305
  summary, 218
Corporate restructuring, 57–58
Cost leadership strategy, 27
  focus and, 28
  organizational life cycle and, 55

440   Index

Cost reduction strategy, 420
Courageous leadership, 224
Creation strategies, domain, 65
  collapse of niche and, 69
  dissolution of niche and, 69
  examples, 66
Creativity, 52
Credentialing, 279–282
Crisis syndrome, 63–64
Critical incidents performance appraisal, 323
Cultural change, 212–217. *See also* Norms
  transformational leadership and, 229
Culture as component of environmental analysis, 33
Culture gap, 207
  closing, 216–217
  identifying, 214–216
  level of hierarchy and, 215
  organizational denial and, 216–217
  patterns, 215
  types, 214–216
Culture rut, 207, 209, 215–217
Culture shock, 207–208

Data collection systems, 143
Decision-making, *see* Management decision-making
Decline phase of organizational life cycle, 54–55, 58–71
  causes, 59
  centralized decision making and, 64
  changes in ecological niches and, 60–61
  constraints in the management of, 70–71
    reasons for, 70–71
    types, 70
  control systems and, 64–65
  crisis syndrome and, 63–64
  as cutback, 58
  decision-making and, 64
  efficiency and conservatism *vs.* effectiveness and innovation, 70–71
  environmental entropy and, 60
  long-run planning and, 64
  motivation of employees and, 64
  organizational atrophy and, 59
  political vulnerability and, 59
  problem depletion and, 60
  references, 79–82
  short-run strategy and, 64
  as stagnation, 58–59
  strategic and tactical responses, 65–70
    collapse, 69–70
    contraction, 68

  dissolution, 68–69
  erosion, 66–68
  table, 67
  types, 65–66
  strategy, human resources management and, 77–79
    reversible decline phase, 77–78
    terminal decline phase, 78–79
    summary, 79
  symptoms, 62–65
    early, 63
    later, 63–65
    table, 62
Decline phase in product (service) life cycle, 40
  human resources management and, 47
Defender organizations, 55
  human resource management systems and, 418–419
Defense strategies, domain, 65
  contraction of niche and, 68
  dissolution of niche and, 69
  examples, 66
Demographics:
  environmental analysis and, 33
  market definition and, 31
Departmental level, assessment at, 310
Development, *see* Training and development of subordinates, 224–225
Diagnosis-related groups (DRGs), 28
  job pricing and, 355
Differentiation strategy, 27–28
  focus and, 28
  organizational life cycle and, 55
Disciplinary procedures, 402–404
  arbitration and, 403
  basic rules, 403–404
  as a contract clause, 432–433
  management rights and, 402–403, 404
  preventive labor-management program and, 382–383
  records and, 404
  union organization and, 378
Discrimination, 246, 253, 255
  performance appraisal and, 321, 335
  selection and, 269, 275–276
    interviews and, 278
    tests and, 277
  sex, 353
Disparate rejection rates, 276
Dissolution of niche, 60–61
  strategic and tactical responses to, 68–69
Distinct competence, 31
Diversification analysis, 25

Divestment strategy, 27
Divisionalized form of organization, 116
Dogs, 36
Domain strategies, 65
  collapse and, 69–70
  contraction and, 68
  dissolution and, 68–69
  erosion and, 66–68
  examples, 66
Drug testing, 287–288
Dysfunctional cultures, 209
  norms, 213–214, 217

Ecological niches, changes in, 60–61
  strategic and tactical responses to, 65–70
Economics as component of environmental analysis, 33
Educational institutions as recruitment sources, 253
Education and organizational commitment, 191
Empathy, 233
Employee attitude assessment, 380
Employee relations, *see* Labor relations
Employee Retirement Income Security Act (ERISA), 362–363
Employment agencies, 252
Employment systems, 159–162. *See also* Recruitment; Selection and placement
  applicant file and, 159–160
  case study, 160–162
Empowerment of workers, 225
Entire agreement contract clause, 434
Entrepreneurship, 52, 53
Environment:
  accommodation with, 56
  analysis, *see* External assessment
  change in, 409–411
    future, 410–411
    human resources implications, 411–414
    recent, 409–410
  compensation and, 349
  as determinant of structure, 114–115
  effect on health practitioners, 4–6
  manipulation, 56
  niche, 60–61
Environmental entropy, 60
Environmental strategy, development based on, 234
Equal Employment Opportunity Commission (EEOC), 246
  appraisal systems and, 321
  recruitment questions that cannot be asked and, 260–261

Erosion of niche, 60, 61
  strategic and tactical responses to, 66–68
Evaluation, *see* Performance appraisal
Exceptional performance awards, 361–362
Executive compensation programs, *see* Incentive compensation programs, executive
Executive search firms, 252–253
  characteristics of reputable, 254–255
Expectancy theory of motivation, 201, 202
External assessment, 31–33
  adaptations and, 91
  competitor analysis, 32
  consumer groups, 32
  industry and environmental analysis, 33
  strategy, technology and, 91–92
External recruitment, 251–255
  advantages and disadvantages, 251
  advertisements, 253, 255
  direct applications, 252
  educational institutions, 253
  employment agencies, 252
  executive search firms, 252–253
    characteristics of reputable, 254–255
  referrals, 252
  special events, 253
Extrinsic rewards, 104
  job performance and, 344–345

Federal Mediation and Conciliation Service (FMCS), 395, 396
Financial information systems, 142, 143
First-line supervisors in a preventive labor-management program, 381
Flexible benefit plans, 365–367
Focus strategy, 28
  organizational life cycle and, 55
Forced choice performance appraisal, 323
Forced distribution performance appraisal, 322
Formalization of structure, 113–114
  influence on human resources management, 116–137
  mature organization and, 53–54
Formal structure, 113
"Free rider", 399
Fringe benefits as reward, 105. *See also* Benefits
Functional grouping of workers, 114

Gender:
  discrimination, 353

Gender (*Continued*)
  female–male employee ratio and union organization, 376–377
  interviews and, 278
  organizational commitment and, 191
General Electric (GE):
  business screen, 36–37
    drawback, 44
    strategy identification and selection and, 40–41
  work planning and review process, 341–342
    defined, 342
    development, 342
    diagrammed, 343
Goals:
  individual, 303–304
  motivation and setting, 203–204, 335
  organizational, 302–303, 304
Government:
  as component of environmental analysis, 33
  HRIS and reporting to, 164–165
  job analysis and, 269
  performance appraisal and, 321
  selection process and, 275–276
Graphic rating performance appraisal, 321, 323
Graphology testing, 287
Grievance procedures, 404–406
  manager's responsibility for, 405
  objectivity and consistency and, 404–405
  preventive labor-management program and, 382–383
  rules for administration, 405–406
  steps in, 404–405
  union organization and, 378
Griggs v. *Duke Power*, 277, 321
Group incentives, 360
Growth phase of organizational life cycle, 52–53
  reinstituting, 55–58
    corporate restructuring and, 57–58
    transformational leadership and, 56–57
  strategy, human resources management and, 76
Growth phase of product (service) life cycle, 39
  human resources management and, 45–46

Halo effect error, 332
Harvest strategy, 27
Health Care Amendments of 1974 (Public Law 93–360), 372, 395–397
  changes brought by, 395–397
  definition of health care institution, 373
  social turmoil and, 374
  strikes and, 395, 396–397
Health maintenance organizations (HMOs), 46
Hedonism, 199
Hierarchical structure, 115
Hierarchy of needs theory of motivation, 200–201
Hold strategy, 26
Holiday contract clause, 435–436
Horizontal complexity (differentiation) of structure, 114
Horizontal integration, 57–58
Hospital Corporation of America (HCA), 210
Hospital information systems, 144–147
Hours of work contract clause, 431
HTLV–III, 286
Human resources information systems (HRIS), 141–177
  applicant file and, 155, 156
  clinical information systems, 141
  data bases for, 154–156
  data collection systems, 143
  definitions, 141
  evolution in hospitals, 142–147
  financial information systems, 142, 143
  fully integrated, 145, 147
  functions, 157–165
    compensation systems, 162–163
    employee relations, 164
    employment systems, 159–162
    government reporting, 164–165
    manpower planning and allocation, 157–159
    performance appraisal, 164
    training and development, 163–164
    work load measure systems, 158
  goal, 152–153
  hospital information systems, 144–147
  implementation, 165–174
    borrowing software and, 172
    building software and, 171–172
    buying software and, 171
    data collection and, 170
    data maintenance, 165–166, 167
    design and, 166, 168, 170–171
    input, 165, 167
    modules for, 168–170
    outputs, 166, 168
    selecting software, 172–174
    three major functional components, 166
  integration of subsystems and, 149–151
  management decision-making level and, 151, 152, 154
  management information systems, 141, 145

medical information systems, 143, 144
model, 149–152
objectives, 166, 168, 170
operational decision-making level and, 151, 152, 154
overview, 9–10
payroll file and, 155, 156
performance appraisal and, 334–335
personnel department and, 148–149
personnel file and, 155, 156
references, 177
role, 152–156
special issues, 174–176
　fit of system with organizational culture, 174–175
　top management involvement, 176
　user or employee resistance, 175–176
state of the art, 145, 146
strategic decision-making level and, 151–152, 153
strategies for coping with the environment and, 148
summary, 176
systems theory and, 149–151
trends in, 145, 147–148
weaknesses in, 147
Human resources model of hospital administration, 412–414
Human resources planning (HRP):
described, 415
integration of strategic planning with, 414–421
　business strategies for, 415, 418–420
　readiness for, 415, 416–417
Human value vs. efficiency, 373
Hygiene factors and job enrichment, 95

Incentive compensation programs, 356–362
bonuses and, 357
defined, 357
exceptional performance awards, 361–362
executive, 357–360
　annual, 358
　benefits from, 359–360
　effectiveness, 358, 359
　health care industry and, 357
　long-term, 358–359
　perquisites and benefits, 358
　targets and, 358
group or team, 360
physician, 361
purposes, 357
rewards, see Rewards
summary, 362

trends in, 360–362
union organization and, 377
Industry analysis, 33
Influence of structure on human resources management, 116–140. See also Organizational structure
in the group medical practice, 122–128
　compensation, 125–126
　differences between solo and, 127–128
　evaluation, 126
　job design, 123–124
　recruitment, 124–125
　training, 125
in the hospital, 128–137
　compensation, 133–134
　evaluation, 134–136
　illustration, 136–137
　job design, 128–130
　recruitment, 130–132
　training, 132–133
overview, 116–117
references, 137–140
in the solo medical practice, 118–122
　compensation, 120–121
　differences between group practice and, 127–128
　evaluation, 121–122
　job design, 118–119
　recruitment, 119
　training, 119–120
summary, 137, 138
Informal structure, 113
Information systems, see Human resources information systems (HRIS)
Information technology, 86
defined, 87–88
Innovative process, 220
Innovation strategy, 420
In-service education, 299. See also Training and development
Intellectual stimulation, 224
Intensive technology, 87
Internal assessment, 33–34
Internal work motivation, 95
Interorganizational structures, 113
Interviews, selection process and, 277–279
Intrinsic rewards, 104
Introduction phase of organizational life cycle, 52
　strategy, human resources management and, 73, 76
Introduction phase of product (service) life cycle, 39
　human resources management and, 45

Job analysis, 268–272
　compensation and, 352–353
　components, 268
　conventional job descriptions, 270–271
　defined, 97, 352
　government guidance in, 269
　information gathering methods, 269
　steps in, 269–272
Job characteristics and organizational commitment, 191
Job descriptions:
　conventional, 270–271
　criteria for selection and, 272
　defined, 268
　job design and, 351
　sections, 269
Job design, 94
　compensation and, 351–352
　organizational structure, human resources management and:
　　in the group medical practice, 123–124
　　in the hospital, 128–130
　　in the solo medical practice, 118–119
　　summary, 137, 138
　technology and, 94–97
　　defined, 94
　　hygiene factors and, 95
　　internal work motivation and, 95, 96
　　job characteristics model and, 95, 96
　　job enlargement, 94
　　job enrichment, 94–95
　　job rotation, 94
　　motivators and, 94–95
　　relationship between, 95–97
Job evaluation, 353–354
　centralization and, 354
　comparable worth and, 353
　point system for, 353
　relative value and, 353–354
Job performance:
　extrinsic rewards and, 344–345
　job design and, 351, 352
　organizational commitment and, 194–195
Job pricing, 354–355
　appearance and reality of fairness and, 354
　complexity, 354–355
　pay range system and, 355
　unusual requirements and, 354
Job rotation, development based on, 234
Job satisfaction and voting for union representation, 374, 375
Job simulation, 279
Job specifications, 97–98
　selection criteria and, 100–102

technological change and, 98–99
Job transfers and union organization, 378
Justification of personnel actions, 335–336

Kilmann–Saxton Culture-Gap Survey, 214–216

Laboratories, staff allocation in, 158–159
Labor law history and trends, 372–373
Labor Law Reporter (LLR), 387, 388
Labor Management Relations (Taft–Hartley) Act, 372
Labor–Management Reporting and Disclosure (Landrum–Griffin) Act, 372
Labor market, 355
Labor relations, 370–408
　causes of labor–management problems, 373–378
　contracts, *see* Labor relations contracts, negotiating and administering
　developing a philosophy and strategy, 371
　HRIS and, 164
　human value *vs.* efficiency and, 373
　labor law history and trends, 372–373
　maintaining nonunion status, 371–372
　management rights and, 373
　management strategy for reactions during union organizing campaigns, 383–384
　overview, 14–15, 370–371
　preventive program, *see* Preventive labor–management program
　references, 384–385
　summary, 384
　what the union organizer looks for, 376–378
　　discipline and grievance procedures, 378
　　employee loyalty by work shift, 376
　　female–male employee ratio, 376–377
　　fringe benefits, 378
　　incentive pay, 377
　　job transfers, 378
　　overtime practices, 377
　　promotion policy, 377–378
　　seniority, 377
　　wage rates, 377
　　work environment, 377
　why employees join unions, 374–376
　　correlation between job satisfaction and voting for unions, 374, 375
　　factors in, 374, 375
　　perception of poor treatment, 376, 379
　　poor communication, 376, 379
Labor relations contracts, negotiating and administering, 386–408
　administering, 402–407
　arbitration, 406–407

discipline, 402–404
  grievance procedure, 404–406
considerations that affect the bargaining milieu, 390–391
contract clauses, see Contract clauses
cornerstones of successful, 386
described, 386
"good faith" bargaining, 387–388
Health Care Amendments of 1974, see Health Care Amendments of 1974 (Public Law 93–360)
HRIS and, 164
legal definition of bargaining, 387–388
management negotiators and, 390
  selecting, 391–392
mandatory bargaining subjects, 388–389
overview, 386
references, 408
strategies for bargaining, 392–395
  dos and don'ts, 393
  information gathering and, 392
  strikes and, see Strikes
summary, 407–408
voluntary bargaining subjects, 389–390
Labor relations model of hospital administration, 412, 413, 414
Landrum–Griffin Act, 372
Leadership and change, 221–231
  empirical evidence on, 222
  references, 235–237
  relationship between, 226–231
  significance, 222
  summary, 234–235
  transactional, see Transactional leadership
  transformational, see Transformational leadership
  types, 221–226
Leadership development, 231–234
  adaptability and, 232
  at all levels, 234
  attitudes and, 232
  behavioral strategy and, 234
  cognitive strategy and, 234
  empathy and, 233
  environmental strategy and, 234
  human resources system and, 233
  identity and, 232
  nature vs. nurture and, 231–232
  network analysis and, 233
  performance and, 232
  planful opportunism and, 233
  symbolic leadership and, 233
  visioning and, 233
Learned-need theory of motivation, 201

Legitimacy, 60
Levels of analysis in organizations, 92–93
Licensing, 282
Life cycles:
  career, 106
  organizational, see Organizational life cycles
  product, see Product life cycle portfolio matrix
Life-long learning characteristic, 224
Long-linked technology, 87
Long-range planning, 23–24
  defined, 21
  described, 23
  flow of steps in, 24
  limitations, 23–24
Loss of control by professionals, 185–186, 189, 196–198, 217, 410, 411
Loyalty by work shift and union organization, 376

Machine bureaucracy, 116
  in the group medical practice, 122–128
Maintenance of membership shop, 398
Management-by-exception, 223
Management-by-objectives (MBO), 102, 336–344
  advantages of a successful program, 341
  applications, 341–344
    General Electric work planning and review process, 341–342, 343
    St. Charles Hospital, 342–344
  as an appraisal system, 339
  common elements, 336
  defined, 338
  described, 324, 336
  guidelines for implementing, 340–341
  health organizations and, 339–340
  objective list for, 337–338
  objective setting process, 337
  origin, 321, 336
  overview of cycles, 338
Management decision-making:
  management level, 151, 152, 154
  operational level, 151, 152, 154
  strategic level, 151–152, 153
  structure and, 115
Management information systems (MIS), 141, 145. See also Human resources information system (HRIS)
Management rights, 373
  as a contract clause, 399–400, 433
  disciplinary procedures and, 402–403, 404
  grievance procedure and, 404–406
Management technology, 86–87

446 Index

Managers, recruiting, 242
  external, 251–252
  issues concerning, 243
  sources for, 252–256
  summary, 263
Manpower planning and allocation, 157–159
Market definition, 31
Market segments, 31
Market structure, 115
Mason Medical Center HRIS case study, 160–162
Matrix model of hospital administration, 412–414
Matrix structure for grouping workers, 114
Maturity phase of organizational life cycle, 53–54
  strategy, human resources management and, 77
Maturity phase of product (service) life cycle, 39–40
  human resource management and, 46–47
Mediating technology, 87
Medical information systems, 143, 144
Medical insurance, 366
Medical technologies, 85–86
Medicare, 196, 355
Mentoring, 234
Minorities, *see* Discrimination
Mission statement, 29–31
  corporate- or business-level, 30
  described, 29
  distinct competence, 31
  market definition, 31
  product definition, 30–31
Modified union shop, 398
Mortality committees, 196
Motivation of professionals, 199–205. *See also* Commitment of professionals
  basics, 199
  content theories, 200–201
    classical theory, 200
    hierarchy of needs, 200–201
    learned-need theory, 201
    two-factor theory, 201
  in decline phase of organizational life cycle, 64
  expectancy theory, 201, 202
  goal setting and, 203–204, 335
  in health services organizations, 202–203
  hedonism and, 199
  importance, 199
  internal work, 95, 96
  job enrichment and, 94–95
  overview, 10–11, 181–182, 199

  performance appraisal and, 335
  process theories, 201, 202
  references, 204–205
  summary, 204
  summary of theories, 202
  work planning and review and, 342
Multihospital systems, 57–58
Multiskilled health practitioners, 422–423

National Labor Relations Act (NLRA), 372
  collective bargaining and:
    "good faith", 387
    mandatory subject, 388
    secondary employers and, 394–395
  Health Care Amendments of 1974 and, 396, 397
National Labor Relations Board (NLRB), 372, 387
  Allied doctrine of, 394–395
  collective bargaining and:
    "good faith", 388
    mandatory subjects, 388–389
    voluntary subjects, 389–390
  Health Care Amendments of 1974 and, 395–397
Network analysis, 233
New developments in human resources management, 409–425
  enhancement of employee productivity and, 421–422
  environmental change, 409–411
  future, 410–411
  human resources implications, 411–414
  recent, 409–410
  integration of strategic and human resources planning, 414–421
    assessing readiness for, 415, 416–417
    business strategies and, 415, 418–420
  multiskilled health practitioners, 422–423
  references, 423–425
  summary, 423
  two models for the future, 412–414
Niches, ecological, 60–61
  strategic and tactical responses to decline of, 65–70
No-lockout clause, 402
Nonunion policy, 379. *See also* Preventive labor–management program
Norms, 211
  assessing and changing cultural, 212–217
    articulating new directions, 213
    closing culture gaps, 216–217
    establishing new norms, 213–214
    identifying culture gaps, 214–216

Index 447

surfacing actual norms, 213
described, 211, 212
need for acceptance and, 211–212
No-strike clause, 402
Nurses:
commitment, *see* Commitment of professionals
job analysis and, 352
job design and, 351–352
motivation, *see* Motivation of professionals
recruiting, 242, 257–258
external, 251
incentives for, 259
issues concerning, 243
sources, 252–256
summary, 263
time frame for, 246

Offense strategies, domain, 65
erosion of niche and, 66–67
examples, 66
Open shop, 398
Operational decision-making level, 151, 152
tasks, 154
Operational planning:
recruitment and, 245
time frame for, 243
Organizational atrophy, 59
Organizational change, 219–237
innovation process and, 220
leadership and, *see* Leadership and change
transformational, *see* Transformational leadership
overview, 11, 219–220
types, 221
Organizational culture, *see* Corporate culture
Organizational development, 296
Organizational domain, 60
Organizational level, assessment at, 310
Organizational life cycle, 51–82. *See also* Product life cycle portfolio matrix
described, 51–55
decline phase, 54–55
growth phase, 52–53
maturity phase, 53–54
start-up phase, 52
summary, 55
matching human resources to, 72
strategic objectives and, 72–73
organizational decline, *see* Decline phase of organizational life cycle
references, 79–82
reinstituting growth, 55–58

corporate restructuring and, 57–58
transformational leadership and, 56–57
strategy, human resources management and, 73–79
growth phase and, 76
maturity phase and, 77
reversible decline phase and, 77–78
start-up phase and, 73, 76
table, 74–75
terminal decline phase and, 78–79
summary, 79
Organizational planning, 243–246
recruitment planning and, 245
reasons for, 245–246
time frames for, 243
Organizational strategy, 20–50. *See also* Strategic management of human resources; Strategic planning
external assessment, 31–33
competitor analysis, 32
consumer groups, 32
industry and environmental analysis, 33
HRIS and, 148
human resources management and, 42–47
appraisal, 44
compensation, 44
defined, 42
development, 43–44
portfolio analysis and, 44–47
recruitment and selection, 43
identifying strategic business units (SBUs), 28–29
internal assessment, 33–34
mission of the organization and, 29–31
distinct competence, 31
market definition, 31
product definition, 30–31
model process, 30
overview, 20
planning methods and, *see* Planning methods
portfolio assessment, 34–40
BCG business grid, 35–36
GE business screen, 36–37
product life cycle matrix, 37–40
references, 49–50
strategy identification and selection, 40–42
summary, 47–49
technology and, 91–92
types of strategy, 26–28
build, 26
competitive advantage and, 27–28
cost leadership, 27
differentiation, 27–28

Organizational strategy (*Continued*)
  divest, 27
  focus, 28
  harvest, 27
  hold, 26
Organizational structure, 112–140
  adhocracy, 116
  bureaucracies, 115–116
  centralization, 114
  commitment and, 191–192
  complexity, 114
  conceptualizing, 113–116
  determinants, 114–115
  dimensions, 113–114
  divisionalized, 116
  environment and, 114–115
  formal, 113
  formalization, 113–114
  functional groupings, 114
  hierarchies, 115
  horizontal differentiation, 114
  human resources management and, *see* Influence of structure on human resources management
  informal, 113
  interorganizational, 113
  machine bureaucracy, 116
  managerial decisions and, 115
  markets, 115
  matrix, 114
  overview, 9, 112–113
  product grouping, 114
  professional bureaucracy, 116
  quasi-firms, 115
  references, 137–140
  simple structure, 116
  size of organization and, 115
  summary, 116, 137
  technology and, 90–91, 115
  vertical differentiation, 114
Orientation training, 298–299. *See also* Training and development
Outcomes approaches to performance appraisal, 102–104
  defined, 102
Outplacement, 47
Overtime contract clause, 431
Overtime practices and union organization, 377

Paired comparison performance appraisal, 322
Paraprofessional technologies, 86
Pay as reward, 105

Payroll file, 155, 156, 162
  performance appraisal and, 164
Peer Review Organization (PRO), 186–187
Pensions, 364
Performance appraisal (PA), 319–347
  absolute standards, 322–324
  alternative ranking, 322
  in autonomous structure, 186–187
  behavioral approaches to, 102–104
  behaviorally anchored rating scale (BARS), 323–324
  clarity and, 328
  common methods, 322–324
  comparative methods, 322
  complexity and, 328
  criteria for choice of techniques for, 326
  criteria specifications and, 102–104
  critical incidents, 323
  developmental function, 327, 335
  evaluation and, 327
  evaluation of, 329–331
  extrinsic rewards and job appraisal, 344–345
  forced choice, 323
  forced distribution, 322
  graphic rating, 321, 323
  historical development, 320–321
  HRIS and, 164
  as linked to other corporate processes, 329
  management by objectives and, *see* Management by objectives (MBO)
  maturity stage of product life cycle and, 46–47
  objectives, 329
  organizational strategy and, 44
  organizational structure, human resources management and:
    in the group medical practice, 126
    in the hospital, 134–136
    in the solo medical practice, 121–122
    summary, 137, 138
  outcomes approaches to, 102–104
  overview, 13–14, 319–320
  paired comparison, 322
  peer review, 186–187
  predictability and, 328–329
  preventive labor–management program and, 381–382
  problems, 102
  purposes, 102, 327–328
  questionnaire to evaluate, 329, 330
  reasons for, 32
  recommendations on techniques for, 326
  references, 346–347

reliability of data and, 329
straight ranking, 322
strategic role, 333–336
  control of personnel and, 335
  development of personnel and, 335
  dual nature, 333–334
  human resources planning and, 334–335
  justification of personnel actions and, 335–336
  summary, 345
system problems, 331–333
  developmental, 331
  implementational, 331–332
  operational, 332–333
technology and, 102–104
trait approaches to, 102
weighted checklists, 322–323
"Performance gap", 56
Perquisites, 358
Personal freedom culture gap, 215
Personnel department. *See also functions of e.g.* Compensation; Training and development
  HRIS and, 148–149
Personnel file, 155, 156
  training and development, 163
Physical examinations, 286
Physicians:
  commitment, *see* Commitment of professionals
  compensation, 361
  credentialing, 279–282
  motivation, *see* Motivation of professionals
  recruiting, 242, 257–258
    analysis of medical staff and, 247–249
    difficulties in, 257–258
    external, 252
    incentives for, 259
    issues concerning, 243
    legal issues and, 261
    sources for, 252–256
    strategic and operational planning and, 247
    summary, 263
    time frame for, 246, 248–249
Placement, *see* Selection
Planful opportunism, 233
Planning problems, 20–25
  budgeting, 21–23
  long-range planning, 23–24
  strategic management, 25
  strategic plannings, 24–25
Political change, 229

Political vulnerability, 59
Polygraph testing, 286–287
Population comparisons, 276
Portfolio analysis, 25
Portfolio assessment, 34–40
  BCG business grid, 35–36
  GE business screen, 36–37
  human resources management and, 44–47
    decline, 47
    growth, 45–46
    integration, 48
    introduction, 45
    maturity, 46–47
  product life cycle matrix, 37–40
Positioning, 30–31
Predictability of task and performance appraisal, 328–329
Predictive validity model, 274
Preventive labor–management program, 378–383. *See also* Labor relations
  disciplinary policies and procedures, 382–383
  employee attitude assessment, 380
  employee training, 380–381
  employee value systems, 381
  first-line supervisors, 381
  nonunion policy, 379
  performance appraisal, 381–382
  personnel selection, 379–380
  reasons for joining unions and, 379
  strategy for reactions during union organizing campaigns, 383–384
  wages, 383
"Primary" responsibilities, 118–119
  compensation and, 121
Private employment agencies, 252
Proactive *vs.* reactive leadership, 225–226
Probationary period contract clause, 429
Problem depletion, 60
Process modifications, 56
Process systems, overview, 11–15. *See also* Compensation; Labor relations; Performance appraisal; Recruitment; Selection; Training and development
Process theories of motivation, 201, 202
Product definition, 30–31
Product grouping of workers, 114
Productivity, enhancement, 421–422
Product life cycle portfolio matrix, 37–40. *See also* Organizational life cycle
  decline stage, 40
  human resources management and, 47
  growth stage, 39

Product life cycle portfolio matrix (*Continued*)
  human resources management and, 45–46
  introductory stage, 39
    human resources management and, 45
  life cycle model, 38
  maturity stage, 39–40
    human resources management and, 46–47
  strategy identification and selection and, 41–42
Professional bureaucracy, 116
  in the group medical practice, 122–128
  in the hospital, 128–137
Professionals in health service organizations, 181–190. *See also* Commitment of professionals; Motivation of professionals
  complex groupings, 182
  conflict for, 184–186
    conditional loyalty and, 184–185
    increasing, 185–186
    resolution of, 189–190
  definition, 183
  loss of control by, 185–186, 189, 196, 198, 217, 410, 411
  organizational forms used by, 186–189
    administrative dominance organizational form, 187–189
    autonomous structure, 186–187
  overview, 182–183
  references, 204–205
  summary, 204
  unions and, 376–377. *See also* Labor relations
  variety, 182–183
Promotion policy and union organization, 377–378
Promotion as reward, 105
Prospector organization, 55
  human resource management systems and, 418–419, 420
Public employment agencies, 252
Public Law 93-360, *see* Health Care Amendments of 1974

Quality enhancement strategy, 420
Quasi-firm structure, 115
Question marks, 36

Raw materials, 88
Reactive *vs.* proactive leadership, 225–226
Reactors, 55
Recognition contract clause, 426–427
Reconfiguration, 56–57

Recruiters, 257
Recruitment, 241–266
  advertising and, 253, 255
  analysis of medical staff and, 247–249
  categories of health care personnel, 242
  demand for new strategies in human resources, 243–246
  economic conditions and, 246–247
  external, *see* External recruitment
  HRIS case study and, 160–162
  implementation, 256–261
    attracting recruits, 256–257
    legal issues and, 260–261
    realistic job previews and, 257
    recruiters and, 257
    recruitment of nurses and physicians, 257–258, 259
    screening recruits, 258, 260
  internal, 249–250
    advantages and disadvantages, 251
  introduction stage of product life cycle and, 45
  law and, 246, 253, 255, 260–261
  linkages between organization-wide planning and, 245
  methods, 250, 253, 255
  new recruiting needs, 241–242
  number and type of positions, 246
  organizational strategy and, 43
  organizational structure, human resources management and:
    in the group medical practice, 124–125
    in the hospital, 130–132
    in the solo medical practice, 119
    summary, 137, 138
  overview, 12, 241–242
  planning, 243–256
  process, 242–243
    evaluating, 261–262
    steps in, 244
  references, 264–266
  replacement chart for, 250
  responsibility for, 249
  sources, 249–256
    evaluating, 255–256
  summary, 262–263
  time frame for, 246
  yield ratios for steps in, 247
Reference checks, 279
Relative ability test, 401
Reliability of selection instruments, 273–274
  specific instruments, 276–288
Replacement chart, 250

Rewards, 104–106
  extrinsic, 104
    job performance and, 344–345
  fringe benefits, 105
  to insure effectiveness, 104–105
  intrinsic, 104
  pay, 105
  promotion, 105
  reasons for, 104
  requirements, 104
  special awards, 105
  status symbols, 105
  technology and, 105–106
Role interchangeability, 119
  compensation and, 120–121
Routineness, 88

St. Charles Hospital work planning and review process, 342–344
Salaries, *see* Compensation
Scientific management, job design and, 94
Scope of services, 30
Selection, 267–293
  adverse impact and, 275–276
  compensatory approach, 288
  defined, 97
  designing the program, 268–275
    choosing criteria, 272
    choosing predictors, 272–273
    job analysis, *see* Job analysis
    relationship between predictors and criteria, 273–275
    reliability and, 273–274
    selection instruments and, 273
    steps in, 268
    validity and, 274–275
  determining the utility of the process, 288–290
    in dollar terms, 289
    errors and, 289, 290
    factors in, 288, 289
  federal regulations and, 275–276
  HRIS case study and, 160–162
  hybrid approach, 288
  introduction stage of product life cycle and, 45
  multiple-hurdles approach, 288
  organizational strategy and, 43
  overview, 12–13, 267–268
  preventive labor–management program and, 379–380
  Principles for the Validation and Use of Personnel Selection Procedures, 275–276
  references, 291–293
  selection instruments, 276–288
    application blanks, 282–283, 284–285
    assessment centers, 283, 286
    credentialing, 279–282
    drug screening, 287–288
    graphology, 287
    interviews, 277–279
    job simulation, 279
    licensing, 282
    physical examination, 286
    polygraphs, 286–287
    reference checks, 279
    table, 273
    tests, 276–277
  summary, 290–291
  technology and, 97–102
    changes in, 98–99
    criteria and, 99–102
    decision process and, 99–102
    job analysis and, 97
    job specifications and, 97–102
    predicting successful performances and, 98–99
    reliability and, 98, 100, 101
    routine *vs.* nonroutine tasks and, 99–100
    validity and, 98, 100, 101
  Uniform Guidelines on Employee Selection Procedures, 268–269, 275–276, 277
Selection ratios, 261–262, 288
Selective retention, 47
Seniority, 401–402
  ability and, 401
  contract clause, 429–431
  noninterchangeable occupational groups and, 401–402
  seniority unit, 401
  union organization and, 377
Seniority plus ability test, 401
Separability contract clause, 433–434
Sex discrimination, 353
Shift and shift differentials contract clause, 432
Sick leave contract clause, 434–435
Similar-to-me error, 332
Simple personnel model of hospital administration, 412, 413, 414
Social relationships culture gap, 214
Software, computer, 170–174
  borrowing, 172
  building, 171–172
  buying, 171
  manpower allocation and, 158

Software, computer (*Continued*)
  packaged, 171
  state of the art, 145, 146
  steps in selecting, 172–174
    contract considerations, 174
    implementation and maintenance options, 174
    justification, 172
    selection criteria, 173
    software evaluation committee, 172–173
    system requirements, 173
    system security, 173
    vendor evaluation, 173–174
Special awards as rewards, 105
Staff development, 295–296. *See also* Training and development
Stars, 36
Start-up phase of organizational life cycle, 52
  strategy, human resources management and, 73, 76
Status symbols as rewards, 105
Straight ranking performance appraisal, 322
Strategic business units (SBUs):
  assessment, *see* Portfolio assessment
  identifying, 28–29
  mission statements and, 30
Strategic decision-making level, 151–152
  tasks, 153
Strategic management, 21, 25
Strategic management of human resources, 3–19. *See also* Organizational strategy
  advances in science and technology and, 4
  behavioral systems and, 10–11. *See also* Commitment of professionals; Corporate culture; Motivation of professionals; Organizational culture
  defined, 3–4, 42
  described, 6–8
  funding for health services and, 6
  health care as a service industry and, 3
  integration with strategic planning, 414–421
  model, 7
  outcomes and, 15–18
    employee, 16–17
    organizational, 16
    strategic management and, 17–18
  patterns of utilization and, 4, 6
  process systems, 11–15. *See also* Compensation; Labor relations; Performance appraisal; Recruitment; Selection and placement; Training and development
  references, 18–19
  staff development programs and, 300–305, 315–316
  strategy and, *see* Organizational strategy
  structural systems and, 8–10. *See also* Information systems; Organizational structure; Technology
  summary, 18
  supply of health personnel and, 4–6
Strategic modifications, 56–57
Strategic planning, 24–25
  defined, 21, 333
  emergence, 24
  integration of human resources planning and, 414–421
  performance appraisal, 333–336
    control of personnel and, 335
    development of personnel and, 335
    dual nature, 333–334
    human resources planning and, 334–335
    justification of personnel actions and, 335–336
    motivation of personnel and, 335
  recruitment and, 245
  steps in, 25
  time frame for, 243
Strategy, *see* Organizational strategy
Strategy-action match, 41–42
Strengths and weaknesses, assessment of, 33–34
Strikes, 393–395
  allied doctrine and, 394–395
  Health Care Amendments of 1974 and, 395, 396–397
  impact of, 393–394
  no-strike clauses, 402
Structural systems, overview, 8–10. *See also* Information systems; Organizational structure; Technology
Substitution strategies, domain, 65
  collapse of niche and, 69–70
  dissolution of niche and, 69
  examples, 66
"Success sequence", 40, 41
Symbolic leadership, 233
Systems, 57–58
Systems theory, 149–151

Tactical planning:
  recruitment and, 245
  time frame for, 243
Taft–Hartley Act, 372, 373, 395
Tardiness and organizational commitment, 193
Task innovation culture gap, 214

Task support culture gap, 214
Taxes and employee benefits, 367–368
Team incentives, 360
Technical change, 221
  problems resulting from, 229
  transactional leadership and, 226
Technical workers, recruiting, 242
  external, 251
  issues concerning, 243
  sources for, 252–256
  summary, 263
  time frame for, 246
Technology, 85–111
  client/staff relationship and, 89–90
  complexity continuum and, 87
  as component of environmental analysis, 33
  decision trees for, 107–108
  defined, 87–88, 89
  degree of diffuseness and, 87
  describing the nature of, 107
  as determinant of structure, 114–115
  in human service organizations, 88–90
  identifying changes in, 107
  impact of human resources management, 107–108
  influences, on the human resources functions, 93–106
    job design, see Job design, technology and
    occupations development, 106
    performance appraisal, 102–104
    personnel selection, see Selection, technology and
    rewards, 105–106
  information, 86, 87–88
  intensive, 87
  isolating changes in, 107
  long-linked, 87
  management, 86–87
  mediating, 87
  medical, 85–86
  overview, 9, 85–87
  paraprofessional, 86
  raw materials and, 88
  references, 109–111
  routineness and, 88, 89
  search process and, 88–89
  summary, 108–109
  technical rationality, 87
  theory of, in organizations, 87–93
    environment, strategy and, 91–92
    levels of analysis and, 92–93
    structure and, 90–91
  transformational processes and, 87–88

Tenure and organizational commitment, 191, 193–194
Tests as selection instruments, 276–277
Threat-rigidity response, 68
Training, 299
Training and development, 294–318
  components of effective programs, 308–313
    decision to institute or modify, 311
    defining curriculum, 311–312
    environmental assessment, 308, 310
    evaluation and control, 312–313
    implementation, 312
    organizational assessment, 310
    preparing materials, 312
    program preparation, 311–312
    setting objectives, 311
  continuum, 296–300
    objectives, 298–299
    skill diversity and personal growth, 299–300
    table, 297
    terms used in, 298
    training site and frequency, 300
  definitions, 295–296
  goal accomplishment and, 302–304
    individual, 303–304
    organizational, 302–303
  growth stage of product life cycle and, 46
    technology and, 106
  HRIS and, 163–164
  implementation of organizational culture and, 304–305
  importance, 300–305
  incentives for upgrading, 305–308
  metamorphosis, 305, 306
  model, 309
  organizational development and, 296
  organizational strategy and, 43–44
  organizational structure, human resource management and:
    in the group medical practice, 125
    in the hospital, 132–133
    in the solo medical practice, 119–120
    summary, 137, 138
  overview, 13, 294–295
  performance appraisal and, 327, 335
  preventive labor–management program and, 380–381
  references, 317–318
  strategic posture and, 315–316
  strategic relevance, 300–305
  summary, 316–317
  synonyms for, 298
  trends in, 313–315

Training and development (*Continued*)
  differentiation among categories of personnel and, 314
  expansion of knowledge and, 315
  multiple careers and, 315
  table, 315
Trait approaches to performance appraisal, 102
Transactional leadership, 222–223
  contingent reward and, 222–223
  defined, 222, 223
  function, 223
  management-by-exception and, 223
  processes, 227
  relationship between transformational and, 225–226
  technical change, transition and, 226
Transformation change, 221
  hierarchical order for, 56
  transformational leadership and, 226–227
Transformational leadership, 56–57, 223–225
  belief in people and, 224–225
  charisma and, 223–224
  courage and, 224
  dealing with complexity and, 224
  defined, 56, 222, 227
  development, *see* Leadership development
  function, 225
  individualized considerations and, 224–225
  intellectual stimulation and, 224
  life-long learning and, 224
  proactive *vs.* reactive leadership and, 225–226
  process, 227–231
    creating commitment to the vision, 230
    creating a felt need for change, 227–229
    creating a vision, 229–230
    environmental triggers for, 56, 227
    illustrated, 228
    institutionalizing the change, 230
    overcoming resistance to change, 56–57, 229
    performance gap and, 56, 227
    summary, 230–231
    types of change and, 229
  references, 235–237
  relationship between transactional and, 225–226
  summary, 234–235
  transformational change and, 226–227
  value-driven, 223–224
  visionaries and, 223
Transformational processes and technology, 87–88
Transitional change, 221
  transactional leadership and, 226
Turnover rates:
  organizational commitment and, 192–193
  realistic job previews and, 257
Two-factor theory of motivation, 201

Unions, *see* Labor relations; Labor relations contracts, negotiating and administering
Union security contract clause, 427–428
Union shop, 398
Unskilled workers, recruiting, 242
  external, 251
  issues concerning, 243
  sources for, 252–256
  summary, 263
  time frame for, 246

Validity of selection instruments, 274–275
  coefficient of, 288
  specific instruments, 276–288
Value-driven leaders, 223–224
Value systems, 381
Vertical complexity (differentiation) of structure, 114
Vertical integration, 57–58
Visionaries, 223
  creating a vision, 229–230
  commitment after, 230
  development, 233
Vision of the firm, 30. *See also* Mission statement
Vocational Rehabilitation Act of 1973, 275

Wage rates and union organization, 377
Wages, *see* Compensation
Wagner Act, 372
Weighted checklists performance appraisal, 322–323
Work environment and union organization, 377
Work experiences and organizational commitment, 191
Work load measure systems, 158
Work planning and review process, 341–342, 343
Work unit level, assessment at, 310